T0213721

Lecture Notes in Business Information Processing 279

Series Editors

Wil M.P. van der Aalst
Eindhoven Technical University, Eindhoven, The Netherlands
John Mylopoulos
University of Trento, Trento, Italy
Michael Rosemann
Queensland University of Technology, Brisbane, QLD, Australia
Michael J. Shaw
University of Illinois, Urbana-Champaign, IL, USA
Clemens Szyperski
Microsoft Research, Redmond, WA, USA

More information about this series at http://www.springer.com/series/7911

Stefano Za · Monica Drăgoicea
Maurizio Cavallari (Eds.)

Exploring
Services Science

8th International Conference, IESS 2017
Rome, Italy, May 24–26, 2017
Proceedings

 Springer

Editors
Stefano Za
Faculty of Economics
eCampus University
Rome
Italy

Maurizio Cavallari
Università Cattolica del Sacro Cuore
Milan
Italy

Monica Drăgoicea
Department of Automatic Control
and System Engineering
University Politehnica of Bucharest
Bucharest
Romania

ISSN 1865-1348   ISSN 1865-1356   (electronic)
Lecture Notes in Business Information Processing
ISBN 978-3-319-56924-6   ISBN 978-3-319-56925-3   (eBook)
DOI 10.1007/978-3-319-56925-3

Library of Congress Control Number: 2017936706

Printed on acid-free paper

This Springer imprint is published by Springer Nature
The registered company is Springer International Publishing AG
The registered company address is: Gewerbestrasse 11, 6330 Cham, Switzerland

Foreword

Exploring Services and Information Technologies: The Investigation of Managerial and Policy Challenges from Different Perspectives

In his brief note in the *Harvard Business Review*, when describing the reasons why the discipline of "services science" seemed to be a promising area of research, Chesbrough [1] stressed the role of intangibility as the specific trait of services with two main consequences: (a) the lack of a tangible artifact to test consumer needs that makes innovation a different endeavor from other sectors; (b) productivity is harder to measure since both inputs and outputs are intangibles. He considers these two areas of research not satisfactorily explored by other disciplines. The same occurs for the question of the transfer of tacit knowledge, which Chesbrough finds particularly relevant since services promote encounters among people that have to learn from each other for an effective service to occur. The accent on tacit knowledge was kept at center stage in the article that Chesbrough co-authored with Spohrer the following year [2]. A successive article by Spohrer and other members of the Almaden IBM Research center [3] points to explicit knowledge ("information" in their wording) as the key issue. Furthermore, given the importance of information technology in services, they highlighted the need to both differentiate and find connections between computational systems and service systems since the components can be modelled and simulated in the former whereas in the latter the presence of human beings complicates matters substantially. They suggest: "*Perhaps*, if we model people as components with stochastic behaviour" existing theories of computational systems can be applied to service systems ([3] p. 76, italics added).

The word "perhaps" used by the authors shows that caution is necessary when proposing computational models to be applied to human behavior. Furthermore, the entanglement of information technology and society as well as the digitization of private and public organizations in a service perspective — i.e., in a perspective where co-production is key to the development of effective e-services — call for multidisciplinary approaches in investigating the extensive and articulated subject matter of service science. Indeed, contributions from different disciplines help shed light on the relevant phenomena. Other authors, besides the ones mentioned here, are often cited, such as the works by Vargo and Lusch in marketing [4] and the seminal work by Ostrom and colleagues back in 1981 [5] concerning co-production in public services. Some authors explicitly refer to the need for interdisciplinary and cross-disciplinary approaches to understand how services should be designed, delivered, and supported [6]. There is a cross-fertilization among disciplines as is the case of ethics where, for example, Floridi [7] has introduced the concept of "infosphere" and the possibility of considering as moral agents not only humans but information technology artifacts with certain traits. Indeed, specific ethical dilemmas concern electronic services, e.g., the privacy versus security issue. The problem, however, is how ethical issues are managed by the business world; sometimes ethicists have been criticized for their difficulty in finding an appropriate way to enter a dialogue with entrepreneurs and managers [8].

The 8th IESS 2017 Conference invited researchers and practitioners to submit contributions and to critically discuss the results of their work by drawing on disciplines not only in management and engineering, but also in social and cognitive sciences, in law, ethics, economics, public administration and other fields to address the theoretical and practical challenges that the services industry and its economy are facing. This book collates interesting contributions that were discussed at the conference and deal with some of these research areas.

References

1. Chesbrough, H.W.: Toward a new science of services. Harv. Bus. Rev. 83, 43–44 (2005)
2. Chesbrough, H., Spohrer, J.I.M.: A research manifesto for services science. Commun. ACM. 49, 35–40 (2006)
3. Spohrer, J., Maglio, P.P., Bailey, J., Gruhl, D.: Steps toward a science of service systems. Computer (Long. Beach. Calif). 40, 71–77 (2007)
4. Vargo, S.L., Lusch, R.F.: Evolving to a New Dominant Logic for Marketing. J. Mark. 68, 1–17 (2004)
5. Parks, R.B., Baker, P.C., Kiser, L., Oakerson, R., Ostrom, E., Ostrom, V., Percy, S.L., Vandivort, M.B., Whitaker, G.P., Wilson, R.: Consumers as Coproducers of Public Services: Some Economic and Institutional Considerations. Policy Stud. J. 9, 1001–1011 (1981)
6. Bardhan, I.R., Demirkan, H., Kannan, P.K., Kauffman, R.J., Sougstad, R.: An Interdisciplinary Perspective on IT Services Management and Service Science. J. Manag. Inf. Syst. 26, 13–64 (2010)
7. Floridi, L.: The Ethics of Information. Oxford University Press, Oxford, UK (2013)
8. Stark, A.: What's the Matter with Business Ethics? Harv. Bus. Rev. 71, 38–40 (1993)

March 2017   Paolo Depaoli

Preface

This volume gathers papers presented at the 8th International Conference on Exploring Service Science, IESS 1.7, organized during May 24–26, 2017, by the Department of Management, University Sapienza of Rome, Italy, in collaboration with the Italian chapter of the Association for Information Systems (http://www.itais.org).

The conference hosts academic scientists and practitioners from the service industry and their worldwide partners in a collegial and stimulating environment. According to his tradition, IESS 1.7 covered major research and development areas related to service science foundations, service engineering and management, service innovation, service orientation of processes, applications in service sectors, and ICT support for services. Services comprise about 75% of mature economies today, being a fast-growing sector in emerging economies, too. This motivates an intense preoccupation to establish the philosophy of a new management and marketing, which highlights a paradigm shift away from the goods-dominant (G–D) logic. This paradigm is the theoretical concept of service-dominant (S–D) logic, fundamental for the service system developments reported in IESS1.7 papers.

The IESS1.7 event collects papers that extend the view on different concepts related to the development of the service science domain of study, applying them to frameworks, advanced technologies, and tools for the design of ICT-based service systems. As IESS 1.7 papers describe, specific items of service business models are analyzed and debated, such as target markets and customers, product offerings or value propositions, distribution channels (activities for services), and constraints and profits, together with the description of case studies and business solutions in various service sectors. All these aspects are covered in the present book, which we hope you will find useful reading.

All the selected papers have been evaluated through a standard blind review process in order to ensure theoretical and methodological rigor. The fourfold structure of the volume reflects four main pillars that have been explored by the included papers.

This publication is the result of a teamwork where many people actively contributed. We are grateful to the authors, the conference chairs and committee members, to the members of the editorial board, and to the reviewers for their competence and commitment.

Theoretical Contributions: Literature Analysis and Conceptual Models

The study of the organizational impact of information technology services on the value generation process in organizations has a story that spans over several decades. In this stream of research, which goes under the name of IT value, scholars debated over years on the potential existence of a positive link between the presence of IT services and the organizational performance.

The contribution of Za and Braccini complements existing studies on the value of IT services, through a literature analysis, focusing on the summarization and systematization of existing literature. Centering on a specific topic, the contribution of Seebacher and Schritz provides a structured literature review investigating the real-world impact and benefit of the Blockchain technology. This specific technology is centered around a peer-to-peer network, enabling collaboration between different parties, where the service system is chosen as unit analysis to examine its potential contribution.

On a different topic is the contribution of Savastano et al. They provide an assessment of the past and current literature on FabLabs (networked platforms for the dissemination of digital culture through the sharing of technological tools and knowledge) from a service perspective, exploring the main research themes and methods associated with this new business model. The workplace health promotion applications are the topic analyzed by Dehkordi at al.; the authors summarize the state of the art and identify the major research streams through a literature review and cluster analysis. A multi-method approach (a systematic literature review and a case study) is used by Reis et al. in order to investigate organizational synergies in the omni-channel service context. In doing so, they disclose new omni-channel trends and discuss the implications for managers and academics.

Focusing on MicroServices architecture, where the legacy architecture is decomposed in micro-components, each one with an independent life cycle but interconnected and correlated, Cavallari et al. analyze in detail the structure and the development of eServices based on this specific architecture. Their paper discusses the new technological tendencies under the lens of an organizational approach.

In the last literature analysis paper, Sorrentino et al. advance the current debate on the co-creation of value in the delivery of health care by treating the informal caregivers as a key organizational resource for the providers, and not as one of the many spokes in the customer wheel.

Results of a different analysis are provided by Morais et al. They examine the relationship between the corruption perception index (CPI) provided by Transparency International and the human development index (HDI) of the United Nations Development Program and its components. The results obtained reinforce the importance of efforts by international politicians and organizations in fighting corruption, especially in highly developed countries.

On the basis of data analysis, Militaru et al. extend research on social networking sites and the role of these tools in the business performance of service firms. Using cross-section data from a sample in Romania, the authors empirically investigated the mediating effects of service innovation on the relationship between social networking sites and business performance. The results indicate that innovation capability played a key role in business performance but its mediating role between social networking sites and business performance of service firms was not confirmed.

Service Systems Analysis and Design

The chapters in this part provide conceptual models and theoretical frameworks for supporting the analysis or the design of a service system.

Verlaine defines and depicts the generic components of service systems as well as the relations between them, proposing a new conceptual model of a generic (simple or complex) service system. Costanza proposes a framework of analysis combining system dynamics and performance management to represent the stock-and-flow structure of the phenomenon of value co-creation through social media marketing and brand communities.

In the current economic scenario, big data are offering considerable economic benefits. Spagnoli and Morelli evaluate the diversity of the creative industries (CI) when related to the use of big data, providing a multi-criteria methodology for assessing their effects on CIs, and a model for implementing collaborative and virtual value chains through its usage. Schmitz et al. develop a method for defining data acquisition strategies to improve uncertainty analyses for industrial service contracting, as well as an approach for ranking acquisition strategies by measuring their acquisition effort and business benefit. The method is applied in an industrial use case to demonstrate its benefit for assessing cost uncertainties in full-service repair contracts.

Meierhofer and Meier show a systematic approach to use data science for the process of service design. The authors develop a structure of data science method-ologies in the dimensions of their potential to create service benefit. This allows one to map the value contribution of the data science tools on the different perspectives and phases of the service design process, establishing a direct link between the outcomes of the data science methodologies and the value drivers for the customer.

Based on a combination of established statistical methods, Hunke at al. propose a systematic approach that allows one to identify different domains of business model patterns. The authors apply it on a dataset of 58 e-mobility projects and, as a result, they identify five distinct and semantically meaningful business model types. One of the main contributions of this paper is to suggest a new approach for identifying different patterns of business models, the second one is to provide valuable insight into the current state of e-mobility service business models that can further drive the adoption.

Kummler presents an approach with which to evaluate and model quality by using requirements from automotive development projects as a practical example. As first results, he provides the development of an assessment tool and an initial analysis of the available dataset. Reuter-Oppermann et al. present the outline of a decision support system for (optimally) locating general practitioners' practices, in the context of pri-mary care services. Silva and Migueis propose a prescriptive analytics solution to enhance the service provided by libraries, by optimizing the layout of libraries. The results of their study corroborate the effectiveness of the method proposed and its potential in supporting library management decisions.

Service Organization Case Studies and Practices

This part gather the contributions describing service organization experiences and case studies. A longitudinal analysis opens this section, followed by two sets of contributions regarding customers' experiences and higher education and training. Two others contributions on different topics close this part.

Tapandjieva et al. present a longitudinal action design research project, describing the transformation of a service-oriented organization. It has slowly matured into the design and development of a visualization tool called Service Cartography. The authors provide the evolution of their conceptualization of how Service Cartography facilitates service-oriented thinking.

Through a content analysis, Molnar and Moraru examine what customers evaluate when they choose a hotel or have an experience with it. At the same time, they study the ability of the hotel to incorporate customers' feedback affecting them and customers' decisions. Considering a similar topic, the contribution of Alcoba et al. compares a wide sample of tourists' numerical scores and verbal assessments, analyzed with a sentiment analysis engine. The objective is to acquire a deeper knowledge of the concept of experience quality to find out what the tourist really values.

Dima analyzes the evolution of convergence in higher education during 2002–2013, based on previously used macroeconomic and transition indicators. Using forecasting techniques for the 2014–2020 period, the author foresees the dynamics of the outlined clusters, and the perspectives of convergence in the near future. Paunescu analyzes the way in which different attributes of community engagement and service are taken into consideration among the indicators used in university rankings.

Faria and Nóvoa describe the design and implementation process of a new BPM platform adopted at the University of Porto.

Menshikova et al. provide an exploratory research of the on-line training courses and initiatives developed by the bank sector institutions and addressed to both their employees and their real and potential clients. The paper provides a preliminary analysis of the programs based on the experience of an Italian bank.

Focusing on fraud risks in telecom services and products, Yesuf et al. describe the outcome of two workshops in which they involved experts from a telecom provider and experts from multidisciplinary areas. They present two exemplary telecom fraud scenarios, analyzing and estimating the impacts of fraud risks qualitatively.

Perna et al. conduct a review of the literature and a comparative analysis of Summer Olympic Game editions during the period 1992–2016, to investigate the crucial role played by the social networks in the mega sporting events improving their social value in the co-creation value perspective.

Sustainability: Service Ecosystems, Environment Control, and Transportation

This part collects contributions discussing topics connected with sustainability through service issues in general.

The public transport service system provides safe and secure urban mobility for all citizens. The contribution of Drăgoicea et al. offers a new perspective on designing sustainable public transport services in times of emergency in order to support the transport service company's operational activities based on sustainable institutions principles. On the basis of Ostrom principles, it is suggested that a common–pool resource (CPR) institutional approach would be able to adapt more responsively to an emergency situation than a centralized or privatized one.

A focus often linked to transportation issues is pollution. Chiru et al. present a perspective related to information service integration for pollution awareness evaluation. Using digital tools, based on indirect information analysis as retrieved from the available literature over time, the paper investigates how pollution events influenced public awareness.

Borangiu et al. introduce a framework for designing flexible environment control services (e-services) based on generic sensing, modelling, and control process specifications that allows the customization of a holonic facility environment control system (HFES). The authors also present and discuss experimental results.

The role of non-profit organization and citizen satisfaction in the service ecosystem is analyzed by the following contributions. Bonomi et al. conduct an in-depth longitudinal study (2013–2016) on an ICT-enabled community of IT professionals, aiming to provide unemployed professionals with employment opportunities while also providing small and micro enterprises and non-profit organizations with affordable, high-level IT services. The authors found that the presence of non-profit organizations in the service ecosystem strongly influences the service ecosystem's institutional logic and worldview and facilitates sustainability-oriented self-organizing throughout the ecosystem.

Zagorie et al. examine the quality of municipal services within inner-city services. They identify the most important service quality dimensions that determine citizen satisfaction. A system dynamics approach is used to model and analyze ways to improve citizen satisfaction. The managing of the queue is often associated with the quality of service delivery.

Fragnière et al. provide an empirical exploratory enquiry to develop a queue's ontology on an ethological basis, taking a concrete example of cable car queues in the Alps, in the Canton of the Valais. This human–machine case is particularly interesting because the results show that a queue's regulation is mostly based on ethological behavior (therefore innate rather than learned) to adjust to the rigid system of the cable cars.

March 2017

Stefano Za
Monica Drăgoicea
Maurizio Cavallari

Organization

IESS 1.7 was organized by the Department of Management, Sapienza University of Rome, Italy, in collaboration with the Italian Chapter of the Association for Information Systems (www.itais.org) during May 24–26, 2017.

Steering Committee

Michel Léonard	University of Geneva, Switzerland
João Falcão e Cunha	University of Porto, Portugal
Eric Dubois	Luxembourg Institute of Science and Technology, Luxembourg
Theodor Borangiu	University Politehnica of Bucharest, Romania
Monica Drăgoicea	University Politehnica of Bucharest, Romania
Marco de Marco	Università Catollica del Sacro Cuore, Italy
Henriqueta Nóvoa	University of Porto, Portugal
Gerhard Satzger	Karlsruhe Service Research Institute, Germany
Mehdi Snene	University of Geneva, Switzerland
Fabrizio D'Ascenzo	Sapienza University of Rome, Italy

Conference Chairs

Mauro Gatti	Sapienza University of Rome, Italy
Fabrizio D'Ascenzo	Sapienza University of Rome, Italy

Program Chairs

Stefano Za	eCampus University, Novedrate (CO), Italy
Monica Drăgoicea	University Politehnica of Bucharest, Romania

Sponsoring Institution

IBM Italia

International Program Committee

Sabrina Bonomi	eCampus University, Italy
Theodor Borangiu	University Politehnica of Bucharest, Romania
António Brito	University of Porto, Portugal
Bettina Campedelli	University of Verona, Italy
Jorge Cardoso	University of Coimbra, Portugal
María Valeria de Castro	Universidad Rey Juan Carlos, Spain

Contents

Service Organizations Case Studies and Practices

Sustainability: Service Ecosystems, Environment Control and Transportation

Theoretical Contributions: Literature Analysis and Conceptual Models

Tracing the Roots of the Organizational Benefits of IT Services

Stefano Za[1,2(✉)] and Alessio Maria Braccini[3]

[1] eCampus University, Novedrate, CO, Italy
stefano.za@uniecampus.it
[2] LUISS Guido Carli University, Rome, Italy
sza@luiss.it
[3] University of Tuscia, Viterbo, Italy
abraccini@unitus.it

Abstract. The study of the organizational impact of information technology services on the value generation process in organizations has a story that spans over several decades. In this stream of research, that goes under the name of IT value, scholars debated over years on the potential existence of a positive link between the presence of IT services and the organizational performance. The field of IT value is ample and continuous. Considering these aspects this paper contributes to the literature by proposing a quantitative analysis of the IT value literature. Our study complements existing studies on the value of IT services, which followed qualitative approaches, focusing on the summarization and systematization of existing literature. This paper presents the results of a longitudinal bibliometric study of the IT value literature based on 435 papers published from 1990. Our analysis shows a growing trend of publications and citations, the existence of a common theoretical foundation for the research stream identified by the most influential and co-cited sources. Our analysis identified the existence of three core research areas inside which most of the IT value discourse is developed. We concluded our research formulating some considerations regarding future IT service value investigations.

Keywords: IT services · IT value · Bibliometric analysis · Co-citation analysis

1 Introduction

While setting the rationale for *services science* as a stand-alone discipline, Chesbrough stressed the role of intangibility as the specificity of services with two main consequences [1]: (i) the lack of a tangible artefact in services to test consumer needs that makes innovation a different endeavour from other sectors; (ii) the difficulty of assessing productivity due to the intangible nature of both inputs and outputs. According to him these two peculiarities are not satisfactorily explored by other disciplines.

Given the importance of Information Technology (IT) in services, Chesbrough and Spohrer [2] highlight the need to study how organizational capabilities and artefacts – including IT – can be combined to generate value. Organizations have invested in IT

S. Za et al. (Eds.): IESS 2017, LNBIP 279, pp. 3–11, 2017.
DOI: 10.1007/978-3-319-56925-3_1

service management as a set of praxes to define, manage, and deliver IT services to ensure the support to business and customers' needs [3]. The costs of IT services accounts to 60%–90% of the total cost of IT ownership [4], making IT service management the core of IT management in organizations.

Researchers have been studying the organizational impact of IT services for several decades [5, 6], to assess positive or negative benefits in terms of organizational performance [7–9]. Over the years different methodologies and theoretical approaches were used to investigate this phenomenon [10] in what is a multi-disciplinary research field [7, 8, 11].

Assessing organizational benefits of IT services is of great importance both for the literature and for the practice [7, 11–14]. This field of research goes under the name of "IT value", and is reputed fundamental for the managerial literature [7, 14]. With the aim of contributing to the ongoing debate in this field of research, previous researchers summarized and systematized the many contributions by way of literature reviews or conceptual frameworks [7, 10, 15]. We complement these works by proposing a quantitative-based analysis of the literature over a longitudinal perspective, seeking to investigate the following research questions:

RQ1.   What are the *foundations* of IT Value in terms of key sources cited in articles discussing it?
RQ2.   How are the *foundations* of the IT Value literature *evolving over time*?
RQ3.   What are the most active *research areas* discussing IT Value?
RQ4.   How are the *research areas* evolving over time?

Our work contributes in tracing the roots and the common theoretical ground of the interdisciplinary discourse on IT services value, and to trace the trends and evolution of such research field over time, to formulate implications on the progression of the discourse.

2   Research Methods

We performed a quantitative bibliography analysis using number of publications and number of citations and co-citation as proxies of the influence of contributions in the IT value organizational discourse [16, 17]. We applied this method, that allows to study the cumulative knowledge generation process in the literature, following the protocol described in Fig. 1.

Literature sources were searched on ISI (Institute for Scientific Information) Web of Science which is a multi-disciplinary literature archive covering more than 12,000 journals with 40 million entries. The conceptualization of the literature is based on our previous knowledge of the discourse complemented by reviews of the IT value literature [7, 9, 18]. We used the following keywords: "organizational performance", "information technology", "IT value", "business value", and "economic value" stemmed and used in combination with wild cards to include both singular and plural expressions. We executed the following query to search for literature:

Dataset Setting   Data Analysis

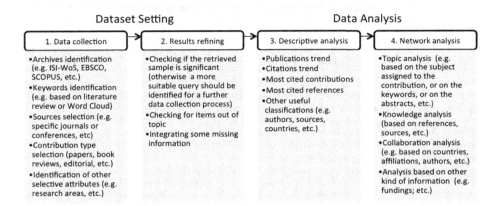

Fig. 1. Research protocol

```
Topic = (("organizational performance" AND "information
technolog*") OR ("IT value" AND "information technolog*")
OR ("business value" AND "information technolog*") OR
("economic value" AND "Information technolog*"))
Refined by: Languages=(ENGLISH) AND Document Types =
(ARTICLE OR REVIEW) AND [excluding] Document Types =
(PROCEEDINGS PAPER OR BOOK CHAPTER)
```

On ISI-WoS, what is indicated in the "TOPIC" field is searched in title, keywords, and abstract of each contribution stored in the databases. To include only the most reliable sources we limited our results only to contributions written in English and published in peer reviewed scientific journals. We made a first selection out of the identified sources by reading titles and abstracts of selected sources. We eventually removed 120 false positives from the original sample. The final set was composed by 435 contributions published after 1990. The number is in line with figures reported by other IT value literature studies [18, 19].

On this final set, following the third and fourth steps of the research protocol, we performed a descriptive analysis, and a network analysis [20, 21].

3   Main Findings

Table 1 lists the top 50 most frequently cited sources by the papers included in the sample. The percentage between brackets is a measure of the relative impact of the specific source in the 435 papers analysed. Rows in grey represent sources that are in the sample analysed. The most relevant sources are the first five, which are the only ones cited more than 100 times, and which represent building blocks shared amongst about one third of the papers included in the sample.

Table 1. Most influential sources in the samples

References	Cit.
Melville N, 2004, Mis Quart, V28, P283	139 (32%)
Bharadwaj As, 2000, Mis Quart, V24, P169, doi:10.2307/3250983	138 (32%)
Barua A, 1995, Inform Syst Res, V6, P3, doi:10.1287/isre.6.1.3	111 (26%)
Brynjolfsson E, 1996, Manage Sci, V42, P541, doi:10.1287/mnsc.42.4.541	106 (24%)
Barney J, 1991, J Manage, V17, P99, doi:10.1177/014920639101700108	106 (24%)
Hitt Lm, 1996, Mis Quart, V20, P121, doi:10.2307/249475	85 (20%)
Mata Fj, 1995, Mis Quart, V19, P487, doi:10.2307/249630	81 (19%)
Wade M, 2004, Mis Quart, V28, P107	78 (18%)
Mukhopadhyay T, 1995, Mis Quart, V19, P137, doi:10.2307/249685	70 (16%)
Tallon Pp, 2000, J Manage Inform Syst, V16, P145	68 (16%)
Powell Tc, 1997, Strategic Manage J, V18, P375, doi:10.1002/(sici)1097-0266(199705)18:5<375::aid-smj876>3.0.co;2-7	67 (15%)
Fornell C, 1981, J Marketing Res, V18, P39, doi:10.2307/3151312	66 (15%)
Weill P., 1992, Information Systems, V3, P307, doi:10.1287/isre.3.4.307	62 (14%)
Sambamurthy V, 2003, Mis Quart, V27, P237	58 (13%)
Devaraj S, 2003, Manage Sci, V49, P273, doi:10.1287/mnsc.49.3.273.12736	56 (13%)
Teece Dj, 1997, Strategic Manage J, V18, P509, doi:10.1002/(sici)1097-0266(199708)18:7<509::aid-smj882>3.0.co;2-z	56 (13%)
Bharadwaj As, 1999, Manage Sci, V45, P1008, doi:10.1287/mnsc.45.7.1008	56 (13%)
Podsakoff Pm, 2003, J Appl Psychol, V88, P879, doi: 10.1037/0021-9101.88.5.879	54 (12%)

The table indicates that these sources are: Melville et al. [18], Bharadwaj [22], Barua [23], Brynjolfsson and Hitt [24], and Barney [25].

The influence of the different sources is shown by the network analysis of the co-citations (Fig. 2). Each node in the figure is a paper cited by the papers in the sample. An arc between two papers indicates a co-citation of the two papers in one of the papers in the sample. Arcs thicker than others indicate co-citation pairs that are more frequent than others. The numbers on the arcs indicate the absolute frequency of the co-citation occurrence. The most evident co-citation pairs link the papers of Melville et al. [18], Bharadwaj [22], Barney [25], Wade [26], Mata et al. [27], and Brynjolfsson and Hitt [24].

Figure 3 shows instead the network analysis of the co-occurrence of the research areas [28]. The figure shows the presence of several research areas. Much part of the IT value discourse takes place among the Business & Economics, Computer Science, and Information Science & Library Science research areas. This triad is by far the most frequent in terms of research areas co-occurrence. Besides this there are also the areas of Engineering and of Operations Research & Management Science, which appear to be also relevant, even though of second order. All the other research areas are instead only mentioned one time, with the only exceptions of Public Administration, Health Care Sciences & Services, and Construction & Building Technology.

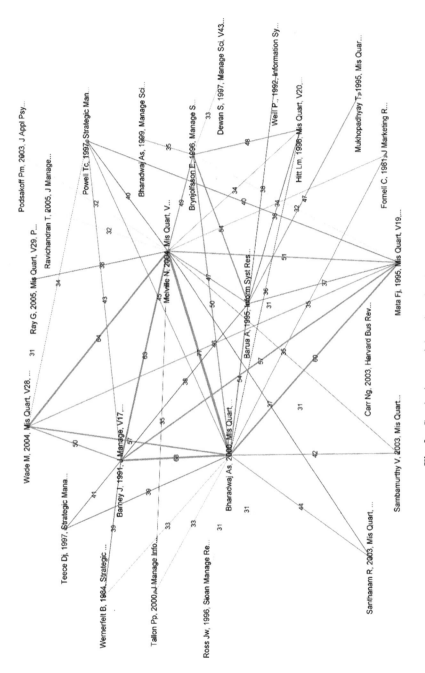

Fig. 2. Co-citation graph based on the cited references

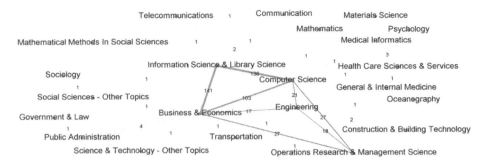

Fig. 3. Research areas co-occurrence

4 Theoretical, Empirical, and Managerial Implications and Contributions

The analysis performed depicts a multi-disciplinary area of research, where the weight of the different disciplines is uneven. Answering to RQ3, the IS discipline is central to the discourse, and the core domain of the IT value research is formed by the intersection of Computer Science, Business & Economics, and Information Science & Library Science.

We performed also the longitudinal analysis. Since the number of publications and citations, it was possible to recognize three main periods: before 2000, 2000–2008, and after 2008.

The first period is the one where the discourse is set up. No contribution emerges as clearly influential in this period but for the work of Barua [23] and Weill [29]. In the second period three sources are influential: Bharadwaj [22], Barua et al. [23], and Brynjolfsson and Hitt [24]. In the third period the work of Barua et al. [23] is confirmed as influential source together with the review of Melville et al. [18] which sets a theoretical framework for IT value investigations.

The most active research area over time are that of: Information Science & Library Science, Business & Economics, and Computer Science. The analysis shows the emergence of secondary areas: Operations Research & Management Science, Engineering (from the first period), and Public Administration (from the third).

As per RQ4 what the longitudinal analysis of the research areas shows instead is that secondary domains flank the core domain. Among these Engineering, and Operations Research & Management Research emerged against the others. We suggest also that Public Administration, Health Care Science & Services, and Medical Informatics could be the next emerging areas. Particularly Healthcare Science & Services and Medical Informatics appear to be more strongly related each other than the rest of the core domain.

Concerning the evolution of the IT value discourse, the longitudinal analysis of the top 10 most influential sources shows that the discourse has the traits of continuity in cumulating research results, as most sources maintain their relevance over time. Except from the first period, there are no clusters in the IT value discourse. This suggests us that there is a stable paradigm informing the IT value research.

Answering RQ1, the most influential sources, which are also quite stable over time, of the IT value are the works of: Melville et al. [18], Bharadwaj [22], Barua et al. [23], Bryinjolfsson and Hitt [24], and Barney [25]. These sources found empirical evidences of value created by IT, but they also contribute on two different aspects: the methodology, and the theory for the IT value assessment.

The paper from Brynjolfsson and Hitt [24] contributes on the theory side. First, it marks an important step in the IT value discourse testifying the overpassing of the productivity paradox, which greatly contributed to animate debate in this field. Secondly the paper builds over economic theory to propose an IT value assessment method based on Cobb-Douglas productivity functions.

Both the papers from Bharadwaj [22], and Barney [30] still contribute on the theory side, but on a different perspective. The paper of Barney [25] is one of the seminal papers which contributed to develop the RBV theory. The paper of Bharadwaj [22] presents empirical evidences of IT value in a firm level assessment resorting on a RBV theoretical perspective to perform the assessment.

Finally the paper from Melville et al. [18] contributes identifying a theoretical framework based against the RBV theory that shows loci and focus of IT adoption and of IT value generation, and proposes guidance for future investigations of the IT value phenomenon.

The paper of Barua et al. [23] instead contributes on the methodology level, proposing a process based approach to assess IT value in a post implementation scenario, and providing empirical evidences of the existence of such value.

Concerning the evolution of these foundations (RQ2), the co-citation analysis showed in the first period the existence of three distinct areas, of different importance, in the literature. The two prevailing ones are focusing on the collection of empirical evidences of IT value, and on the investigation of the strategic importance of IT implementation. The third, and less frequent one, concerns instead the investigation of IT impacts at inter-organizational level. Out of these three aspects only the collection of empirical evidences survives in the second and in the third period, and over time most influential sources focus instead on the theoretical foundations of IT value investigation (the already mentioned RBV, and economic productivity function), on methodological guidance for assessments of IT value at the intra-organizational level, and on the provision of integrative frameworks to synthesize an extensive and articulated research area.

Looking at the results of our analysis we posit that the RBV theory constitutes a foundation of the IT value discourse. The IT value phenomenon has been studied mainly at three different levels of analysis: the business process level [18, 23, 31], the firm level [18, 24], and the network/inter-organizational level [18, 32].

One challenge for the IT value discourse is the study of the benefits achieved by ICT adoption at the inter-organizational level. We believe this to be a challenge as the underpinning theoretical approach, the RBV, does not necessarily matches with inter-organizational cooperation scenarios where resources are of value if they are exchanged, and not if they are protected or made scarce. This is of importance in a service centred scenario where different organizations combine their resources, pieces of knowledge, and competences to deliver services of value for customers. In this setting the value of IT could also be explained in terms of shared IT resources rather than in terms of private and protected resources [33]. That would imply a shift in the underpinning theoretical approach to study the value delivered by IT services.

Another aspect we believe to be important for the IT value research is the emergence of the domain of public administration studies in the literature of IT value. Concerning this we also see a potential future challenge as in such domain the RBV theory is again not fit in explaining context of public value [9, 34, 35].

References

1. Chesbrough, H.W.: Toward a new science of services. Harv. Bus. Rev. **83**, 43–44 (2005)
2. Chesbrough, H., Spohrer, J.: A research manifesto for service science. Commun. ACM **49**, 35–40 (2006)
3. Winniford, M., Conger, S., Erickson-Harris, L.: Confusion in the ranks: IT service management practice and terminology. Inf. Syst. Manag. **26**, 153–163 (2009)
4. Galup, S.D., Dattero, R., Quan, J.J., Conger, S.: An overview of IT service management. Commun. ACM **52**, 124 (2009)
5. Nevo, S., Wade, M.R.M.: The formation and value of IT-enabled resources: antecedents and consequences of synergistic relationships. MIS Q. **34**, 163–183 (2010)
6. Grover, V., Kohli, R.: Co-creating IT value: new capabilities and metrics for multifirm environments. MIS Q. **36**, 225–232 (2012)
7. Kohli, R., Grover, V.: Business value of IT: an essay on expanding research directions to keep up with the times. J. Assoc. Inf. Syst. **9**, 23–39 (2008)
8. Braccini, A.M.: Value Generation in Organisations. LAMBERT Academic Publishing, Saarbrücken (2011)
9. Wilkin, C., Campbell, J., Moore, S., Van Grembergen, W.: Co-creating value from IT in a contracted public sector service environment: perspectives on COBIT and Val IT. J. Inf. Syst. **27**, 283–306 (2013)
10. Oh, W., Pinsonneault, A.: On the assessment of the strategic value of information technologies: conceptual and analytical approaches. MIS Q. **31**, 239–265 (2007)
11. Gable, G.G., Sedera, D., Chan, T.: Re-conceptualizing information system success: the IS-impact measurement model. J. Assoc. Inf. Syst. **9**, 377–408 (2008)
12. Irani, Z., Love, P.E.D.: The propagation of technology management taxonomies for evaluating investments in information systems. J. Manag. Inf. Syst. **17**, 161–177 (2000)
13. Thatcher, M.E., Oliver, J.R.: The impact of technology investments on a firm's production efficiency, product quality, and productivity. J. Manag. Inf. Syst. **18**, 17–45 (2001)
14. Agarwal, R., Lucas, H.C.: The information systems identity crisis: focusing on high-visibility and high-impact research. MIS Q. **29**, 381–398 (2005)
15. Kohli, R., Sherer, S.A., Baron, A.: Editorial - IT investment payoff in e-business environments: research issues. Inf. Syst. Front. **5**, 239–247 (2003)
16. Pritchard, A.: Statistical bibliography or bibliometrics? J. Doc. **24**, 348–349 (1969)
17. Culnan, M.J.: The intellectual development of management information systems, 1972–1982: a co-citation analysis. Manag. Sci. **32**, 156–172 (1986)
18. Melville, N., Kraemer, K.L., Gurbaxani, V.: Review - information technology and organizational performance: an integrative model of IT business value. MIS Q. **28**, 283–322 (2004)

19. Chau, P.Y.K., Kuan, K.K.Y., Liang, T.-P.: Research on IT value: what we have done in Asia and Europe. Eur. J. Inf. Syst. **16**, 196–201 (2007)
20. Za, S., Spagnoletti, P.: Knowledge creation processes in information systems and management: lessons from simulation studies. In: Spagnoletti, P. (ed.) Organization Change and Information Systems. LNISO, vol. 2, pp. 191–204. Springer, Heidelberg (2013)
21. Brumana, M., Decastri, M., Scarozza, D., Za, S.: A bibliometric study of the literature on technological innovation: an analysis of 60 international academic journals. In: Baglieri, D., Metallo, C., Rossignoli, C., Iacono, M.P. (eds.) Information Systems, Management, Organization and Control, vol. 6, pp. 141–152. Springer, Cham (2014)
22. Bharadwaj, A.S.: A resource-based perspective on information technology capability and firm performance: an empirical investigation. MIS Q. **24**, 169–196 (2000)
23. Barua, A., Kriebel, C.H., Mukhopadhyay, T.: Information technologies and business value: an analytic and empirical investigation. Inf. Syst. Res. **6**, 3–23 (1995)
24. Brynjolfsson, E., Hitt, L.: Paradox lost? Firm-level evidence on the returns to information systems spending. Manag. Sci. **42**, 541–558 (1996)
25. Barney, J.: Firm resources and sustained competitive advantage. J. Manag. **17**, 99–120 (1991)
26. Wade, M., Hulland, J.: Review: the resource-based view and information systems research: review, extension, and suggestions for future research. MIS Q. **28**, 107–142 (2004)
27. Mata, F.J., Fuerst, W.L., Barney, J.B., Mata, J.: Information technology and sustained competitive advantage: a resource-based analysis. MIS Q. **19**, 487–505 (1995)
28. Ricciardi, F., Za, S.: Smart city research as an interdisciplinary crossroads: a challenge for management and organization studies. In: Mola, L., Pennarola, F., Za, S. (eds.) From Information to Smart Society. LNISO, vol. 5, pp. 163–171. Springer, Cham (2015)
29. Weill, P.: The relationship between investment in information technology and firm performance: a study of the valve manufacturing sector. Inf. Syst. Res. **3**, 307–333 (1992)
30. Barney, J.B.: The resource based view of strategy: origins, implications, and prospects. J. Manag. **17**, 97–211 (1991)
31. vom Brocke, J., Braccini, A.M., Sonnenberg, C., Spagnoletti, P.: Living IT infrastructures— an ontology-based approach to aligning IT infrastructure capacity and business needs. Int. J. Account. Inf. Syst. **15**, 246–274 (2014)
32. Lane, O., Howison, J., Wiggins, A., Rowston, K., Crowston, K., Mukhopadhyay, T., Kekre, S., Kalathur, S.: Business value of information technology: a study of electronic data interchange. MIS Q. **19**, 137 (1995)
33. Amit, R., Zott, C.: Value creation in e-business. Strateg. Manag. J. **22**, 493–520 (2001)
34. Bannister, F.: Dismantling the silos: extracting new value from IT investments in public administration. Inf. Syst. J. **11**, 65–84 (2001)
35. Braccini, A.M., Federici, T.: IT value in public administrations: a model proposal for e-procurement. In: D'Atri, A. Saccà, D. (eds.) Information Systems: People, Organizations, Institutions and Technologies, pp. 121–129. Physica-Verlag, a Springer Company, Heidelberg (2009)

Blockchain Technology as an Enabler of Service Systems: A Structured Literature Review

Stefan Seebacher$^{(\boxtimes)}$ and Ronny Schüritz

Karlsruhe Institute of Technology, Kaiserstr. 89, 76131 Karlsruhe, Germany
{stefan.seebacher,ronny.schueritz}@kit.edu

Abstract. Blockchain technology is expected to revolutionize the way trans-actions are performed, thereby affecting a vast variety of potential areas of application. While expectations are high, real world impact and benefit are still unclear. To be able to assess its impact, the first structured literature review of peer-reviewed articles is conducted. As blockchain technology is centered around a peer-to-peer network, enabling collaboration between different parties, the service system is chosen as unit analysis to examine its potential contribu-tion. We have identified a set of characteristics that enable trust and decen-tralization, facilitating the formation and coordination of a service system.

Keywords: Blockchain technology · Service system · Technology impact

1 Introduction

Blockchain technology is known as the underlying basis of Bitcoin [1]. Apart from its utilization in the Bitcoin network, many researchers and practitioners expect it to generally revolutionize the way we interact and transact over the Internet, resulting in the dawn of a new economy (e.g. [2, 3]). A vast potential for its application is pre-dicted, for example affecting the way governments [4], public notary services [5] or contracts in an online environment [6] work. Expectations towards the potential of this new technology are rising, which can be seen in Gartner's Hype Cycle, where blockchain technology has already reached the peak of inflated expectations [7]. But as the term *inflated expectations* indicates, there is a difference between expectations and experienced real world impact [4]. In that context, Gideon Greenspan, the CEO of Multichain a blockchain provider, is stating that businesses are still "waiting to gain a clearer understanding of where blockchains genuinely add value in enterprise IT" [8]. While, there are several startups, that already offer blockchain solutions to their cus-tomers, no application has yet achieved large scale recognition, as they face compe-tition of existing and well-established systems [9]. Therefore, additional and pervasive use cases are needed to foster the adoption of blockchain technology [2] and to reveal real world benefits for its users [10].

In order to facilitate the identification of practical use cases, it is necessary to be aware of potential impacts, which result from the application of blockchain technology. As it is built upon interaction in networks or systems, we investigate its implications in

© Springer International Publishing AG 2017
S. Za et al. (Eds.): IESS 2017, LNBIP 279, pp. 12–23, 2017.
DOI: 10.1007/978-3-319-56925-3_2

the context of service systems, which themselves are characterized by collaborative processes and, therefore, serve as an excellent unit of analysis [11].

Performing the first structured literature review on blockchain technology, which is entirely based on peer-reviewed literature, we derive a distinct set of characteristics that we illustrate in a concept matrix and interpret. The characteristics are then assessed concerning their contribution to service systems, developing a better understanding of the potential of blockchain technology.

The structure of this paper is as follows: Sect. 2 presents the methodological approach for conducting the literature review. Section 3 lays the theoretical foundations concerning blockchain technology by synthesizing a definition for the concept as well as presenting its inherent characteristics. Subsequently, Sect. 4 gives an overview over the concept of service systems and discusses the implications of applying blockchain technology in service systems. Section 5 closes with a conclusion and gives an overview over the research agenda.

2 Research Methodology

Since blockchain technology is a rather new field of study [12], publications have based their research on available white papers and practitioner-oriented sources, such as related forums (e.g. [2]). Until now, the extent of peer-reviewed publications was very limited and therefore an analysis of peer-reviewed articles has not yet been conducted. With rising academic interest, more and more publications ensuring scientific rigor are surfacing. Therefore, this work intends to focus on peer-reviewed publications as principal source of information. Non-peer-reviewed literature is used to support and underline the derived results.

In order to fully explore the concept of blockchain technology and its underlying characteristics, a structured and systematic literature review is conducted. Google Scholar is used as search engine to retrieve relevant literature.

Table 1. Overview over keywords and hits

Keyword	Number of hits
Blockchain	6.790
Block chain	4.570
Keyword combined with "blockchain"	*Number of hits*
Peer-to-peer database	1.110
Immutable database	213
Consensus database	1.430
Consensus protocol	1.180
Distributed ledger	1.170

As a first step, keywords, covering the field of blockchain technology, have to be identified. The terms "Blockchain" and "Block chain" are used as starting points for a database search, as they are treated synonymously throughout the blockchain community.

Furthermore, they do not describe unrelated concepts or technologies and are therefore suited as initial set of keywords. In order to incorporate additional perspectives on the technology, the list of keywords is gradually and iteratively extended through the analysis of the identified results. The applied set of keywords as well as their hit counts are presented in Table 1. For each of the presented keywords, the first 50 search results are analyzed and examined for relevance. Results that do not fulfil the following filtering requirements are discarded: Publications are written in English and have passed a peer-review process.

As a means to uncover different characteristics of blockchain technology, the resulting 31 peer-reviewed articles are used to develop a concept matrix [13], thereby synthesizing the literature at hand.

3   Review on Blockchain Technology

In this section, we present the results of our literature review on blockchain technology. We start by formulating a definition for the basic concept, which is followed by a presentation of the technology's inherent characteristics.

3.1   The Concept of Blockchain Technology

Although blockchain technology was first introduced in the year 2008 in Nakamoto's whitepaper as the underlying technology of Bitcoin [1], a generally accepted definition of the concept has not been established. Therefore, this section, provides a definition of the concept based on peer-reviewed literature.

While some authors refer to a blockchain as a distributed data structure, database or system [4, 9, 12, 14–17], others call it a decentralized network [18, 19]. Serving as a log or ledger to document all transactions and activities that took place within the construct [12, 14, 15, 19–24], it is a linked sequence of transactions [9, 25], in which time-stamped transactions [26] are broadcasted to and shared with participating entities, located in its belonging peer-to-peer network [12, 16]. Transactions are secured through public-key cryptography and verified by the participants for correctness [9, 12, 17, 23, 26]. Once a transaction is verified by the participatory community, it is added to an unpublished block. Amongst others, a block serves as storage unit for transactions and contains a reference to the settled and verified chain of blocks. Through the use of a consensus mechanism new blocks are added to the blockchain in an append-only manner and then cannot be altered anymore [20, 21, 25, 27].

Based on the presented statements, we synthesize the following definition for a blockchain:

A blockchain is a distributed database, which is shared among and agreed upon a peer-to-peer network. It consists of a linked sequence of blocks, holding timestamped transactions that are secured by public-key cryptography and verified by the network community. Once an element is appended to the blockchain, it can not be altered, turning a blockchain into an immutable record of past activity.

Furthermore, a distinction can be made between public and private blockchains. Public blockchains are not restricted in terms of access rights and allow all participants to append new blocks, whereas private blockchains may be used in a stricter setting in which it is important to limit who enters and contributes to the network [25].

3.2 Characteristics of Blockchain Technology

Although blockchain technology can be regarded as an emerging technology [28] and therefore still has room for improvement in terms of efficiency and technical aspects [12], its underlying characteristics can already be discussed. To assess these characteristics in a structured and systematic manner, the identified peer-reviewed articles and the respectively mentioned attributes are presented in the concept matrix in Table 2.

Our analysis shows that blockchain technology brings to bear a variety of characteristics, which are, in the following, analyzed concerning their interrelations, deriving a set of key characteristics. For example, it is assumed that the characteristics "shared and public" as well as "low friction" lead to increased transparency in a system, since information is made publicly available between participants without being influenced by a third party. An overview over the resulting key characteristics and their underlying elements is presented in Fig. 1 and is further elaborated in the following.

Two principal characteristics are to be identified when looking at blockchain technology, namely its trust evoking and decentralized nature.

Its decentralization facilitates the creation of a private, reliable and versatile environment, which is further described below.

As blockchain technology is based on a peer-to-peer network [9], which combined with the technology's ability to secure interactions between two individuals by using public-key cryptography, and the fact that identities are covered by pseudonyms, a high degree of privacy for its participants is enabled [37].

Reliability within the system is established through use of two factors. On the one hand, information on transactions is shared and stored throughout the network and is therefore treated in a redundant way [25] and on the other hand, since the technology is based on data and code, the introduction of automated measures is facilitated [40], which in turn may reduce individual mistakes as there is little need for manual intervention [34].

By enabling its participants to integrate their own programs, develop and distribute their own code, thereby shaping their own environment, blockchain technology facilitates the creation of an open and versatile system [4]. A popular example for this characteristic is a so-called smart contract, which is a piece of code that serves as programmed contractual agreement between two parties [2].

While some authors explicitly mention blockchain technology's trust enabling notion (e.g. [4, 20, 24, 41]), others describe it in an indirect manner as through the establishment of transparency via a shared and public view on occurring transactions throughout the peer-to-peer network (e.g. [27, 36]), through ensuring the integrity of data in the blockchain (e.g. [23, 42]), or its immutable architecture (e.g. [9, 39, 40]).

Table 2. Concept matrix of the reviewed literature

Author(s)	Trust	Shared and public	Low friction	Peer verification	Cryptography	Immutability	Decentralization	Pseudonymity	Redundancy	Versatility	Automation
Barber et al. [29]	✓		✓	✓		✓	✓			✓	✓
Beck et al. [20]	✓	✓	✓	✓	✓	✓	✓		✓	✓	✓
Böhme et al. [9]		✓				✓	✓	✓		✓	✓
Bonneau et al. [19]	✓			✓			✓	✓			
Cai and Zhu [30]	✓	✓		✓	✓	✓					
Cucurull and Puiggalí [23]	✓	✓		✓	✓	✓	✓				
Delmolino et al. [31]		✓		✓	✓					✓	✓
Eyal et al. [32]				✓			✓				
Garay et al. [15]		✓				✓	✓	✓			
Garman et al. [33]	✓	✓		✓		✓	✓	✓			
Garay et al. [15]		✓				✓	✓	✓			
Gerstl [24]	✓	✓		✓		✓	✓	✓	✓		
Guo and Liang [34]							✓				✓
Heilman et al. [35]							✓	✓			
Herrera-Joancomartí and Pérez-Solà [36]		✓		✓			✓	✓			
Hull et al. [37]	✓	✓			✓		✓	✓	✓	✓	✓
Idelberger et al. [27]		✓		✓		✓	✓			✓	✓
Kosba et al. [18]	✓			✓	✓		✓				
Kraft [21]	✓	✓			✓		✓				
Lewenberg et al. [14]							✓				
McCorry et al. [38]	✓	✓		✓		✓		✓			
McCorry et al. [22]					✓	✓		✓			
Ølnes [4]	✓		✓		✓		✓			✓	
Sharples and Domingue [25]		✓				✓	✓		✓		✓
Sun et al. [39]	✓	✓	✓	✓	✓		✓	✓			✓
Tschorsch and Scheuermann [17]	✓	✓		✓			✓	✓			
Wang et al. [16]	✓	✓		✓			✓				
Weber et al. [40]	✓	✓		✓	✓	✓	✓				✓
Wilson and Ateniese [41]	✓			✓							
Xu [42]	✓	✓		✓	✓	✓	✓			✓	✓
Zhao et al. [12]	✓	✓		✓	✓	✓	✓	✓			✓
Zyskind et al. [26]	✓	✓		✓	✓		✓	✓			

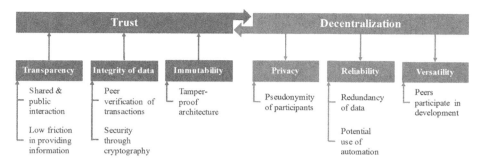

Fig. 1. Characteristics of blockchain technology

Using blockchain technology enables its participants to establish a shared and publicly unfolded relationship. As there is a shared view on all passed and current transactions, participants have full disclosure on activities of the system [33]. New transactions are broadcasted through the entire network [31] and as there is no single intermediary who controls the system, users can interact directly, resulting in a reduction of friction [20, 39].

Trust may also be facilitated through the technology's inherent characteristic of ensuring the integrity of data, which is stored in the database itself, since direct interaction is secured through public-key cryptography and the fact, that through its transparent nature every user is able to verify broadcasted transactions based on pre-defined rules [31].

Another factor that contributes to establishing trust is the immutable design of the database, meaning that once a transaction is added to a block, which in turn is added to the blockchain, this transaction cannot be altered [23]. This process is facilitated by applying a so-called consensus mechanism, which e.g. require the calculation of a proof-of-work. A proof-of-work may be regarded as a computational puzzle, which takes a lot of effort to solve, but whose solution is easily verifiable by others. In case a user finds the solution, it is shared with the remaining participants in the network, who in turn can verify its correctness, thereby reaching a consensus on the solution. One crucial aspect of the proof-of-work is that the puzzle a user is solving, depends on the previously accepted and agreed upon blocks of the blockchain. Since a variety of participants is trying to form and append new blocks to the blockchain, changes in the blockchain would result in varying solutions, revealing misuse or manipulation [33]. Both trust and decentralization are closely connected and interrelated in case of blockchain technology. On the one hand, the mechanisms used to establish trust, such as transparency, integrity and immutability of data, are needed for the creation of a decentralized network, in which reliable and dependable transactions can take place without a trusted third party. On the other hand, decentralization provides the mean for users to get involved in the network, establishing the foundations for consensus mechanism thereby rendering the necessity of a trusted third party obsolete.

4   Impact on Service Systems

This section lays the theoretical foundations for the concept of service systems and elaborates the way they might be influenced utilizing blockchain technology. Therefore, the first subsection deals with presenting the notion of service systems as well as their inherent characteristics. The second part of this section discusses the results of Sect. 3 and applies them to the context of a service system.

4.1   Service Systems

For decades no common basis has been established concerning services [11]. Even at the beginning of the 20th century, in which services have already accounted for a remarkable share of economic performance, service still remained on a residual place of the economic worldview [43].

With the introduction of the Service-Dominant (S-D) logic by Vargo and Lush [44], this worldview changed, shifting the overall perspective on services. They define a service "as the application of specialized competences (knowledge and skills) through deeds, processes, and performances for the benefit of another entity or the entity itself". Thereby they introduce a truly inclusive notion of the term service.

Knowledge-intensive as well as customized services call for a closer integration of customers [45]. In this context, S-D logic is embracing the thought that value creation takes part through the involvement of service providers and beneficiaries in a co-creating manner. To be more precise, they argue that operant resources, which might be machinery or employees, act upon operand resources, increasing their value [44].

Through the integrated reflection of at least two participating parties, S-D logic motivates the creation of the service system abstraction. Instead of calling the involved parties provider or beneficiary, Maglio et al. [11] express the need to regard them in a more generic and conceptual way, as both entities are needed for the process of value creation, therefore calling each of them as well as their combination "service systems".

In general, there exists a variety of definitions for the term *system*, since a system may incorporate different characteristics depending on its underlying purpose [46].

Spohrer et al. define a service system as a "value-coproduction configuration of people, technology, other internal and external service systems, and shared information (such as language, process, metrics, prices, policies, and laws)" [47].

By separating between internal and external service systems, the concept allows for describing interactions between "unique identities", which are "instances of a type or class of service systems". The collaboration between two service systems may be installed in two different ways. The composition may be based upon a hierarchical structure in which only the decision maker needs to addressed or market-based structure, in which an immediate collaboration between participating service systems is established. [11].

Furthermore, a distinction can be made between formal service systems, which are bound to a set of legal and economic rules in order to fulfill pre-defined contracts, obligations and expectations, and informal service systems, in which cultural and behavioral norms play a predominant role [48].

An important aspect for the functioning of a service system is the availability and distribution of information, as collaboration requires a shared basis of information as a mean for coordination. Language, laws and measures are the principal types of shared information in a service system setting. Since service systems are subject to both change within their system as well as their environment, the characteristics of shared information may change over time [47].

Although service systems are to be characterized by their complexity, adaptive nature [47], openness and dynamic composition of operant and operand resources as well as their expandability with other service systems, the integration of two individual service systems does not necessarily end in the development of a greater service system. Generally, in order to facilitate co-creation of value, there has to be at least one operant entity who delivers a proposal to the other operant resource, who settles a mutual agreement concerning the aspired result and who further promotes the realization of value. Given these prerequisites, a service system has the ability to improve both the partner system's as well as its own state [11].

4.2 Understanding the Impact of Blockchain Technology

As we have shown in Sect. 3, both the establishment of an environment for trusted interactions as well as the formation of a decentralized network constitute the core of blockchain technology. Both of which appear to be important aspects for a service system, as it is a configuration of different entities or resources, relying on trusted and shared information (see Sect. 4.1).

Since value is co-created between the involved partners of a service system, trust is an essential aspect that has to be ensured in order to facilitate collaborative processes [40]. Therefore, a typical interaction in a service system involves a governing authority, whose task is to verify and ensure that the involved parties follow shared agreements and laws [47]. The introduction of a blockchain would render the use of a third party unnecessary, as it would evoke a trusted and transparent environment, where all participating entities have full insights into ongoing processes and can rely on the integrity of immutable data. An example for this would be provenance tracking of a good, as every participant of the network would be able to reconstruct the origin of a given good [49].

As blockchain technology facilitates the exchange of information in a way that all involved parties have access to a transparent and shared database, thereby establishing a common basis of information for all users, an important prerequisite for the functioning of a service system is satisfied. Even involved parties who are located at the edges of a service system would gain access to current and direct information, thereby solving problems, which are caused by insufficient or inadequate information. An example for this may be seen in a supply chain setting, where it may seem beneficial for an individual to keep information for themselves, but sharing information would lead to an improvement of the overall system (see Bullwhip effect) [50].

In this context, Weber et al. [40] use a blockchain in two different ways as part of a collaborative setting. They call the first one "choreography monitor", as it serves as a storage unit for joint and individual data. The second one is called "active mediator", where it is used to oversee and initiate the execution of joint processes.

Regarding the latter, the blockchain technology enables a great potential for standardization and automation, as it is a transparent system, which relies on formal code and data. The implementation of both standardization and automation in service systems often releases bound productive capacity, while also reducing transaction costs and having a beneficial impact on coordination [47]. Standardization and automation might also have a favorable impact on minimizing manual mistakes and accelerating interaction processes.

As we have presented, the co-creation of value depends on making a proposal, sharing a common understanding of an interaction's outcome, which is stipulated in an agreement and whose realization is consequently monitored. A blockchain might facilitate all of these activities, as it provides a platform, in which interacting parties can transparently and precisely interact with each other, for example through the definition of coded contracts. The blockchain platform Ethereum may serve as an example for this, as it delivers a toolset for the design of coded contracts [5].

This might also have an impact on the formation of formal service systems, which are determined and regulated by rules [48]. Since interactions in a blockchain are per definition precise and pre-defined, this might facilitate the accelerated creation of such service systems.

If information, time and cost can be managed in a more effective way, a blockchain will even enable the establishment of new service systems that were not possible before. An example for this would be the Bitcoin, which eliminated the need for a trusted third party as information is shared among Bitcoin users, potentially reducing the time needed for the execution of a transaction, drastically reducing transaction cost, and therefore is not limit the minimum practical transaction size [1].

As interacting in such a system depends on strict conditions, leaving no room for vague formulations which might result in conflict, blockchain technology could even help at solving one of the key research objectives in service science, which is to understand how disputes are to be settled effectively [11].

5   Conclusion

To be able to discuss the impact of blockchain technology on service systems, the first structured literature review on the technology, based entirely on peer-reviewed literature, was performed. Thereby, a set of characteristics was revealed, enabling trust and decentralization in a collaborative setting. Blockchain technology creates a trusted environment through its transparent nature, making information publicly available thought out its entire network, while also assuring the integrity and immutability of data. Decentralization allows for the protection of privacy, through pseudonymization, and creates a reliable and versatile setting. The identified characteristics were subsequently assessed in the context of a service system. Blockchain technology addresses many important aspects, which support the functioning of a service system, such as facilitating co-creation of value, ensuring availability of information and offering mechanisms of coordination. Therefore, the technology is expected to have an extensive impact on current and contribute to the formation of new service systems.

As for further research, it would be of interest to explore blockchain technology's contribution within real world use cases. Hence, insights are to be generated by performing a large-scale empirical analysis on existing areas of application.

References

1. Nakamoto, S.: Bitcoin: A Peer-to-Peer Electronic Cash System (2008)
2. Swan, M.: Blockchain: Blueprint for a New Economy. O'Reilly Media, Inc., Sebastopol (2015)
3. Tapscott, D., Tapscott, A.: Blockchain Revolution. Penguine Random House, New York (2016)
4. Ølnes, S.: Beyond bitcoin enabling smart government using blockchain technology. In: Scholl, H.J., et al. (eds.) EGOVIS 2016. LNCS, vol. 9820, pp. 253–264. Springer, Cham (2016). doi:10.1007/978-3-319-44421-5_20
5. Nachiappan, Crosby, M., Pattanayak, P., Verma, S., Kalyanaraman, V.: BlockChain technology: beyond bitcoin. Applied Innovation Review, pp. 6–19 (2016). http://scet.berkeley.edu/wp-content/uploads/AIR-2016-Final-version-Int.pdf. (Accessed 10 Oct 2016)
6. Szabo, N.: Formalizing and securing relationships on public networks. First Monday 2 (1997). http://pear.accc.uic.edu/ojs/index.php/fm/article/view/548/469. (Accessed 10 Oct 2016)
7. Gartner: Hype Cycle for Emerging Technologies. http://www.gartner.com/newsroom/id/3412017. Accessed 21 Nov 2016
8. Greenspan, G.: Avoiding the pointless blockchain project. http://www.multichain.com/blog/2015/11/avoiding-pointless-blockchain-project/. Accessed 13 Nov 2016
9. Böhme, R., Christin, N., Edelman, B., Moore, T.: Bitcoin: economics, technology, and governance. J. Econ. Perspect. **29**, 213–238 (2015)
10. Warburg, B.: How the blockchain will radically transform the economy. https://www.ted.com/talks/bettina_warburg_how_the_blockchain_will_radically_transform_the_economy/transcript?language=en. Accessed 21 Nov 2016
11. Maglio, P.P., Vargo, S.L., Caswell, N., Spohrer, J., Vargo, S.L., Caswell, N., Maglio, P.P.: The service system is the basic abstraction of service science. Inf. Syst. E-bus. Manag. **7**, 395–406 (2009)
12. Zhao, J.L., Fan, S., Yan, J.: Overview of business innovations and research opportunities in blockchain and introduction to the special issue. Financ. Innov. **2**, 28 (2016)
13. Webster, J., Watson, R.T.: Analyzing the past to prepare for the future: writing a literature review. MIS Q. **26**, xiii–xxiii (2002)
14. Lewenberg, Y., Sompolinsky, Y., Zohar, A.: Inclusive block chain protocols. In: Böhme, R., Okamoto, T. (eds.) FC 2015. LNCS, vol. 8975, pp. 528–547. Springer, Heidelberg (2015). doi:10.1007/978-3-662-47854-7_33
15. Garay, J., Kiayias, A., Leonardos, N.: The bitcoin backbone protocol: analysis and applications. In: Oswald, E., Fischlin, M. (eds.) EUROCRYPT 2015. LNCS, vol. 9057, pp. 281–310. Springer, Heidelberg (2015). doi:10.1007/978-3-662-46803-6_10
16. Wang, H., Chen, K., Xu, D.: A maturity model for blockchain adoption. Financ. Innov. **2**, 12 (2016)
17. Tschorsch, F., Scheuermann, B.: Bitcoin and beyond: a technical survey on decentralized digital currencies. IEEE Commun. Surv. Tutorials **18**, 2084–2123 (2016)
18. Kosba, A., Miller, A., Shi, E., Wen, Z., Papamanthou, C.: Hawk: the blockchain model of cryptography and privacy-preserving smart contracts. In: 2016 IEEE Symposium on Security and Privacy, pp. 839–858 (2016)

19. Bonneau, J., Miller, A., Clark, J., Narayanan, A., Kroll, J.A., Felten, E.W.: Research perspectives and challenges for bitcoin and cryptocurrencies. In: 2015 IEEE Symposium on Security and Privacy, pp. 104–121 (2015)
20. Beck, R., Stenum Czepluch, J., Lollike, N., Malone, S.: Blockchain - the gateway to trust-free cryptographic transactions. In: Twenty-Fourth European Conference on Information Systems (ECIS), pp. 1–14 (2016)
21. Kraft, D.: Difficulty control for blockchain-based consensus systems. Peer-to-Peer Netw. Appl. **9**, 397–413 (2016)
22. McCorry, P., Möser, M., Shahandasti, S.F., Hao, F.: Towards bitcoin payment networks. In: Liu, J.K.K., Steinfeld, R. (eds.) ACISP 2016. LNCS, vol. 9722, pp. 57–76. Springer, Cham (2016). doi:10.1007/978-3-319-40253-6_4
23. Cucurull, J., Puiggalí, J.: Distributed immutabilization of secure logs. In: Barthe, G., Markatos, E., Samarati, P. (eds.) STM 2016. LNCS, vol. 9871, pp. 122–137. Springer, Cham (2016). doi:10.1007/978-3-319-46598-2_9
24. Gerstl, David S.: Leveraging bitcoin blockchain technology to modernize security perfection under the uniform commercial code. In: Maglyas, A., Lamprecht, A.-L. (eds.) Software Business. LNBIP, vol. 240, pp. 109–123. Springer, Cham (2016). doi:10.1007/978-3-319-40515-5_8
25. Sharples, M., Domingue, J.: The blockchain and kudos: a distributed system for educational record, reputation and reward. In: Verbert, K., Sharples, M., Klobučar, T. (eds.) Adaptive and Adaptable Learning: 11th European Conference on Technology Enhanced Learning, EC-TEL 2016, pp. 490–496. Springer International Publishing, Cham (2016)
26. Zyskind, G., Nathan, O., Pentland, A.S.: Decentralizing privacy: using blockchain to protect personal data. In: Proceedings of 2015 IEEE Security and Privacy Workshop, SPW 2015, pp. 180–184 (2015)
27. Idelberger, F., Governatori, G., Riveret, R., Sartor, G.: Evaluation of logic-based smart contracts for blockchain systems. In: Alferes, J.J.J., Bertossi, L., Governatori, G., Fodor, P., Roman, D. (eds.) RuleML 2016. LNCS, vol. 9718, pp. 167–183. Springer, Cham (2016). doi:10.1007/978-3-319-42019-6_11
28. Wright, A., De Filippi, P.: Decentralized blockchain technology and the rise of lex cryptographia. SSRN Scholarly Paper ID 2580664. Social Science Research Network, Rochester, NY (2015). https://papers.ssrn.com/sol3/papers.cfm?abstract_id=2580664. (Accessed 10 Oct 2016)
29. Barber, S., Boyen, X., Shi, E., Uzun, E.: Bitter to better - how to make bitcoin a better currency. In: Keromytis, A.D. (ed.) Financial Cryptography and Data Security: 16th International Conference, FC 2012, pp. 399–414. Springer, Heidelberg (2012)
30. Cai, Y., Zhu, D.: Fraud detections for online businesses: a perspective from blockchain technology. Financ. Innov. **2**, 20 (2016)
31. Delmolino, K., Arnett, M., Kosba, A., Miller, A., Shi, E.: Step by step towards creating a safe smart contract: lessons and insights from a cryptocurrency lab. In: Clark, J., Meiklejohn, S., Ryan, P.Y.A., Wallach, D., Brenner, M., Rohloff, K. (eds.) FC 2016. LNCS, vol. 9604, pp. 79–94. Springer, Heidelberg (2016). doi:10.1007/978-3-662-53357-4_6
32. Eyal, I., Gencer, A.E., Sirer, E.G., van Renesse, R.: Bitcoin-NG: a scalable blockchain protocol. In: Proceeding of 13th USENIX Symposium Networked Systems Design and Implementation (NSDI 2016), pp. 45–59 (2016)
33. Garman, C., Green, M., Miers, I.: Decentralized anonymous credentials. In: Network and Distributed System Security (NDSS) Symposium 2014, pp. 23–26 (2014)
34. Guo, Y., Liang, C.: Blockchain application and outlook in the banking industry. Financ. Innov. **2**, 24 (2016)

35. Heilman, E., Baldimtsi, F., Goldberg, S.: Blindly signed contracts: anonymous on-blockchain and off-blockchain bitcoin transactions. In: Clark, J., Meiklejohn, S., Ryan, P.Y.A., Wallach, D., Brenner, M., Rohloff, K. (eds.) FC 2016. LNCS, vol. 9604, pp. 43–60. Springer, Heidelberg (2016). doi:10.1007/978-3-662-53357-4_4

36. Herrera-Joancomartí, J., Pérez-Solà, C.: Privacy in bitcoin transactions: new challenges from blockchain scalability solutions. In: Torra, V., Narukawa, Y., Navarro-Arribas, G., Yañez, C. (eds.) MDAI 2016. LNCS (LNAI), vol. 9880, pp. 26–44. Springer, Cham (2016). doi:10. 1007/978-3-319-45656-0_3

37. Hull, R., Batra, V.S., Chen, Y.-M., Deutsch, A., Heath III, F.F.T., Vianu, V.: Towards a shared ledger business collaboration language based on data-aware processes. In: Sheng, Q. Z., Stroulia, E., Tata, S., Bhiri, S. (eds.) ICSOC 2016. LNCS, vol. 9936, pp. 18–36. Springer, Cham (2016). doi:10.1007/978-3-319-46295-0_2

38. McCorry, P., Shahandashti, S.F., Clarke, D., Hao, F.: Authenticated key exchange over bitcoin. In: Chen, L., Matsuo, S. (eds.) SSR 2015. LNCS, vol. 9497, pp. 3–20. Springer, Cham (2015). doi:10.1007/978-3-319-27152-1_1

39. Sun, J., Yan, J., Zhang, K.Z.K.: Blockchain-based sharing services: what blockchain technology can contribute to smart cities. Financ. Innov. 2, 26 (2016)

40. Weber, I., Xu, X., Riveret, R., Governatori, G., Ponomarev, A., Mendling, J.: Untrusted business process monitoring and execution using blockchain. In: La Rosa, M., Loos, P., Pastor, O. (eds.) BPM 2016. LNCS, vol. 9850, pp. 329–347. Springer, Cham (2016). doi:10. 1007/978-3-319-45348-4_19

41. Wilson, D., Ateniese, G.: From pretty good to great: enhancing PGP using bitcoin and the blockchain. In: Qiu, M., Xu, S., Yung, M., Zhang, H. (eds.) SSR 2015. LNCS, vol. 9497, pp. 368–375. Springer, Cham (2015). doi:10.1007/978-3-319-25645-0_25

42. Xu, J.J.: Are blockchains immune to all malicious attacks? Financ. Innov. 2, 25 (2016)

43. Chesbrough, H., Spohrer, J.: A research manifesto for services science. Commun. ACM 49, 33–40 (2006)

44. Vargo, S.L., Lusch, R.F.: Evolving to a new dominant logic. J. Mark. 68, 1–17 (2004)

45. Sampson, S.E., Froehle, C.M.: Foundations and implications of a proposed unified services theory. Prod. Oper. Manag. 15, 329–343 (2006)

46. Backlund, A.: The definition of system. Kybernetes 29, 444–451 (2000)

47. Spohrer, J., Maglio, P.P., Bailey, J., Gruhl, D.: Steps toward a science of service systems. Comput. (Long. Beach. Calif.) 40, 71–77 (2007)

48. Spohrer, J., Kwan, S.K.: Service Science, Management, Engineering, and Design (SSMED). Int. J. Inf. Syst. Serv. Sect. 1, 1–31 (2009)

49. Greenspan, G.: Four Genuine Blockchain Use Cases. http://www.coindesk.com/four-genuine-blockchain-use-cases/. Accessed 13 Nov 2016

50. Bray, R.L., Mendelson, H.: Information transmission and the bullwhip effect: an empirical investigation. Manag. Sci. 58, 860–875 (2012)

FabLabs as Platforms for Digital Fabrication Services: A Literature Analysis

Marco Savastano[1(✉)], Francesco Bellini[1], Fabrizio D'Ascenzo[1], and Eusebio Scornavacca[2]

[1] Management Department, Sapienza University, Rome, Italy
{marco.savastano,francesco.bellini,
fabrizio.dascenzo}@uniroma1.it
[2] Information Systems and Decision Science Department,
University of Baltimore, Baltimore, USA
escornavacca@ubalt.edu

Abstract. Digital fabrication is contributing to the paradigm shift that is determining a new way to design, produce and consume goods and services. In order to understand the role of FabLabs, networked platforms for the dissemination of digital culture through the sharing of technological tools and knowledge, this paper explores the main research themes and methods associated with this new business model. Through a systematic literature analysis, it provides an assessment of the state of art of the past and current literature about FabLabs from a service perspective. Based on research papers published exclusively on scholarly journals, the study describes the emergence of this research area and characterize its current status. A critical analysis of the existing research as well as some recommendations for future studies in this field are also offered.

Keywords: Literature analysis · FabLab · Digital fabrication · Services · Digital transformation · Digital manufacturing · 3D printing · Knowledge · Innovation

1 Introduction

The digital transformation of manufacturing is contributing to the ongoing paradigm shift of the way to design, produce and consume goods and services. These now happen in cooperative and shared manner allowing the user-consumer to be actively part of the process, or becoming itself the producer. "Digital fabrication" can be defined as the making of physical digitally enhanced artifacts as well as the production of materialized objects by means of digital models. Following this definition, technologies for digital fabrication encompass physical computing technologies as well as digital production machines for printing three-dimensional objects (i.e. 3D printers) and for cutting, shaping or milling material [1]. This revolution, that finds its roots both in the fields of knowledge creation/sharing and technological development has become more than a simple formal exploration in architecture and design, or a set of tools exclusive to advanced industries. Innovative applications, devices and processes are becoming accessible to the masses through physical spaces as FabLabs and makerspaces, while

© Springer International Publishing AG 2017
S. Za et al. (Eds.): IESS 2017, LNBIP 279, pp. 24–37, 2017.
DOI: 10.1007/978-3-319-56925-3_3

being shared all over the world through Internet platforms with an open source philosophy both for hardware and software [2].

It is hard to define digital fabrication and tools associated with this field as "new technology". Indeed the profession of landscape architecture has been articulating its ideas using these tools for decades now. What is new are the means by which individuals gain access to fabrication tools and services as well as the way in which they develop their capabilities to digitally and directly manipulate them.

In a short time, makerspaces and fabrication laboratories (i.e. FabLabs) have emerged, popping up in cities, universities and garages around the world. When designers and entrepreneurs granted direct access to a wide range of digital fabrication tools, the workshop or the lab becomes a proving ground for shaping ideas and developing innovative projects [3].

FabLabs are a worldwide networked platform for the development and diffusion of digital knowledge and skills through services inspired by the values of social innovation and value co-creation. This model can represent an effective research platform to study the phenomenon of distributed digital manufacturing and have an important role in the diffusion of digital knowledge and skills, as well as the emergence of startups and new business models. As defined by Neil Gershenfeld, the MIT professor (head of MIT's Center for Bits and Atoms) who initiated this concept, it is a global network of locally integrated laboratories that allow to transform ideas into innovative projects and products, in a collaborative process of value creation, providing public access to digital fabrication tools and knowledge [4]. According to data released by the Fab Foundation, today there are about 600 FabLabs globally connected (i.e. FabLab Global Network), in over 70 countries worldwide. Although the maker and FabLab movements find their origins in the United States, after a few years also Europe experienced a fast diffusion, initially finding fertile ground especially in countries where the technological culture is deeply rooted, as Germany and the Netherlands. More recently, they have become popular also in many other European countries. For instance, one of the most active and famous FabLab in Europe was founded in Barcelona.

In this context, Italy is the second country in the world by number of FabLabs: recently the Country has experienced a rapid growth, passing from 0 to 100 in the last three years. This extremely positive trend is largely due to cultural and social traits, peculiar of the nation. Indeed, Italy has always been recognized worldwide as a country of inventors, artisans, pioneering small businesses and entrepreneurs able to combine precision engineering, high quality and design [5, 6]. This background has always been behind the "Made in Italy", that the digital revolution can further strengthen [7]. Moreover, since 2013 the most important maker exhibition in Europe, the Maker Faire-European Edition, is organized in Rome. The third edition of 2015, hosted within the Sapienza University Campus, recorded numbers similar to the World Maker Faire New York, with more than 100,000 visitors, 452 projects and inventions, 511 conferences and courses on digital manufacturing applied to different industries.

Considering the important role that FabLabs are assuming for the digital transformation of industry and society, it results crucial to carry out an in-depth analysis of the literature regarding this phenomenon.

The main purpose of this literature analysis is to obtain a "big picture" representing the actual state of art of the scholarly research published on FabLabs. From this, the present paper seeks to understand the extent to which this model can be considered as a platform to provide services to the network of stakeholders surrounding it (i.e. businesses, students, consumers, makers, etc.) as well as the eventual gaps existing in the literature.

Furthermore, this work aims to expand the results of a previous study based on a multiple-case study design [7], where the authors described how a FabLab creates its own local network as a community of users and partners, by promoting certain principles and values of peer production and knowledge sharing, and at the same time provides a boost to local entrepreneurship.

This paper is structured as follows. Next section describes the methodology used to gather and analyze the data for the literature analysis on the FabLabs phenomenon. The results of the analysis are presented and discussed in Sect. 3. The paper concludes with a summary and recommendations for future research on this topic.

2   Research Method

To better understand the state of art of research in this field, we started examining the existing relevant literature exploring the phenomenon of FabLab and its relevant attributes. This is an established tradition in social science research of assessing the research literature for creating a solid base of knowledge about the "state of play" of the research in the field [8]. One of the aim of this paper, in fact, is to discern patterns in the development of the field by analyzing the research published to date.

Initially, it was required to establish the primary parameters to carry out our literature source:

- **Keyword:** *FabLab*, in order to search only for specific research contributions associated with the phenomenon we are investigating;
- **Relevant Research:** we selected only peer-reviewed/scholarly journal articles, more than two-pages in length. Conference proceedings and industry reports/specialized publications on magazines were excluded;
- **Research Fields:** we didn't limit this criterion to specific research disciplines/fields;
- **Time period:** the last 10 years (2006–2016), since the phenomenon originated in the mid 2000s;
- **Language:** English.

The literature search was conducted through comprehensive bibliographic databases in order to cover a broad range of journals. The sources explored are shown in Table 1.

Table 1.   Selected online databases for the literature search.

Online research platform	Databases
EBSCOhost	Academic Search Complete; Business Source Premier; Ebsco Discovery Service (EDS); etc.
ProQuest	ABI/INFORM Global; Emerald Insight, etc.

As abovementioned, a keyword search was conducted on these databases in order to identify relevant papers, following the search strategy outlined above. From this first selection resulted a total of 54 papers.

The next step was to examine the abstract of every paper selected in the previous phase. Thus, we exported the results of our literature search in a citation management software and carefully analyzed this element together with the title, keywords chosen by the author/s and the journal. Any article considered pertinent to the topic was selected for further analysis. The general guideline was as follows:

- the central theme should be FabLab, including its activities and services connected to digital fabrication;
- papers should focus on the theoretical and practical implications of this phenomenon.

Following these criteria, we excluded all the papers not relevant to our purpose. Furthermore, we compared the articles resulting from the different databases and found eventual duplicates to be excluded. In addition, we also evaluated the quality of the research published, by examining the rigorousness of the articles. Thus, we decided to focus only on articles which met or exceeded a quality baseline typical of established journals. For this reason, a number of articles were eliminated from further consideration, resulting in a sample of 24 papers.

3   Key Results and Discussion

In order to guide the analysis of the resulting literature, the following questions were posed:

- What is the main focus of the research (i.e. technology, services, users, etc.)?
- What research design/methods were used?
- What type of data the research relies on (primary or secondary data)?
- What were the key contributions of the study?

To answer these questions, the selected papers were read in their entirety, carefully categorized and subsequently analyzed in detail. At the end of this iterative process we

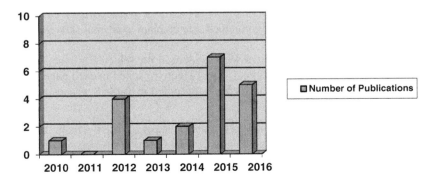

Fig. 1. Number of papers per year

excluded other 4 papers we found not responding to our inclusion criteria (i.e. FabLab was not their main focus), obtaining a final sample of 20 publications (the complete list of references is provided in Table 7, in the A Appendix).

Table 2 lists the articles on the base of their source (scholarly journal) and year of publication. Observing the distribution it is clear that research in this area is relatively recent (it covers the last six years) and is expanding more rapidly in the last two years (12 publications in the period 2015–2016, corresponding to 60% of the total - see also Fig. 1).

Table 2. Selected articles by years and source of publication.

Source	2010	2011	2012	2013	2014	2015	2016	Tot
Foreign Affairs			1					1
IEMJ					1			1
IJCCI						1		1
IJ Design				1				1
JOCM							1	1
JSET	1							1
Library Hi Tech					1	3	1	5
New Library World							1	1
Nexus Network Journal			2					2
Physics Education							1	1
Rapid Prototyping Journal						1		1
Science as Culture							1	1
Symphonya			1			1		2
TFSC						1		1
Total	1	0	4	1	2	7	5	20

The limited number of publications can be explained by the fact that, as for 3D printing and digital manufacturing phenomena, at this time most of the publications are held in the domain of popular publications (blogs, magazines, etc.) and specialized reports [9], which we excluded from our search. In addition, also paper proceedings haven't been included in this study.

The in-depth analysis of the papers allows the authors to categorize them based on four main dimensions (focus; design/method; data; main contribution), in order to answer the guiding questions.

Indeed, a range of research focuses were evident in the selected articles. Based on the objective (variously described as goal, aim, objective) stated in each paper or interpreted through their analysis, the classification presented in Table 3 was developed.

A. Focus/orientation

One of the goals of this literature analysis is to understand what are the main themes and activities connected to the FabLab model and to what extent it has been studied from a service perspective.

By deeply analyzing our final sample, we found out some recurring patterns and focus that we connected to the following four categories (see Table 3).

Table 3. Main themes from the FabLab Literature.

Category	Description	N (%)
Education	Digital Knowledge and technological skills enhancement (through creativity, problem solving, and collaboration); Intellectual Capital; Research; School/University/library programs for Digital fabrication and diffusion-access to Digital culture and tools	9 (45%)
End user/ consumer	3D printing/Digital Fabrication tools usage by FabLab users and consumers: to what extent it enhances their life, experiences, capabilities and/or collaborative participation (technological implications)	4 (20%)
Organization	How digital technologies such as 3D Printing modify organization structures and strategies, and/or originate innovative business models or ecosystems	4 (20%)
Technology	Implementation of technological tools for the digital fabrication and creation of innovative environments through their adoption. Exploration of the FabLab/makerspace general characteristics, enabling tools and knowledge, requirements and possibilities	3 (15%)

The predominant theme in the studies published to date was *Education* (45% of the total), intended as the enhancement of digital knowledge and technological skills through specific workshops, classes and university programs organized or held within FabLabs and Makerspaces. For instance, Beyers (2010) in his paper describes the "FabKids" case study, which relates to the promotion of science and technology literacy among the public, generally, and the youth in particular; and secondly, to increase the enrolment of a cohort of demographically representative youth with talent and potential into science, engineering and technology [10]. More recently, Fonda and Canessa (2016), review their experiences within the Scientific Fabrication Laboratory (SciFabLab) of the ICTP in Trieste (Italy), providing evidences of how it can be an "open place to learn, and get notions of science, beyond the traditional classrooms and without limits" [11]. Starting from the activities that are of interest to scientists as well as the formation of new communities interested in science and development around FabLabs, the paper reports insights for an effective development of a Fabrication Laboratory inserted in a scientific framework, describing the extent to which it can open new dimensions to science and education, inspire curiosity and offer powerful new ways to facilitate the development of innovative impactful ideas.

Following with the resulting themes from the literature analysis, 20% of the publications were focused on the technological implications of digital fabrication on users and consumers' lives, experiences and capabilities (see "End User/Consumer" category in Table 3). On this topic Bosqué (2015) carried out a large-scale field survey on personal digital fabrication as practiced in more than 30 FabLabs, hackerspaces and makerspaces from all over the world. A general picture of the limits and promises of the adoption of entry-level 3D printers by FabLabs and their user-friendliness was obtained [12]. Moreover, Katterfeldt et al. (2015) by building on principles rooted in constructionist learning tradition, explore the growing demand and need of learning environments for digital fabrication. The latter, considering mainly programmable construction kits, contribute not only to the acquisition of skills but also to a deep and sustainable learning process (i.e.

"*Bildung*", german word that can be translated as ''learning-to-be'' instead of the traditional ''learning about'') [1].

Concerning the third category, 20% of the publications were found to be focused on how digital fabrication technologies such as 3D printing modify organization structures and strategies or create innovative business models. Cautela et al. (2014) through a qualitative analysis of three cases demonstrate that the value of 3-D printing technology depends on new business models based on the ability to structure and integrate creative inputs, crowd-sourcing processes, and market distribution networks. The authors make a difference between the use of 3-D printing technology in different company types: companies already specialized in prototyping services employ this technology as an additional service to client firms; FabLabs use it both as advanced technology means to keep offering prototyping services to manufacturing companies and to create new business services for digital plat-form consumers, where the final consumers and/or designers can create their own concepts and designs with the intention of using and/or selling them [13].

The remaining 15% of the analyzed publications is represented by the last category: *Technology*. This category is about the exploration of FabLab/makerspace broad character-istics, studied as innovative environments enabled by digital fabrication tools and knowl-edge. Neil Gershenfeld (2012) in his paper describes how digital fabrication allows individ-uals to design and produce tangible objects on demand, wherever and whenever they need them. "Widespread access to these technologies will challenge traditional models of busi-ness, aid, and education". This "new industrial revolution" started in 2001, with the opening of the MIT's Center for Bits and Atoms and the beginning of the class called "How to Make (almost) Anything". The Center was developed to study the boundary between computer and physical science. Due to the enthusiasm of the students for digital fabrication, the first fabri-cation laboratory (i.e. FabLab) was then set up in 2003 by adopting digital manufacturing tools connected by custom software [4].

Moorefield-Lang (2015) in their multiple–case study research describes the implemen-tation of makerspaces in various library settings, exploring how through the adoption of 3D printing technologies the resulting innovative environments could expand traditional library services for engaging curiosity, creativity, and collaboration in enhanced learning spaces [14].

As common evidence across the various categories described emerges that in many papers from our sample (30% of the total) the main topic was the implementation of digital fabrication (i.e. through the adoption of 3D printers and scanners) and makerspaces within library settings. The strong relationship between these two fields could be explained by the common mission of knowledge democratization and sharing of spaces and tools (i.e. digital technologies, books, etc.) that inherently characterize the two environments. This datum is also emphasized by the relative high frequency shown by the publications on journals speci-alized in the field of library science (i.e. *Library Hi Tech* and *New Library World*). Although this interest for the emergence of makerspaces in the library community continues to grow as more libraries of all kinds continue to create spaces that foster making and active learning [15], being a relatively new phenomenon research on digital fabrication and makerspaces in libraries is still modest [9, 16].

Moreover, it is important to point out the evident lack of a specific focus on services within the analyzed research area.

B. Research Design/Method

To identify the research designs or methods used in the literature about FabLabs, all articles were classified according to the approach stated by the authors or resulting from the analysis. Table 4 presents the distribution found in the sample.

Table 4. Research methods used in the sample.

Design/Method	N	Percentage
Multiple-case study	7	35%
Case study	6	30%
Narrative study	3	15%
Experiment	2	10%
Mixed method	2	10%

Due to the limited literature existing on these new service platforms based on the sharing of digital fabrication tools and skills, the current phenomenon of interest is an emerging topic requiring exploratory research. Indeed, empirical *case study* research resulted the most common approach used by the authors (65% overall, by adding the two categories of multiple and single case studies). Usually focused on specific applications and spaces developed within universities, libraries or research centers, case studies allowed investigators to retain the holistic and meaningful characteristics of real-life events. The distinctive need for case studies arise out of the desire to understand complex social phenomena [17]. Building theories from case studies is a research strategy that involves using one or more cases to build theoretical constructs, propositions and/or midrange theories from case-based, empirical evidence [18]. They are rich, empirical descriptions of particular instances of a phenomenon that are typically based on a variety of data sources [19].

In our sample, cases are mainly used as the basis from which a theory can be developed inductively. The theory emerges from a practical case and is developed by recognizing patterns of relationships in constructs and cases. The use of an inductive theory building approach from cases is relevant especially in the first stage of an analysis, because it can produce new theories that are accurate, interesting and testable. In addition, as Eisenhardt and Graebner (2007) highlight, publications using multiple cases (which represent the 35% of our sample) can delineate constructs and relationships more precisely because it is easier to determine accurate definitions and appropriate levels of construct abstraction [20]. Therefore, theory building from multiple cases typically yields more robust, generalizable, and testable theories than single-case research [13, 17, 21].

A significantly lower number of publications within our sample present a different research design: 3 articles (15%) have been classified as *narrative studies*, since they are descriptive studies based mainly on intuition-based reasoning and qualitative data collected through direct experience, observation or different sources to create awareness of benefits, challenges and strategies for developing and managing makerspaces and FabLabs; 2 articles (10%) for both the categories of *experimental* and *mixed method* researches, based respectively on field experiments carried out with workshop participants and FabLab users, as well as studies based on a combination of different types of data and sources (e.g. qualitative from case studies and quantitative from surveys).

It is also remarkable that among the analyzed papers no one could be classified as *literature review*, emphasizing that the sample is mainly characterized by empirical research, which rarely applies this research approach (this aspect will be further explained in the next section).

C. Data
In addition to the analysis of the research methodology, to investigate whether the academic publications regarding the FabLab phenomenon are dominated by conceptual analysis and intuition-based reasoning rather than empirical investigations, a further categorization was needed to classify the selected articles [8, 22]. This study considers "*empirical research*" all research originating in or based on observation or direct experience, independently of whether the researcher gathered data through primary or secondary data collection. For instance, case studies based on information collected from secondary sources (such as websites, databases and practitioner reports) were considered empirical. Publications characterized by intuition-based reasoning and conceptual analysis were classified as "*conceptual research*". Following this classification, 16 articles (80%) were found to be empirical research and 4 (20%) conceptual.

Some interesting insights result from the analysis of the relations among research approach and data collection type (see Table 5 below).

Table 5. Research approach in relationship with data collection.

	Primary - N (%)	Secondary - N (%)	Mixed - N (%)	Total - N (%)
Empirical	14 (70%)	0	2 (10%)	16 (80%)
Conceptual	0	3 (15%)	1 (5%)	4 (20%)
Total	14 (70%)	3 (15%)	3 (15%)	20 (100%)

In most cases (3 studies over 4, 75% of the category) conceptual research resulted based on secondary data collection, associated with qualitative data. On the other hand, it is important to note that the majority of empirical research publications (14 articles, corresponding to 70% of the sample) was found to be characterized by data gathered from primary sources. Within these papers, 11 were based on qualitative data. Moreover, among the 20 publications 3 (15%) showed the combination of qualitative data gathered both from primary and secondary sources. Figure 2 offers a representation of the abovementioned analysis.

Finally, it is necessary to notice that in 10 over 13 case study researches (including both single and multiple case-study publications) the authors gathered data through primary data collection.

D. Key Contribution
An analysis of each article's main contributions was also conducted. Not all the authors clearly highlighted the main contributions of their articles. Consequently, in many cases due to the lack of information given by the authors, this classification required a reviewer judgment. Findings are presented in Table 6.

The fact that "Multiple-Case Study" and "Case Study" were the most common research methods undoubtedly resulted in *insights* emerging as the most common type of contribution of the articles reviewed (50% of the sample). To a lesser extent and precisely in the 30% (6) of the publications, a *research agenda* was the main contribution/orientation of the work.

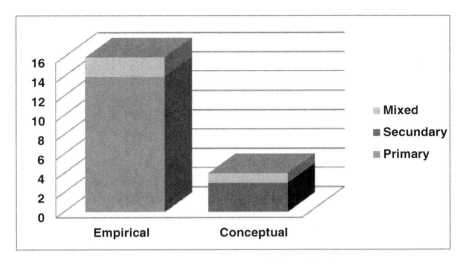

Fig. 2. Research approach and data collection

The 15% (3) of the papers offered a *framework* as their main contribution, and they result all case study researches. Finally, only 1 paper was found to contribute with the creation and testing of a *model*.

Table 6. Primary contribution of papers.

Contribution	N	Percentage
Insights	10	50%
Research agenda	6	30%
Framework	3	15%
Model	1	5%

4   Conclusion and Future Directions

This article has provided an overview of the existing literature on the FabLab phenomenon, adopting a service perspective.

The development of digital manufacturing in modern industrial and manufacturing economies is promoting the rise of necessary digital fabrication capabilities and culture and the need for their widespread diffusion. In particular, it seems that a new competitive arena is emerging in services connected with design and creativity, based on different forms of networked structures that rely on the sharing of digital technologies, knowledge and skills. The proliferation of instruments and software open to design, the spread of cultures linked to "making" and advanced self-production, together with the potential of innovative web applications and social networks, are the key factors and the background for the development of these new forms of creativity and manufacturing [23].

In order to better understand the state of research in this domain, a systematic and comprehensive literature analysis consisting of only peer-reviewed articles was carried out. Our findings support future theory development in this research area.

In addition to provide a summary of the insights that have been gained to date in the studies about the development and management of FabLabs and makerspaces in several different environments, a number of research gaps were identified, suggesting opportunities for profitable future research. Indeed, by classifying the existing publications on this topic of interest, this study has highlighted some recurring patterns as well as important gaps to be filled.

The findings presented here shows that:

- The majority of research in this field is connected to the theme of "*education*", meaning the development of digital fabrication capabilities and the creation of new projects and services through dedicated university programs or workshops organized inside FabLabs and makerspaces. A moderate connection with the library settings was also noted;
- More than 60% of the publications were classified as *case study research*, most of them relying on multiple case-studies;
- Almost the entire sample (80%) was dominated by *empirical research*, based on observation or direct experience in the context under investigation;
- A very high number of articles (corresponding to the 70% of the sample) resulted to be based on primary data collection, mostly of qualitative data;
- Concerning the primary contribution of papers, the most common types resulting from the review were "*insights*" and "*research agenda*".

Although this is a recent research field, from the above mentioned results is evident the lack of studies published on scholarly journals providing conceptual frameworks and models that would allow to create a solid theoretical foundation to this body of literature. Indeed, it is unlikely to become fully recognized as a research area on its own until specific theories will be developed, in addition to key theories applied from reference disciplines. At the same time, an important gap regarding the service perspective on this phenomenon was found in the analyzed literature and needs to be addressed by future research.

Interested researchers should be able to identify additional research opportunities through the information provided here. In particular, the connections between the innovative concept and services of FabLab with the business and entrepreneurial settings need further investigation.

This article also faces some limitations. While the present literature analysis was extensive and spanned a number of different research domains, it is possible that some articles were missed due to the strict criteria adopted for the sample selection. Indeed, the study took into consideration only scholarly peer-reviewed publications; future studies could extend the analysis to conference papers and specialized publications in this field.

Furthermore, the research analysis was focused on four major research questions. Future studies should examine additional and more specific issues to extend the findings of the present work.

Lastly, starting from the above mentioned evidences and literature gaps, researches can focus their efforts more carefully to enhance the body of research in this area and make it becoming more mature by developing its own research tradition.

A Appendix

See Table 7

Table 7. List of the 20 publications analyzed.

Author/s	Title	Year	Journal
Beyers, R.N.	Nurturing Creativity and Innovation Through FabKids: A Case Study	2010	Journal of Science Education & Technology
Birtchnell et al.	3D printing and the third mission: The university in the materialization of intellectual capital	2015	Technological Forecasting and Social Change
Bosqué, C.	What are you printing? Ambivalent emancipation by 3D printing	2015	Rapid Prototyping Journal
Cautela et al.	The emergence of new networked business models from technology innovation: an analysis of 3-D printing design enterprises	2014	International Entrepreneurship and Management Journal
Craddock, I.L.	Makers on the move: a mobile makerspace at a comprehensive public high school	2015	Library Hi Tech
De Boer, J.	The business case of FryskLab, Europe's first mobile library FabLab	2015	Library Hi Tech
De Couvreur et al.	The Role of Subjective Well-Being in Co-Designing Open-Design Assistive Devices	2013	International Journal of Design
Diez, T.	Personal Fabrication: Fab Labs as Platforms for Citizen-Based Innovation, from Microcontrollers to Cities	2012	Nexus Network Journal
Fonda, C., Canessa, E.	Making ideas at scientific fabrication laboratories	2016	Physics Education
Gershenfeld, N.	How to Make Almost Anything: The Digital Fabrication Revolution	2012	Foreign Affairs
Katterfeldt et al.	Designing digital fabrication learning environments for Bildung: Implications from ten years of physical computing workshops	2015	International Journal of Child-Computer Interaction
Moorefield-Lang, H.M.	Makers in the library: case studies of 3D printers and maker spaces in library settings	2014	Library Hi Tech
Moorefield-Lang, H.M.	When makerspaces go mobile: case studies of transportable maker locations	2015	Library Hi Tech
Nascimento et al.	Sustainable Technologies and Transdisciplinary Futures: From Collaborative Design to Digital Fabrication	2016	Science as Culture
Okpala, H.N.	Making a makerspace case for academic libraries in Nigeria	2016	New Library World
Paio et al.	Prototyping Vitruvius, New Challenges: Digital Education, Research and Practice	2012	Nexus Network Journal
Purpur et al.	Refocusing mobile makerspace outreach efforts internally as professional development	2016	Library Hi Tech
Rieple et al.	Business Network Dynamics and Diffusion of Innovation	2012	Symphonya
Rieple, A., Pisano, P.	Business Models in a New Digital Culture: The Open Long Tail Model	2015	Symphonya
Seravalli, A., Simeone, L.	Performing Hackathons As A Way Of Positioning Boundary Organizations	2016	Journal of Organizational Change Management

References

1. Katterfeldt, E.S., Dittert, N., Schelhowe, H.: Designing digital fabrication learning environments for bildung: implications from ten years of physical computing workshops. Int. J. Child-Comput. Interact. **5**, 3–10 (2015). Digital Fabrication in Education
2. Diez, T.: Personal fabrication: fab labs as platforms for citizen-based innovation, from microcontrollers to cities. Nexus Netw. J. **14**(3), 457–468 (2012)
3. Hare, J.: Getting digitally dirty. Landscapes/Paysages **18**(2), 48–49 (2016)
4. Gershenfeld, N.: How to make almost anything: the digital fabrication revolution. Foreign Aff. **91**(6), 58 (2012)
5. Banzi, M., De Benedetti, C., Luna, R.: Il Primo Rapporto sull'Impatto delle Tecnologie Digitali nel Sistema Manifatturiero Italiano. Make in Italy CDB (2015)
6. Menichinelli, M., Ranellucci, A.: Censimento dei Laboratori di Fabbricazione Digitale in Italia. Make in Italy CDB (2014)
7. Savastano, M., Bellini, F., D'Ascenzo, F.: FabLab and digital manufacturing: innovative tools for the social innovation and value co-creation. In: Baglieri, D., Metallo, C., Rossignoli, C., Iacono, M.P. (eds.) The Social Relevance of the Organisation of Information Systems and ICT. Lecture Notes in Information Systems and Organisation. Springer, Cham (2016)
8. Scornavacca, E., Barnes, S.J., Huff, S.L.: Mobile business research published in 2000–2004: emergence, current status, and future opportunities. Commun. Assoc. Inf. Syst. **17**(1), 28 (2006)
9. Moorefield-Lang, H.M.: Makers in the library: case studies of 3D printers and maker spaces in library settings. Libr. Hi Tech. **32**(4), 583 (2014)
10. Beyers, R.N.: Nurturing creativity and innovation through fabkids: a case study. J. Sci. Educ. Technol. **19**(5), 447–455 (2010)
11. Fonda, C., Canessa, E.: Making ideas at scientific fabrication laboratories. Phys. Educ. **51**(6), 1 (2016)
12. Bosqué, C.: What are you printing? Ambivalent emancipation by 3D printing. Rapid Prototyping J. **21**(5), 572–581 (2015)
13. Cautela, C., Pisano, P., Pironti, M.: The emergence of new networked business models from technology innovation: an analysis of 3D printing design enterprises. Int. Entrepreneurship Manag. J. **10**(3), 487–501 (2014)
14. Moorefield-Lang, H.M.: When makerspaces go mobile: case studies of transportable maker locations. Libr. Hi Tech. **33**(4), 462–471 (2015)
15. Purpur, E., Radniecki, T., Colegrove, P.T., Klenke, C.: Refocusing mobile makerspace outreach efforts internally as professional development. Libr. Hi Tech. **34**(1), 130–142 (2016)
16. Craddock, I.L.: Makers on the move: a mobile makerspace at a comprehensive public high school. Libr. Hi Tech. **33**(4), 497–504 (2015)
17. Yin, R.K.: Case Study Research: Design and Methods. SAGE Publications, Thousand Oaks (2003)
18. Eisenhardt, K.M.: Building theories from case study research. Acad. Manag. Rev. **14**(4), 532–550 (1989). SRC
19. Yin, R.K.: Case Study Research: Design and Methods, vol. 5, 2nd edn. Applied Social Research Methods Series. SAGE, Los Angeles (1994)
20. Eisenhardt, K.M., Graebner, M.E.: Theory building from cases: opportunity and challenges. Acad. Manag. J. **50**(1), 25–32 (2007)

21. Rieple, A., Pisano, P.: Business models in a new digital culture: the open long tail model. Symphonya **2**, 75–88 (2015)
22. Hirschheim, R.A.: Information systems epistemology: an historical perspective. Comput. Inf. Sci., 13–35 (1985)
23. Rieple, A., Pironti, M., Pisano, P.: Business network dynamics and diffusion of innovation. Symphonya **2**, 13–25 (2012)

IT-Support in Workplace Health Promotion: Mobile Apps on the Rise

Fujan Nuryan Dehkordi[1]([✉]), Rüdiger Breitschwerdt[2], and Michael Fellmann[1]

[1] Institute of Computer Science, University of Rostock,
Albert-Einstein-Straße 22, 18059 Rostock, Germany
{fujan.dehkordi,michael.fellmann}@uni-rostock.de
[2] Institute for eHealth and Management in Health Care,
Flensburg University of Applied Sciences, Kanzleistraße 91-93,
24943 Flensburg, Germany
Breitschwerdt@hs-flensburg.de

Abstract. The purpose of this paper is to analyze the landscape of workplace health promotion applications. We analyze the contributions that consider this type of apps and their use. Therefore, the published research is summarized by a literature review presenting an overview. The findings are categorized in various clusters. As a result, it was possible to see major research streams described in literature. Furthermore, the results show a need for future research on app development for workplace health promotion.

Keywords: Workplace health promotion · Literature review · Application · Mobile app · mHealth

1 Introduction

Service science is supposed to be practical or theoretical investigation allowing for improved effectiveness and/or costs of services where we have a lack of knowledge to build a sustaining solution [1]. This has particularly been proven in healthcare regarding matters of (re-) organization or digitalization [2, 3] often also called eHealth or more recently mHealth. In this context, a known trend is to use apps on mobile devices [4]. Those are supposed to assist users with improving their health in some way. Correspondingly, the variety of applications is large and has meanwhile been nurtured by health insurers as novel means of health care, such as digital health coaches for individual diet and exercise program services, e.g. in Germany by the largest health insurance provider, AOK [5]. One of the more successful applications of that kind represents the app of Weight Watchers with some 10,000,000 downloads from Google's Play Store [6].

While this trend has so far rather been consumer-driven, enterprises and their employees in all industries have also an interest in enhanced health of individuals for productivity reasons: recently, so-called corporate or workplace health promotion initiatives have been rising. The German Ministry for Health funds a program called 'Unternehmen unternehmen Gesundheit' (translation approximated by 'Enterprise of Health') to support health promotion particularly in small and medium sized enterprises.

© Springer International Publishing AG 2017
S. Za et al. (Eds.): IESS 2017, LNBIP 279, pp. 38–50, 2017.
DOI: 10.1007/978-3-319-56925-3_4

According to § 3 No. 34 German Income Tax Law EStG, companies going for certified external or internal corporate health promotion activities have been motivated to invest 500 € annually per employee on a tax-free basis since 2008 [7].

Corporate health promotion has been boosted by its recognition as a strategic element (e.g. as a potential key differentiator between employers) of corporate endeavours facing

- both difficult recruitment and retention of qualified labor in a competitive environment,
- increasing shortage of workers in Western industrialized countries in times of demographic change (an ageing society with fewer young or fit for work people) as well as
- rising expenses because of sick employees.

The latter was confirmed by 65% of 436 surveyed companies in Austria [8]. Booz & Company estimate that employees on sick-leave cost German companies some 130 bn. Euro per year whereas resulting economic annual loss totals 225 bn. Euro – an equivalent of 10% of Germany's GDP [9]. According to a 2015 report from the German 'Initiative Gesundheit und Arbeit' (Initiative for Health and Work, in short *iga*) evaluating 2,400 studies, sick days can be reduced up to 25% by health promotion [10] – in terms of Booz & Company a benefit of appr. 32 bn. Euro only for Germany. Reduced absenteeism is one alleged outcome of some dedicated health apps. Mobile and smart healthcare applications provide patients with (expected) benefits in terms of better quality of life [11]. So, combining both trends, apps and health promotion, might be of interest, here, since health promotion has been evaluated, accepted and used by employees at higher rates when web- or IT-based [12]. Advantages could on one hand be that companies provide more individual and flexible programs to motivate a healthy lifestyle via work smartphones or tablet-PCs. Also, a more comprehensive view of employees' health status (e.g. important in stressful periods) can be reported automatically. Additionally, [13] quoted different studies for 300–500% Return on Investment for "electronic health" promotion. On the other hand, employees can decide freely when and where to use the advice, even at home or in their leisure time.

Considering our goal and for better preparation, we went through exploratory research for an initial practice-oriented overview and could find such international service offerings of commercial apps for corporate use such as Moove App [14] or go4health.com. From the authors' experience, their implementation in a medium-sized enterprise can be difficult or might eventually be abandoned. There have also been a few apps targeting groups of employees of the same companies as users, e.g. 'HealthyTeam – A Team Challenge'. Moreover, several health insurances offer corresponding prevention apps to individuals, e.g. regarding loss of weight during work like AOK's 'Abnehmen mit Genuss im Job'.

For successful implementation of such an initiative, perspectives and influencing factors for users as well as requirements of employees, employers and – on national bases – at least law need to be considered. In this review, the target groups are employers and employees collaborating within their working environments. To check practicability of workplace health promotion by means of applications, our main goal is to survey acceptance and requirements by looking for corresponding demands and evaluations in scientific contributions by means of a systematic literature review (Sect. 3). This shall result in a comprehensive overview of research (Sect. 4) to be eventually discussed (Sect. 5). This chapter also provides a practice-oriented market overview. Subsequently, Sect. 2 presents related research, first.

2   Related Work

Most reviews related to this topic highlight the increasing costs of the health system in most countries as a general point of view [10]. Even though earlier research targeted similar aspects [13, 15, 16] like our approach, both technology and behavior with health and/or mobile applications presumably changed. The systematic review at hand shares the motivation but adopts to nowadays' situation and also checks if mobile applications have gained acceptance amongst their target groups [17]. One of the clear differences represents the definition of targeted user groups: Often regular consumers or lay people are analyzed [13, 15]. Other meta-analyses emphasize the studies on work health promotion in general and analyze the benefits [13, 18, 19]. In fact, there are systematic reviews of health promotion using digital media and similar methodologies. They lead to the discovery of interesting details about the need of digital workplace prevention, such as the Healthy Employee Lifestyle Program (HELP) [20, 21]. However, there are only very few reviews available addressing the combination of the aforementioned topics of interest [22, 23] inspiring our work at the same time.

3   Method

Hempe et al. propose initial literature reviews to achieve a 'vision' when designing new services in healthcare [2] thus meeting the aforementioned strategic demands in corporate healthcare. At their beginning, defined databases have to be explored: SpringerLink (including European Journal of Information Systems, Journal of Information Technology), Pubmed, IEEE Xplore, ISI Web of Science, Medline, Ebscohost (incl. Information Systems Journal, Information Systems Research, Journal of the Association for Information Systems, Journal of Information Technology, Journal of Management Information Systems, MIS Quarterly), ACM Digital Library (Association for Computing Machinery), Journal of the Association for Information Systems (AIS) and ScienceDirect (Journal of Strategic Information Systems) thus ensuring also coverage of the AIS *senior scholar's journal basket*. For a structured review, we started to collect keywords to be looked for. For instance, more than one definition describing the topic of workplace health promotion exists. Therefore, we searched for equivalent and metonym terms.

Table 1 provides an overview of all the terms observed in literature. We searched the databases with all those terms to figure out whether there is a major keyword for this topic. Whenever a new term appeared, the databases were reviewed once more with this keyword following an iterative approach. Additionally, it is necessary to consider the relevant literature by combining the aforementioned keywords with IT terms such as application, app, or smartphone. The result of this merging was a list of keywords. We added samples of the combination of search terms to Table 1 as well.

Also, we used the instructions of Webster and Watson [24] for facilitated structuring of the review and summarized various papers in form of using the filters, e.g. by years. The existing literature has been restricted to the last 10 years (2006–2016). In Fig. 1, we describe the complete process of data collection following Mohan and Ahlemann [25].

Table 1. Health promotion terms and sample search combinations.

Health promotion terms specified	Sample combinations of search terms
Workplace health promotion	Workplace health promotion AND smartphone;…
Company health promotion	Company health promotion AND application; …
Occupational health promotion	Occupational health promotion AND app; …
Corporate health promotion	…

1. Database selection
2. Search execution
3. 1st filtration
4. Refined Search Execution
5. 2nd filtration
6. References review
7. Definition of clusters and categories
8. Asorting & evaluating papers accordingly

Paper relevance

Fig. 1. Process of data collection

In a first step, we started with database selection and detected suitable ones as per above. In step 2 we used a few defined keywords and started to search for them. This resulted in some promising contribution titles like "Using smartphone-based application (apps) in workplace health promotion: The opinion of German and Austrian leaders" [17], or research on workplace health promotion in the Nordic countries [19]. In the third step, we set the filters to limit the literature (the number of papers were still too high). Therefore, subsequently in step 4 we searched for the keywords and combined terms. After confining the results, we re-applied a filter in step 5 and in step 6 started to read titles and abstracts. This step helped us to find the common denominator of the relevant papers. Last but not least, we defined the variables and evaluate the findings considering these clusters in step 7. In step 8 we finally recognized new keywords connected to the topic so that we were able to iterate the whole process.

Following Orwat et al. [26] we examined our findings for comparison reasons and created clusters and within them categories to identify similarities. In Table 2 there is an overview of the categorization applied starting with the cluster *system type*. That defines the kind of analyzed information system in three subsets such as an app for a mobile device, a web- or browser-based application or a 'usual' computer application (standalone or client-server). A combination of the subsets represents a possible outcome, too. Moreover, we found helpful to know about the *life-cycle stage* of the system. That shows whether it is in an initial Requirements Engineering stage (where specifications are defined or a blueprint exists derived from various sources of information) or if it is in a Test phase after the prototyping or in development and is studied for its performance and quality. Eventually, rather an Evaluation stage applies where the system at hand is rated by users for fulfilment of its purposes like reduction of health-related risks.

Table 2. Clusters and categories.

Cluster name	Cluster categories
System type	Mobile app
	Application
	Web-based
Life-cycle stage	Requirements Engineering
	Test
	Evaluation
Functionality	Alert
	Support
	Information
Target users	Employers
	Employees

The authors aligned those categories from a software engineering point of view: we therefore used Boehm's iterative *spiral model* particularly focusing analysis of *requirements*, *evaluation*, and, as a major part of realization, *testing* [27]. Planning as a project management activity was skipped since too general and not in our core interest. Our focus categories can also be found in Royce's *waterfall model* (*evaluation* there called revalidation) [28] or in the *software life-cycle* of Pomberger and Blaschek [29].

Furthermore, we decided to consider the described functionalities of the system. As in Orwat et al. [26], we summarize the goal of the system as the most significant functionalities such as information, support or alert (so, an application might inform, guide through a workflow or generate warnings). In detail, the purposes of an application are defined and separated into:

- Information: The application aims to deliver information based on the studies available (like facts, diagrams, videos, animations etc.) for healthcare control or management of the user.
- Support: This aims at providing support by further processing (e.g. to medical practitioners for reporting) and/or evaluating user's data to reach some health promotion goals regarding diagnosis, treatment or monitoring.
- Alert: Warnings, advice or alarms for user assistance proactively make the user aware of activities performed and give a functionality for improving health outcomes based on application's user data analyses, E.g. the application monitors sleeping time and alerts a user if not slept enough and giving advice about recommended duration of sleep.

The last criterion is then to investigate and analyze the target audience of the information system thus also facilitating requirements identification in later stages of research. The two suitable groups are on the one hand, of course, employees but on the other hand also employers. They represent the shareholders of the company or the management level. The employers "[…] have found that providing incentives […] is a cost-effective way to improve employee health status" [30] – therefore, they are mostly willing and ready to launch health initiatives because they recognize mainly positive impacts on their businesses. After developing the research funnel to detect the relevant literature, we discovered a few contributions matching our topic. Therefore, the research was continued by explorative research in app stores, at corresponding organizations like health insurances and associations as well as in different kinds of practitioner journals such as the German magazine of corporate health management ('Magazin für Betrie-bliches Gesundheitsmanagement').

4   Results

In the first step, we figure out that there is no clear definition respectively term for this topic yet – Table 3 provides an overview of all the searched keywords and number of papers searched in the database. We looked for 30 different keywords combinations. They are explained in Sect. 3. Subsequently, the major keywords are abbreviated as:

- CHP = Corporate health promotion
- CHM = Corporate health management
- CO = Company
- COHP = Company health promotion
- OHP = Occupational health promotion
- POH = Promotion of Health
- WHM = Workplace health management
- WHP = Workplace health promotion
- A = App
- AP = Application
- S = Smartphone

To be able to consider corresponding papers and identify trends, the data were filtered by the research discipline 'Health Informatics'. Again, we tried to reduce the high number of the papers and to find the relevant papers containing both informatics and healthcare aspects. There are several papers showing the huge importance of health promotion independent from the terms app, application or smartphone. The importance of this topic is also documented in a contribution reviewing 33 studies and their outcomes as well as representing the impact of worksite wellness programs [15] and generally evaluating health promotion programs at workplaces [31, 32]. After combining the terms with app, application, smartphone or alike, the results approach a minimum of 19 contri-butions for further inspection.

On the content side, we could observe that most of the researched papers are about analyzing the purpose and benefits of healthcare information systems at worksite. Various characteristics have been studied to evaluate their effect on employee perform-ance improvement. One study relates to behavioral prevention: some German health insurances evaluated four corresponding features that are physical activity, nutrition, stress management and addiction prevention.

Table 3. An overview of the total results (where any).

Keywords	Number of results
CHM	4,243
CHM + A	3,310
CHM + AP	6,962
CHM + S	3,287
CHP	2,332
CHP + AP	5,355
CHP + S	141
COHP	520
COHP + A	935
COHP + AP	115
COHP + S	300
OHP	5,059
OHP + AP	5,677
OHP + S	341
WHP	11,641
WHP + A	535
WHP + AP	74,939
WHP + S	3,150
WHM	221
WHM + A	21
WHM + AP	1,539

However, a majority of papers describes the effect of healthcare and wellness programs in leveraging physical activity [33, 34]. Other contributions mention the reduced risk of stress and its control including topics such as relief, relaxation or burnout prevention. In [23], a computer-based system has been examined for stress management training for Japanese workers. In addition, results include effects on nutrition [18, 35] respectively focus on diet and improving weight or control of blood values.

We can conclude that most of literature researched was about the evaluation of web-based systems. There are papers introducing various health-based programs that were a mixture of an online application along with physical training, counseling several employees based on data analyzed. Two of the papers contributed evaluations of programs, called *Virtual Fitness Centre* (VFC) [36] and *A healthier You* (AHY) [37], studying their potential means and impact on reduction of risk of physical inactivity, high body weight, helping with smoking cessation and improvements in blood pressure and cholesterol level respectively. The outcome of AHY was a support-based application storing all information and results further evaluating by use of a biometrical screening of the aforementioned values. Such programs have been improving the employee performance, have reduced health risks and employees were found to be more efficient. The study of app-based programs yielded limited results, but showing similar positive results for both companies and employees.

Although a large number of results is listed in Table 3, the funnel and iterative research papers were filtered out and had to be dismissed. Table 4 gives an overview of the processed results per categories and Table 5 shows the geographic distribution of the articles found.

Table 4. Results by Cluster Categorization.

Cluster name	Cluster categories	Contributions referenced
System type	Mobile app	[17, 22, 35]
	Application	[16, 21, 23, 34, 38]
	Web-based	[13, 18, 21, 22, 33–43]
Life-cycle stage	Requirements Engineering	[13, 16, 17, 38, 40]
	Test	[43]
	Evaluation	[15, 18, 21–23, 33–37, 39, 41, 42]
Functionality	Alert	[21]
	Support	[22, 23, 33–36, 39, 41–43]
	Information	[13, 16–18, 35, 37, 38, 40]
Target users	Employers	[16, 17, 19, 38, 40, 41]
	Employees	[13, 15–19, 21–23, 33–37, 39, 42, 43]

Table 5. International distribution of initiatives described.

Country/region	Contributions referenced
Australia	[22]
Austria	[17] (analyzed also German cases)
Germany	[17, 33, 43]
Japan	[23]
Netherlands	[35]
Scandinavia	[19]
South Korea	[38]
UK	[34]
USA	[13, 15, 16, 18, 21, 36, 37, 39, 40, 42]

Most of the papers about evaluation of these applications were about initiatives from the USA, followed by European countries such as Germany and Austria. A small number of reviews were stemming from Asian countries such as Japan or South Korea. Basically all papers studied and evaluated were from industrialized countries.

In Table 3 is also an overview of the total findings. There are several papers showing the huge importance of health promotion independent from the terms app, application or smartphone. For example, in SpringerLink, there were 752 articles, or 232 in ISI web of science generated by searching the keyword WHP. However, in the databases Pubmed, ACM, and Ebscohost we could find the most relevant results. The term WHP alone is mentioned in a very large number of contributions (11,641 papers) in the afore-mentioned databases.

Also interesting is to inspect the distribution of industries described in the significant papers, cp. Figure 2. Generally, there is a good share of industry-specific contributions available. To highlight one of them, education is one of the more investigated industries where targeted users have been school teachers [39] or university employees [34]. Nevertheless, some reviews do not provide industry specifics since only meta-studies and only few reviews focus on more than one industry. They consider several ones such as manufacturing, service, trade/transportation, government and healthcare [15] at once. We finally found interesting that 6% of the papers investigated the IT industry.

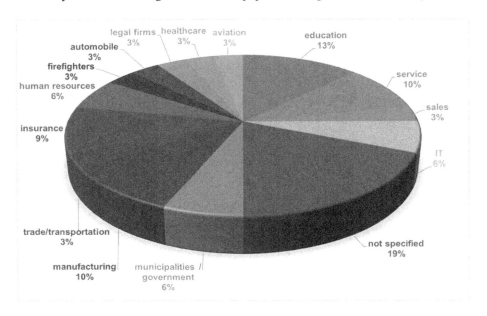

Fig. 2. Distribution of contributions by industry

5   Discussion and Outlook

The review provides an overview of existing work about workplace health promotion and corresponding applications. Although we could find apps tailor-made for WHP, there is still a lack of corresponding IT solutions in the literature. The practical examples such as Moove show new initiatives and a potential need of the companies.

The ideas of employees are an important factor for gaining new insights, comprehending their requirements and therefore the contribution for their general health. A few works point out the general difficulty of healthcare and emphasize the motivation of its rising costs as well as smartphones' increasing influence on our lives [17, 36, 40] proving the relevance of this subject. Two of the most insightful papers are the ones from Guertler et al. 2015 [22] and Dunkl and Jiménez 2016 [17]. Guertler et al.'s focus is more on employees and they investigate a high number of workers while Dunkl and Jiménez are regarding employers such as German managers as their target group. It would be of interest to consider the perspectives of both target groups on one app or in one research approach.

Moreover, Guertler et al. compare different cluster categories such as mobile app and web-based for logging physical activity. The results show a higher engagement on using the app compared to others and a reduced risk of attrition. Dunkl and Jiménez measured the three aspects physical fitness, mental fitness, and nutrition. Furthermore, their work considers the life-cycle category of requirements. They try to "identify individual and organizational aspects which might influence" the decision of using an app in WHP and evaluating the feedbacks. The cluster functionality is also noticed by these two contributions. Guertler et al. use a supportive app to increase the engagement with an intervention. Dunkl and Jiménez use the information functionality to give an instant feedback to the users and motivate them to participate. The results show that younger leaders and leaders with an interest in WHP are more enthusiastic about using the app.

At the same time, our research shows a lack of information regarding the requirement specifications and proper requirement engineering before applications can be developed. It is highly significant to deliver web-, hybrid or native mobile applications grounded on complete requirements identification and to ensure high quality (such as minimized errors and efficient performance) in subsequent software development.

The industry distribution according to own experiences shows that IT employees usually are seated at a desk most of time and sitting is considered one major health threat. Therefore, this industry could be more aware of health promotion (or researchers could focus more on this industry). The geographic distribution of contributions' origin countries is statistically not representative, but shows that United States is somewhat leading in this topic. However, it is possible that there are some more novel workplace healthcare apps not described or not found in the literature. One example where clinical trials are missing is 'Upright' [44], a combination of an app and wearables monitoring an employee's bearing and alerts for moving when sitting incorrectly. However, large companies like SAP and Siemens have been using it [45].

Nevertheless, we should also consider a certain semantic bias because of the limitations in searching the literature in terms of filters and confined research disciplines. With help of additional terms, there is a chance to identify contributions from other disciplines such as public health, medicine or social science. In the future, we will also extend the search terms to work*site* health promotion, as eventually read in one of the papers identified.

As next steps, we will explore more about the existing apps of service providers and compare their contents with expected requirements especially focusing the requirements of employees. Moreover, it will be interesting to consider various industries using these apps and see if they are continuously and successfully used by employees, e.g. in the IT industry. It is also important to consider the requirements of other stakeholders like employers or management more deeply and to comprehend their needs. In general, it should be analyzed if there were legal certificates or support by law available for workplace health promotion apps thus facilitating potential introduction. E.g. in Germany, public health insurance must support workplace health promotion abiding to social law § 20a (1) SGB V. Another matter to be examined will be applying data privacy law and ethics, e.g. regarding big brother situations where employee activity is supervised electronically. Especially regarding aforementioned benefits and profitability margins, service providers and their clients might still be interested to tackle existing challenges.

Acknowledgment. Parts of this contribution originate from the project Baltic Sea Campus on eHealth (BSCeH), an excellence grant from the state of Schleswig-Holstein to support the development of the Institute for eHealth and Management in Healthcare (IEMG) at Flensburg University of Applied Sciences. The authors are pleased to acknowledge all supporting partners.

References

1. Zaballa, R.: Services science: filling the gap between knowledge and needs. In: Snene, M., Ralyté, J., Morin, J.-H. (eds.) IESS 2011. LNBIP, vol. 82, pp. 136–145. Springer, Heidelberg (2011). doi:10.1007/978-3-642-21547-6_11
2. Hempe, E.-M., Dickerson, T., Holland, A., Clarkson, P.J.: Framework for design research in health and care services. In: Morin, J.-H., Ralyté, J., Snene, M. (eds.) IESS 2010. LNBIP, vol. 53, pp. 125–135. Springer, Heidelberg (2010). doi:10.1007/978-3-642-14319-9_10
3. Bonomi, S., Zardini, A., Rossignoli, C., Dameri, P.R.: E-health and value co-creation: the case of electronic medical record in an italian academic integrated hospital. In: Nóvoa, H., Drăgoicea, M. (eds.) IESS 2015. LNBIP, vol. 201, pp. 166–175. Springer, Cham (2015). doi: 10.1007/978-3-319-14980-6_13
4. mHealth Economics Research Program: mHealth App Developer Economics 2015. The current status and trends of the mHealth app market. research2guidance, Berlin (2015)
5. Heller, L.: Gesundheitskontrolle durch Apps und Algorithmen. Engl.: Health control by apps and algorithms, http://bit.ly/1K6j3RR
6. Google Inc.: Weight Watchers Mobile. http://bit.ly/2lS8TfL
7. Bundesgesundheitsministerium | German Federal Ministry of Health: Steuerliche Vorteile (Engl.: Tax advantages). http://bit.ly/2lYQ3Wz
8. Austrian BGF-Network: Gründe und Motive für die Umsetzung von BGF-Projekten (Engl.: Reasons and motives for the implementation of WHP-projects). Netzwerk BGF, Linz (2015)
9. Maar, C., Fricker, R., Hildebrandt, N., Drechsler, M.: Die Rolle der betrieblichen Gesundheitsvorsorge für die Zukunftsfähigkeit des Wirtschaftsstandortes Deutschland (Engl.: The role of workplace health promotion for the future German economy). Booz & Co., Munich (2011)
10. BKK Dachverband, Dt. Gesetzl. Unfallversicherung, AOK-Bundesverband, Verband der Ersatzkassen: Betriebliche Gesundheitsförderung kann krankheitsbedingte Fehlzeiten um ein Viertel senken (WHP can lower sick-leave by 25%). https://www.iga-info.de/veroeffentlichungen/pressemitteilungen/wirksamkeit-bgf/
11. Hommersom, A., Lucas, P.J.F., Velikova, M., Dal, G., Bastos, J., Rodriguez, J., Germs, M., Schwietert, H.: MoSHCA - My mobile and smart health care assistant. In: IEEE 15th International Conference on e-Health Networking, Applications and Services, Healthcom, pp. 188–192 (2013)
12. Cook, R.F., Billings, D.W., Hersch, R.K., Back, A.S., Hendrickson, A.: A field test of a web-based workplace health promotion program to improve dietary practices, reduce stress, and increase physical activity: randomized controlled trial. J. Med. Internet Res. **9**, e17 (2007)
13. Coulter, C.H.: The employer's case for health management. Benefits Q. **22**, 23–33 (2006)
14. vitaliberty GmbH: moove | Innovatives Betriebliches Gesundheitsmanagement (Engl. Innovative WHP). https://www.corporate-moove.de/

15. Osilla, K.C., Van Busum, K., Schnyer, C., Larkin, J.W., Eibner, C., Mattke, S.: Systematic review of the impact of worksite wellness programs. Am. J. Manag. Care. **18**, e68–e81 (2012)
16. Goetzel, R.Z., Shechter, D., Ozminkowski, R.J., Marmet, P.F., Tabrizi, M.J., Roemer, E.C.: Promising practices in employer health and productivity management efforts: findings from a benchmarking study. J. Occup. Environ. Med. **49**, 111–130 (2007)
17. Dunkl, A., Jiménez, P.: Using smartphone-based applications (apps) in workplace health promotion: the opinion of German and Austrian leaders. Health Informatics J. (2016). http://journals.sagepub.com/doi/abs/10.1177/146045
18. MacKinnon, D.P., Elliot, D.L., Thoemmes, F., Kuehl, K.S., Moe, E.L., Goldberg, L., Burrell, G.L., Ranby, K.W.: Long-term effects of a worksite health promotion program for firefighters. Am. J. Health Behav. **34**, 695–706 (2010)
19. Torp, S., Eklund, L., Thorpenberg, S.: Research on workplace health promotion in the Nordic countries: a literature review, 1986–2008. Glob. Health Promot. **18**, 15–22 (2011)
20. Perez, A.P., Phillips, M.M., Cornell, C.E., Mays, G., Adams, B.: Promoting dietary change among state health employees in Arkansas through a worksite wellness program: the Healthy Employee Lifestyle Program (HELP). Prev. Chronic Dis. **6**, A123 (2009). https://www.cdc.gov/pcd/issues/2009/oct/08
21. Aneni, E.C., Roberson, L.L., Maziak, W., Agatston, A.S., Feldman, T., Rouseff, M., Tran, T.H., Blumenthal, R.S., Blaha, M.J., Blankstein, R., Al-Mallah, M.H., Budoff, M.J., Nasir, K.: A systematic review of internet-based worksite wellness approaches for cardiovascular disease risk management: outcomes, challenges & opportunities. PLoS ONE **9**, e83594 (2014)
22. Guertler, D., Vandelanotte, C., Kirwan, M., Duncan, M.J.: Engagement and nonusage attrition with a free physical activity promotion program: the case of 10,000 steps Australia. J. Med. Internet Res. **17**, e176 (2015)
23. Umanodan, R., Shimazu, A., Minami, M., Kawakami, N.: Effects of computer-based stress management training on psychological well-being and work performance in japanese employees: a cluster randomized controlled trial. Ind. Health **52**, 480–491 (2014)
24. Webster, J., Watson, R.T.: Analyzing the past to prepare for the future: writing a literature review. MIS Q. **26**, xiii–xxiii (2002)
25. Mohan, K., Ahlemann, F.: Understanding acceptance of information system development and management methodologies by actual users: a review and assessment of existing literature. In: Bernstein, A., Schwabe, G. (eds.) 10th International Conference on Wirtschaftsinformatik, vol. II, pp. 734–744. AIS, Zurich (2011)
26. Orwat, C., Graefe, A., Faulwasser, T.: Towards pervasive computing in health care - a literature review. BMC Med. Inform. Decis. Mak. **8**, 26 (2008)
27. Boehm, B.: A spiral model of software development. IEEE Comput. (1988)
28. Royce, D.W.W.: Managing the development of large software systems. IEEE Wescon, pp. 1–9 (1970)
29. Pomberger, G.P.: Wolfgang: Software Engineering (1993)
30. Goetzel, R.Z., Ozminkowski, R.J.: Health and productivity management: emerging opportunities for health promotion professionals for the 21st century. Am. J. Heal. Promot. **14**, 211–214 (2000)
31. Viester, L., Verhagen, E.A.L.M., Proper, K.I., van Dongen, J.M., Bongers, P.M., van der Beek, A.J.: VIP in construction: systematic development and evaluation of a multifaceted health programme aiming to improve physical activity levels and dietary patterns among construction workers. BMC Public Health. **12**, 89 (2012)
32. Goldgruber, J., Ahrens, D.: Effectiveness of workplace health promotion and primary prevention interventions: a review. J. Public Health (Bangkok) **18**, 75–88 (2010)

33. Pressler, A., Knebel, U., Esch, S., Kölbl, D., Esefeld, K., Scherr, J., Haller, B., Schmidt-Trucksäss, A., Krcmar, H., Halle, M., Leimeister Jan Marco, J.M.: An internet-delivered exercise intervention for workplace health promotion in overweight sedentary employees: a randomized trial. Prev. Med. (Baltim) **51**, 234–239 (2010)

34. Mackenzie, K., Goyder, E., Eves, F.: Acceptability and feasibility of a low-cost, theory-based and co-produced intervention to reduce workplace sitting time in desk-based university employees. BMC Public Health. **15**, 1294 (2015)

35. van Drongelen, A., Boot, C.R.L., Hlobil, H., Twisk, J.W.R., Smid, T., van der Beek, A.J.: Evaluation of an mHealth intervention aiming to improve health-related behavior and sleep and reduce fatigue among airline pilots. Scand. J. Work. Environ. Heal. **40**, 557–568 (2014)

36. Herman, C.W., Musich, S., Lu, C., Sill, S., Young, J.M., Edington, D.W.: Effectiveness of an incentive-based online physical activity intervention on employee health status. J. Occup. Environ. Med. **48**, 889–895 (2006)

37. Hochart, C., Lang, M.: Impact of a comprehensive worksite wellness program on health risk, utilization, and health care costs. Popul. Health Manag. **14**, 111–116 (2011)

38. Ho, S.H., Chae, Y.M., Kim, S.B., Kang, Y.S., Choi, B.S., Park, Y.W., Lee, M.S., Jin, D.L.: Health information system based on EHR for workplace health promotion. In: AMIA Annual Symposium Proceedings, p. 972 (2008)

39. Aldana, S.G., Merrill, R.M., Price, K., Hardy, A., Hager, R.: Financial impact of a comprehensive multisite workplace health promotion program. Prev. Med. (Baltim) **40**, 131–137 (2005)

40. Naydeck, B.L., Pearson, J.A., Ozminkowski, R.J., Day, B.T., Goetzel, R.Z.: The impact of the highmark employee wellness programs on 4-year healthcare costs. J. Occup. Environ. Med. **50**, 146–156 (2008)

41. Kawakami, N., Takao, S., Kobayashi, Y., Tsutsumi, A.: Effects of web-based supervisor training on job stressors and psychological distress among workers: a workplace-based randomized controlled trial. J. Occup. Health. **48**, 28–34 (2006)

42. Cook, R.R., Billings, D.W., Hersch, R.K., Back, A.S., Hendrickson, A.: A field test of a web-based workplace health promotion program to improve dietary practices, reduce stress, and increase physical activity: Randomized controlled trial. J. Med. Internet Res. **9**, e17 (2007)

43. Jung, J., Nitzsche, A., Neumann, M., Wirtz, M., Kowalski, C., Wasem, J., Stieler-Lorenz, B., Pfaff, H.: The Worksite Health Promotion Capacity Instrument (WHPCI): development, validation and approaches for determining companies' levels of health promotion capacity. BMC Public Health. **10**, 550 (2010)

44. Upright Tech Ltd.: Posture trainer: back pain relief starts with good posture. http://www.uprightpose.com/

45. Medica Tradefair: The winning app from the MEDICA App COMPETITION 2016 helps you maintain correct posture (2016). http://bit.ly/2myPNKB

New Ways to Deal with Omni-Channel Services: Opening the Door to Synergies, or Problems in the Horizon?

João Reis[1(✉)], Marlene Amorim[2], and Nuno Melão[3]

[1] Department of Economics, Management and Industrial Engineering and Tourism,
Aveiro University, Aveiro, Portugal
reis.joao@ua.pt
[2] Department of Economics, Management and Industrial Engineering and Tourism,
and GOVCOPP, Aveiro University, Aveiro, Portugal
mamorim@ua.pt
[3] Department of Management and CI&DETS, School of Technology and Management of Viseu,
Polytechnic Institute of Viseu, Viseu, Portugal
nmelao@estgv.ipv.pt

Abstract. This article aims to investigate organizational synergies in the omni-channel service context. In doing so, it discloses new omni-channel trends and discusses its implications for managers and academics. It uses a qualitative multi-method approach, which includes more than one method of collecting data to generate comprehensiveness and rich knowledge, namely: a systematic literature review and a case study. The transition to an omni-channel service requires companies to overcome many organizational challenges and is compelling academics and practitioners to focus on its operations management. The results indicate that organizational synergies are changing the omni-channel landscape and may provide several opportunities for gaining competitive advantages by implementing new technologies (e.g. m-payments), and anticipating customer needs (e.g. multi-brand experience). It is possible that these organizational synergies are transcending the omni-channel concept, creating new trends, but to confirm this hypothesis further investigation is needed.

Keywords: Organizational synergies · Omni-channel services · Qualitative multi-method approach · Systematic literature review · Case study · Competitive advantage · Operations management

1 Introduction

Recent times have seen an increasing interest in omni-channel services. Whereas traditional retail players are ramping up their Internet presence, online-first retailers are complementing their service delivery systems by opening stores and showrooms [1]. Customers are becoming more self-assured in employing electronic devices (e.g. laptops, tablets, mobile phones…) both for product search and order placement [2]. Moreover, the service delivery arena is now evolving from the adoption of multi-channel approaches, where customers are offered alternative channels, towards the pursuit of omni-channel strategies that aim at leveraging on integration, and the potential synergies

© Springer International Publishing AG 2017
S. Za et al. (Eds.): IESS 2017, LNBIP 279, pp. 51–63, 2017.
DOI: 10.1007/978-3-319-56925-3_5

from combining various interfaces for customer interaction. Moving from multi– to omni-channel service systems is an important opportunity, but also a major challenge for companies [3]. It allows customers to undertake the buying process on their terms and convenience across all channels [4] and as a brand experience [5]. Key challenges are related to the management of service operations so as to allow a seamless integration of decentralized structures [6], as it happens when orders are placed online and have to be collected in store [7]. What underlies is that we have witnessed a continued evolution from single, to multi–, cross–, and recently to omni-channel services in a relatively short period of time. So far, we ask ourselves, what is the next step? Are we consolidating or opening the doors to new strategies? The next sections aim to provide some answers to these questions.

2   Literature Review

The first step is to delimit the omni-channel concept. The multi–, cross–, and omni-channel concepts are commonly used indistinctively in the academic literature. Rigby's [8] first mentioned the word, defining omni-channel retailing as an integrated sales experience that melds the advantages of physical stores with the information-rich experience of online shopping, but is difficult to find a consensual definition. Beck and Rygl [7, p. 175] defined omni-channel retailing as the "set of activities entailed in selling merchandise or services through all widespread channels, whereby the customer can trigger full channel interaction and/or the retailer controls full channel integration". Picot-Coupey et al. [5] performed a systematic literature review to describe omni-channel services as a seamless and integrated shopping experience across all channels that blurs the distinctions between physical and online stores, and culminates in an integrated brand experience. The emergence of Internet and new technologies have changed significantly the foundations of customer-company interactions; self-service technology is a classic example, where no interpersonal contact is required between buyer and seller [9]. The availability of these new channels has drastically changed the way companies interact with customers by introducing substantial degrees of freedom in the way customers can employ different channels for each service activity [10]. Moreover, it extends the possibilities for providers to facilitate customers' direct engagement with specialized intermediaries for specific service delivery activities (e.g. payment, logistics). Choi and Wu [11] predicted, to a certain extent, the emergence of new dynamics in service delivery involving triadic relationships, i.e. buyer-supplier-supplier. More recently, Wynstra et al. [12] extended the supply networks relations to service management, while they suggested a service triad as a business model (buyer-supplier-customer). For instance, if a software company outsources its helpdesk services to a third-party call-center, the primary service interaction is between the customer and the call-center, not between the customer and the software company, even though the customer has a contractual relationship with the software company [12]. What is relevant here is the integration of more than one company to provide a service experience, and for that reason engaging in direct interactions with the final customer. Triadic relationships take place when a company contracts with a supplier to deliver services directly to its final customer. Adding a technological interaction layer to these triadic service delivery approaches leads to the need to update the components involved in the operations of omni-channel services, i.e. evolving the service channel systems that support the traditional dyadic service exchanges between providers and customers towards a network of companies-channels-customers (Fig. 1). To adequately conceptualize these chances, we need to resort to three major interrelated and dynamic components of service delivery systems [13]: (1) Strategic service design choices (SSDC) (companies), (2) Service delivery systems execution (SDSE) (the channels), and (3) Customer-perceived value for the total service concept (CVTSC) (customers).

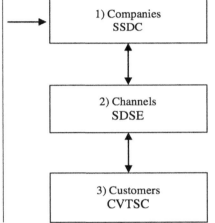

Service Operation Management

Fig. 1. Triad of service management in an omni-channel context (adapted from [2, 12, 13])

According to Roth and Menor [13] the setup of a service delivery system requires a set of company's decisions related to (a) Structural choices, concerning key decisions about physical elements of the delivery system, namely facilities, technology, equipment, and capacity; (b) Infrastructural choices, concerning programs, policies, and behavioural aspects that command service operations strategy; and (c) Integration choices, that refer to the issues of external integration, internal integration and adaptive mechanisms. The interface between customers and companies' service system is performed by means of service channels that result from the aforementioned decisions. The developments in information and communication technologies has increased the number of means by which customers are able to interact with service providers [14–16]. Sousa and Voss [17] distinguish among two types of channels: (a) Virtual channels, consisting of means of interaction using advanced telecommunications, information, and multimedia technologies (e.g. ATMs); and (b) Physical channels, consisting of a means of communication with the customer employing a physical (bricks-and-mortar) infrastructure (e.g. warehouses) and resorting to customer-employee personal interactions. Froehle and Roth [16] offered a classification for banking channels according to the type of customer interface: "face-to-face" or "face-to-screen". In financial services companies, face-to-face contact, for example, occurs at the physical branches [15]. This taxonomy has led to a profusion of definitions for service delivery models; namely Sousa and Voss [17, p. 357] defined virtual service (face-to-screen) as "the pure information component of a customer's service experience provided in an automated fashion through a given virtual channel" and physical service (face-to-face) "as the portion of a customer's service experience provided in a non-automated fashion, requiring some degree of human intervention, either through a virtual or physical channel". Virtual services have grown in number and sophistication with the recent emergence of mobile payment technologies (m-payment). These are solutions that have been anticipated since the early 2000s, but it was only in recent years that their roll out has gathered strength, particularly in the USA, Europe and some parts of Asia [18]. As the availability and reliability of these systems increases, and customers get more acquainted and qualified, they increasingly employ electronic devices for diverse service operations, ranging from information search to order placement and payments. M-payment possibilities came into the retail sector and dramatically altered the process by which products pass from retailer to consumer, but it has received surprisingly little scholarly attention [19, 20]. After all, mobile payment will become an uncontested mode for paying for goods in the near future [21, p. 188]. Such technological innovations lay the ground for service systems where the co-creation of value will be become increasingly common [22]. In this sense, co-creation, is a new paradigm in the management literature that allows companies and customers to create value through interaction [23]. The growing technological developments give academics enough confidence to argue that omni-channel services will continue to evolve and at the same time will create new trends.

3   Methodology

This article follows a qualitative multi-method approach. Consistent with prior definitions, we define multi-method research as one that includes more than one method of collecting data and or more than one method of analyzing the data [24]. Such methods can be supported by qualitative techniques, quantitative techniques, or a mix of both, in what is called mixed-methods approach [24, 25]. Dividing this study into two independent articles could make the methodological approach more suitable to the reader, however, recent work advocates that combining two or more methods of collecting data generates comprehensiveness and rich knowledge [24], counterbalancing the weaknesses that are inherent to individual methods [26]. The first method consisted in conducting a systematic literature review, justified by the fact that omni-channel services are a relatively new area of study [27]. Its purpose was to identify recent trends in the utilization of channels for service delivery. Building on the literature review, a case study was then conducted to empirically validate the theoretical insights. The literature search was performed on Scopus.com, one of the largest abstract and citation peer-reviewed literature databases, using the word *omni-channels* for article title, abstract and keywords, on December 14, 2016.

Table 1. Systematic literature review.

Scopus Search		
Criteria	Filters	Documents
Keyword	Omni-channels	67 documents
Restrictions		
Document type	Article, book chapter and conference paper	57 documents
Source type	Journals, books, conference proceedings	45 documents
Language	English	44 documents

The systematic literature review offers a qualitative overview on the trends of omni-channel services in the management literature. The results support the view that the literature is still incipient (see Table 1), with only 67 documents emerging on the Scopus database. The subsequent case study-based research aimed at further understanding and corroborating the findings from the literature review. It used multiple data collection methods, including 5 semi-structured interviews, direct observations and analysis of official documents from a large private retail bank. The number of participants selected for the interviews is justified by theoretical saturation. Saunders and Townsend [28] consider saturation as a plausible justification for the number of participants, and comment that saturation is being considered the gold standard by some [29]. Participants were chosen according to the employees' different functional areas and different levels of responsibility at a bank's physical branch. Employees tend to follow very similar rules and procedures across branches, for which the research team had reason to believe that data collection from one branch would probably not be substantially different on a different branch. Observation, as a data collection method, involves systematically seeing and listening [30] in order to enable learning and analytical interpretation [31]. During the direct observation field notes were taken. These field notes came from the analysis of the real life phenomenon, and from informal conversations with the interviewees. At the end, the data from direct observation was confronted with the interviews for triangulation purposes. The choice to carry out the case study in the context of banking was driven by the available academic and empirical evidence about the pioneering role that financial services have been taking in the adoption of new channels in service delivery [32], contributing to pave the way for new trends (e.g. multi-channel services).

4   Findings

This section provides a theoretical overview and its empirical validation derived from the case study. The focus of this study is on new trends in omni-channel services, notably its synergies and possible (dis)advantages. Data analysis and discussion integrates statements collected from employees, direct observations and documental analysis.

4.1   Moving from a Marketing to an Operations Management Perspective

During the latter part of the 20th century, the service sector grew significantly in virtually every developed country, leading service operations management to recall for a legitimate field of its own [33]. This rapid growth was determined by several factors, including the application of information technology/information systems (IT/IS), which have significantly altered the landscape of operations management [34]. The IT/IS enables the combination of competences, capabilities, and knowledge [35] that underpin value co-creation in collaborative relationships [36]. Cabiddu *et al.* [22] exemplifies with the airline sector, where information technology has influenced operations (e.g. from paper-based ticketing to e-ticket) as well as service delivery [37]. Contemporaneous evidence suggests that omni-channel services literature has contributed to reinforce earlier arguments. However, Reis *et al.* [38] noted that preliminary insights from the multi-channel services came from the marketing literature. They alerted scholars for the apparently need to carry out studies in the operations management sphere. *De facto*, the introduction of new technologies and the "shifting towards omni-channel strategies were so complex and engaging that it is impossible to evolve directly from a multi-channel, siloed strategy to an omni-channel strategy without any transition" [5, p. 347]. This transition has clearly compelled academics to increase their focus on operations management, as shown by the significant increase of academic articles published in scientific journals. If we analyze the systematic review, we verify that 47% of the current literature is based on operations management, being a higher percentage when compared with a similar study by Reis *et al.* [39] on multi-channel services (only 2.5%). The roots of this phenomenon are probably linked with the calls for organizational adaptation. Companies have found that they have advantages in serving their customers using an integrated network of channels, but these changes normally require process change. Hübner *et al.* [40] corroborates this, stating that the transition from multi-channel to omni-channel requires the redesign of logistic structures and, concomitantly, the optimization of processes [13]. In this context customer participation has an important role. Process optimization needs to align the transition to omni-channel services with customer requirements, as the company and the customers' roles converge [41]. Customers generate value through interaction [42], as their participation is an important element in the value co-creation process [43]. When customer involvement into the company's operations takes place, joint value creation occurs, which means that the customer is engaged in, for example, the design or in front-office operations [44]. However, relatively little is known about how customers engage in the co-creation of value [36], especially in an omni-channel context. The future of omni-channel services and its implications to operations management are still uncertain, but will certainly be promising.

4.2   Synergies Between Companies: Systematic Literature Overview

The results of the systematic review showed a plethora of articles referring to companies' integration and adaptation to the omni-channel strategy [2, 40], but few referred to synergies. An exception is Picot-Coupey *et al.* [5] who investigated internal synergies between clicks and bricks, and how successful the transformative process to become omni-channel can be. Other authors pointed out the need to focus less in products or services, and more in consumer-centric approaches [45]. Melero *et al.* [45] exemplified how Decathlon and Zara companies introduced new communication channels to interact with their customers: Decathlon with a mobile application (App) to facilitate contact between customers who practice the same sport and Zara achieved a leading position on social networks, where customers can interact with the company. Also in the omni-channel context, Notomi *et al.* [46, p. 38] stated that "retailers have found themselves forced to compete for customer attention like never before". The underlying question is: do organizations have to do this on their own? Maybe not. Notomi *et al.* [46, p. 38] remarked that "at the same time, online retailers are expanding their reach by partnering with companies that already have stores and service establishments", because today's consumers no longer go to stores merely to shop; they pursue the optimal purchasing experiences, e.g. best prices, best information. This seems a sensible thing to do because, theoretically, we already have companies that collaborate with each other to best serve their customers. In this regard, Verhoef *et al.* [3] gives the examples of Booking.com and Tripadvisor.com, which shook-out traditional travel intermediaries. These are just a few examples that emerged from the systematic literature review. Companies can also interact with each other to optimize their services. For example, Rumbo.com and Edreams.com have understood the advantages of adding several airline agents to their search engines so as to provide the cheapest available prices and the lowest waiting time to their customers. Indeed, working in partnership can open up new opportunities. Other companies are also adopting this strategy internally. Starbucks Canada is implementing mobile payments through an app [47], enabling customers to make virtual payments of their purchases in any store of its network. These synergies between companies, either within or between groups of companies, can provide a number of opportunities still to investigate. In this sense, it is necessary to address the possible interactions that may arise from the synergies between organizations not only to respond to customer's needs and expectations, but also to promote their retention. From the literature review, we conclude that there is a need to: (i) address operations management issues related with integrated service experience; and (ii) evolve from synergies between channels (bricks and clicks) to synergies between companies and thereby introducing new degrees of freedom in how customers can interact with different companies for each service activity.

4.3   Basic and Complex Synergies: Case Study Analysis

Data analysis from the case study highlighted, in the strategic plan (2016–2018), the importance of channel management, notably by referring to the implementation of more digital and technological tools for customers and workers (explicitly mentioned in the bank's official documents). Similarly, a contact sales employee emphasized that the bank was pursuing a transition from multiple channels to an omni-channel strategy, e.g. is investing and making available to its contact sales employees tablets to facilitate inter-actions with customers. The same contact sales employee reinforces that the required organizational effort is high, because the bank has to invest on: (1) technology and, concomitantly, on restructuring existing processes as new channels are introduced (i.e. revisiting its structural decisions); (2) training its employees (i.e. revisiting infrastructural decisions) and; (3) disseminating the information regarding the availability of new channels to the customers (i.e. acting on the integration of the components of the service system). Cook *et al.* [4] discusses the nature of the omni-channel customer and the associated changes required from the physical retail spaces, referring to the case of Argos, a brick-and-click store that started to use iPad-based kiosks. These kiosks and the empowered staff were helping their customers in selecting the best suited products and purchasing. In the case study, the concept behind this approach is to simplify and speed up the sales process, removing the traditional queuing approach in service provision [4]. When companies do not have capacity to add new technology to their portfolio (structural limitation), or do not intend to invest in additional physical stores or web environment, a viable strategy is establishing partnerships. Examples of basic synergies can also be found in the literature. Ebay.com proposed pickup points at Argos stores in the United Kingdom [7], giving support to Notomi *et al.*'s [46] arguments mentioned before. Amazon, originally a pure online player, already opened brick-and-mortar bookstores and is planning to venture is retail operations to 100 pop-up stores in the United States. These examples show that an omni-channel strategy can be achieved in several ways. On this subject, the contact sales employees argued that the bank was already preparing complex channel synergies, building on the establishment of synergies with other companies from different specialties. Companies that complement each other create value networks [42] in which resources of the partners are orchestrated into a novel value proposition that, in turn, is offered to customers. This co-created value exists when several firms interact with one another, e.g. by means of technological innovation to create a value proposition that can generate greater value for customers compared to a value proposition offered by any single company [22]. For example, the bank already participated in the MB Way service, a functionality that allows customers to connect the bank to several retail companies. This solution allows customers to combine an act of physical purchase and virtual payment, by making a mobile payment for a service or product purchase in a retail store. This kind of synergy is complex because customers can use the payment function across a network of companies, and also combines several types of services (physical and virtual). In return, customers provide value to the network of companies in the form of profits [48], although they also may compete for the extraction of economic value [41]. The roles of producer and consumer are becoming indistinct, as joint interactions lead to the development of new business opportunities [23] and reciprocally co-creating value through the integration of resources (e.g., channels) and customer skills [42]. We believe that this typology of service delivery is beyond the omni-channel capabilities, but to determine the implications of these synergies and the co-creation of value further investigation is needed.

4.4 Opening the Door to Synergies or Problems in the Horizon?

Mobile devices have a number of characteristics, such as ultra-portability and location sensitivity, assisting consumers in a number of shopping activities: search, comparison, purchase and post-purchase [49, 50]. Interviews from the case study revealed that m-payment technology created new prospects in omni-channel services, notably by "increasing the possibility of choosing simultaneously other channels to perform a purchase" (sales employee statement). This is consistent with the literature in that m-payments may facilitate the showrooming practice, which consists of "using mobile technology while in-store to compare products for potential purchase via any number of channels" [51, p. 360]. For instance, a customer can access information and opinions from a variety of sources, including friends, competitors, consumer-to-consumer reviews, and even other channels at the focal retailer (virtual channels and/or physical channels). Additionally, the interviews revealed that m-payments are actually opening the door to synergies, as these technologies are bringing together companies that are using or intending to use the same means of payment. A sales person added that in the case of partnerships, the free-riding phenomenon (when a consumer uses a retailer's channel to prepare a purchase and then switches to another retailer's channel to purchase [52]) may be mitigated, although this not yet been corroborated in the literature. Nevertheless, we know from the literature that consumers can visit a retail website via a mobile device, even in a competing retail store, and even purchase at a competitor's on-line shop without leaving the brick-and-mortar store [50]. Allegedly, when it comes to partnerships, the free-riding phenomenon is not applicable. Direct observation confirms that a customer may choose the Supermarket B that has a partnership with Bank A, which allows the customer to pay for a product with her mobile device. This process comprises simultaneously a physical purchase and virtual payment, involving two different companies. On the other hand, if the customer wishes to add another purchase to the shopping cart, but does not like the wine offers of the Supermarket B, she can alternatively buy that product online from Supermarket C, using her mobile device to pay the purchase (Fig. 2).

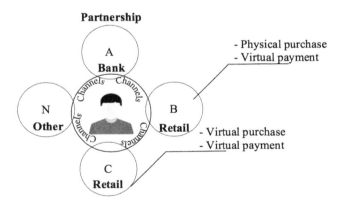

Fig. 2. Omni-channel service synergies

This process connects three different companies; it encompasses, simultaneously, a physical and a virtual purchase with a virtual payment (m-payment) to deliver a service to a customer. What is new here is that we believe this strategy goes beyond the omni-channel experience that originally reflects the articulation of different channels in the context of a single service provider. Picot-Coupey *et al.* [5, p. 339] refers to omni-channel as an integrated "brand experience", but the empirical insights reflect more a multi-brand experience, since it entails several companies. There are, however, limitations. The network of channels of a partnership transcends the channels of a single organization. Thus, customers may have to choose over a portfolio of channels of different organizations, which will certainly bring new operations management challenges. But these challenges can also bring problems on the horizon, the transition from multi- to omni-channel services requires process change, but in this new development stage a paradigm shift is also needed - from a single company to a network of companies, and an overall portfolio of channels.

5   Conclusions

The omni-channel service transition typically requires reengineering of processes and, thus, has led academics to focus more in operations management issues. When companies do not have the capacity to add new technologies to their portfolio or, for some other reason, do not have such interest, they may seek synergies (partnerships). These synergies provide a number of opportunities to customers and organizations: e.g. a multi-brand experience, and the end of the free-riding phenomenon. In the current service delivery contexts, where companies can rely on multiple channels to support different interactions with customers, these organizational synergies (i.e. different companies coordinating to provide distinct activities in service delivery) call for a conceptualization that is beyond the omni-channel concept, as they bring together a mix of channels and providers that need to be articulated in a seamless interaction with the customer. Since this is a recent layer in the service channels landscape, the full understanding of its implications requires further investigation. This article is important for practitioners because it attempts to identify new trends that may be relevant to organizations so as to gain competitive advantages - especially with regard to value co-creation, as synergies between firms lead to the emergence of value networks, making them more competitive. In return, customers collaborate in co-creating value but also compete for the extraction of economic value. Academically, this article shows that the omni-channel concept may need to be adapted to be in line with developments in real-world practice. This paper is not free of limitations. Some relevant articles may be missing since the search is restricted to a selected keyword. This work is also limited because of its exploratory nature, but we hope that it can encourage future investigations at the level of the omni-channel services. On the other hand, by integrating a conceptual and empirical study, the qualitative multi-method approach provided a balanced design, paying due attention to the dimensions of development, triangulation and complementarity, as well as contributing to an overall understanding of the subject under investigation. Due to confidentiality reasons we have not provided any information about key informants and the respective organization. According to Mills *et al.* [53] the removal of identifying information and suppression of confidential information can lead to the removal of the contextual information that is of greatest interest and value to the researcher. To maintain the scientific rigor, the list of all documents of the systematic literature review can be provided on request by the first author. In line with the main results, it may be interesting to conduct a mixed-method investigation in other geographical areas. It seems also relevant to explore some issues (e.g. new trends and synergies) that still remain unclear and require further investigation. With this timely contribution we expect to instigate other investigators to contribute to the operations management discipline, and to advance knowledge in the omni-channel service arena.

References

1. Bell, D., Gallino, S., Moreno, A.: Showrooms and information provision in omni-channel retail. Prod. Oper. Manag. **24**(3), 360–362 (2015)
2. Bernon, M., Cullen, J., Gorst, J.: Online retail returns management. Integration within an omni-channel distribution context. Int. J. Phys. Distrib. Logistics Manag. **46**(6/7), 584–605 (2016)
3. Verhoef, P., Kannan, P., Inman, J.: From multi-channel retailing to omni-channel retailing. Introduction to the special issue on multi-channel retailing. J. Retail. **91**(2), 174–181 (2015)
4. Cook, G.: Customer experience in the omni-channel world and the challenges and opportunities this presents. J. Direct, Data Digital Mark. Pract. **15**(4), 262–266 (2014)
5. Picot-Coupey, K., Huré, E., Piveteau, L.: Channel design to enrich customers' shopping experiences. Synchronizing clicks with bricks in an omni-channel perspective – the Direct Optic case. Int. J. Retail Distrib. Manag. **44**(3), 336–368 (2016)
6. Zhang, J., Farris, P., Irvin, J., Kushwaha, T., Steenburgh, T., Weitz, B.: Crafting integrated multichannel retailing strategies. J. Interact. Mark. **24**(2), 168–180 (2010)
7. Beck, N., Rygl, D.: Categorization of multiple channel retailing in multi-, cross-, and omni-channel retailing for retailers and retailing. J. Retail. Consum. Serv. **27**, 170–178 (2015)
8. Rigby, D.: The future of shopping. Harvard Business Review (2011). https://hbr.org/2011/12/the-future-of-shopping
9. Meuter, M., Ostrom, A., Roundtree, R., Bitner, M.: Self-service technologies: understanding customer satisfaction with technology-based service encounters. J. Mark. **64**(3), 50–64 (2000)
10. Patrício, L., Fisk, R., Cunha, J.: Designing multi-interface service experiences. J. Serv. Res. **10**(4), 318–334 (2008)
11. Choi, T., Wu, Z.: Triads in supply networks: theorizing buyer-supplier-supplier relationships. J. Supply Chain Manag. **45**(1), 8–25 (2009)
12. Wynstra, F., Spring, M., Schoenherr, T.: Service triads: a research agenda for byer- supplier-customer triads in business services. J. Oper. Manag. **35**, 1–20 (2015)
13. Roth, A., Menor, L.: Insights into service operations management: a research agenda. Prod. Oper. Manag. **12**(2), 145–164 (2003)
14. Cassab, H., MacLachlan, D.: A consumer-based view of multi-channel service. J. Serv. Manag. **20**(1), 52–75 (2009)
15. Cortiñas, M., Chocarro, R., Villanueva, M.: Understanding multi-channel banking customers. J. Bus. Res. **63**(11), 1215–1221 (2010)
16. Froehle, C., Roth, A.: New measurement scales for evaluating perceptions of the technology-mediated customer service experience. J. Oper. Manag. **22**(1), 1–21 (2004)
17. Sousa, R., Voss, C.: Service quality in multichannel services employing virtual channels. J. Serv. Res. **8**(4), 356–371 (2006)
18. Mallat, N., Tuunainen, V.: Exploring merchant adoption of mobile payment systems: an empirical study. E-Serv. J. **6**(2), 24–57 (2008)
19. Groß, M.: Mobile shopping: a classification framework and literature review. Int. J. Retail Distrib. Manag. **43**(3), 221–241 (2015)

20. Taylor, E.: Mobile payment technologies in retail: a review of potential benefits and risks. Int. J. Retail Distrib. Manag. **44**(2), 159–177 (2016)
21. Raina, V.: Overview of mobile payment: technologies and security. In: Banking, Finance, and Accounting: Concepts, Methodologies, Tools, and Applications, pp. 186–222 (2014)
22. Cabiddu, F., Lui, T.-W., Piccoli, G.: Managing value co-creation in the tourism industry. Ann. Tourism Res. **42**, 86–107 (2013)
23. Galvagno, M., Dalli, D.: Theory of value co-creation: a systematic literature review. Managing Serv. Qual. **24**(6), 643–683 (2014)
24. Mills, A., Eurepos, G., Wiebe, E.: Encyclopedia of case study research. Sage Publications, California (2010)
25. Davis, D., Golicic, S., Boerstler, C.: Benefits and challenges of conducting multiple methods research in marketing. J. Acad. Mark. Sci. **39**(3), 467–479 (2011)
26. Wood, M., Daly, J., Miller, J., Roper, M.: Multi-method research: an empirical investigation of object-oriented technology. J. Syst. Softw. **48**(1), 13–26 (1999)
27. Thorpe, R., Holt, R.: The Sage dictionary qualitative management research. Sage, London (2008)
28. Saunders, M., Townsend, K.: Reporting and justifying the number interview participants in organizational and workplace research. J. Manage. **27**, 836–852 (2016)
29. Guest, G., Bunce, A., Johnson, L.: How many interviews are enough? Field Methods **18**, 59–82 (2006)
30. Taylor-Powell, E., Steele, S.: Collecting Evaluation Data: Direct Observation. University of Wisconsin, Madison (1996)
31. Saunders, M., Lewis, P.: Research Methods for Business Students, 4th edn. Prentice Hall, London (2007)
32. Sousa, R., Amorim, M.: A framework for the design of multichannel services. Project for the Foundation for Science and Technology, under grant number PTDC/GES/68139/2006 (2009)
33. Heineke, J., Davis, M.: The emergence of service operations management as an academic discipline. J. Oper. Manag. **25**(2), 364–374 (2007)
34. Gunasekaran, A., Ngai, E.: The future of operations management: an outlook and analysis. Int. J. Prod. Econ. **135**(2), 687–701 (2012)
35. Srivastava, M., Gnyawali, D.: When do relational resources matter? Leveraging portfolio technological resources for breakthrough innovation. Acad. Manag. J. **54**(4), 797–810 (2011)
36. Payne, A., Storbacka, K., Frow, F.: Managing the co-creation of value. J. Acad. Mark. Sci. **36**(1), 83–96 (2008)
37. Basole, R., Rouse, W.: Complexity of service value networks: conceptualization and empirical investigation. IBM Syst. J. **47**(1), 53–70 (2008)
38. Reis, J., Amorim, M., Melão, N.: Disclosing paths for multi-channel service research: a contemporaneous phenomenon and guidelines for future investigations. In: Nóvoa, H., Drăgoicea, M. (eds.) IESS 2015. LNBIP, vol. 201, pp. 289–300. Springer, Cham (2015). doi: 10.1007/978-3-319-14980-6_23
39. Reis, J., Amorim, M., Melão, N.: Research opportunities in multi-channel services: a systematic review. In: 21st EurOMA Conference on Operations Management in an Innovation Economy, Palermo, 20–25th of June (2014)
40. Hübner, A., Wollenburg, J., Holzapfel, A.: Retail logistics in the transition from multi-channel to omni-channel. Int. J. Phys. Distrib. Logistics Manag. **46**(06Jul), 562–583 (2016)
41. Prahalad, C., Ramaswamy, V.: Co-creation experiences: the next practice in value creation. J. Interact. Mark. **18**(3), 5–14 (2004)
42. Vargo, S., Lusch, R.: Service-dominant logic: continuing the evolution. J. Acad. Mark. Sci. **36**(1), 1–10 (2008)

43. Vega-Vazquez, M., Revilla-Camacho, M., Cossío-Silva, F.: The value co-creation process as a determinant of customer satisfaction. Manag. Decis. **51**(10), 1945–1953 (2013)
44. Grönroos, C.: Value co-creation in service logic: a critical analysis. Mark. Theory **11**(3), 279–301 (2011)
45. Melero, I., Javier Sese, F., Verhoef, P.: Recasting the customer experience in today's omni-channel environment [Redefiniendo la experiencia del cliente en el entorno omnicanal]. Universia Bus. Rev. **50**, 18–37 (2016)
46. Notomi, N., Tsukamoto, M., Kimura, M., Yamamoto, S.: ICT and the future of the retail industry - Consumer-centric retailing. NEC Tech. J. **10**(1), 38–41 (2015)
47. Pastoll, C., Rochwerg, T., Vlaar, B., Compeau, D.: Starbucks canada: the mobile payments decision. In: 35th International Conference on Information Systems. Build a better world through information systems, ICIS, New Zealand (2014)
48. Gupta, C., Lehman, D.: Managing customers as investments: the strategic value of customers in the long run. Pearson Prentice Hall, Upper Saddle River (2005)
49. Shankar, V., Balasubramanian, S.: Mobile marketing: a synthesis and prognosis. J. Interact. Mark. **23**(2), 118–129 (2009)
50. Voropanova, E.: Conceptualizing smart shopping with a smartphone, implications of the use of mobile devices for shopping productivity and value. Int. Rev. Retail Distrib. Consum. Res. **25**(5), 529–550 (2015)
51. Rapp, A., Baker, T., Bachrach, D., Ogilvie, J., Beitelspacher, L.: Perceived customer showrooming behavior and the effect on retail salesperson self-efficacy and performance. J. Retail. **91**(2), 358–369 (2015)
52. Heitz-Spahn, S.: Cross-channel free-riding consumer behaviour in a multichannel environment: an investigation of shopping motives, sociodemographics and product categories. J. Retail. Consum. Serv. **20**(6), 570–578 (2013)
53. Mills, A., Durepos, G., Wiebe, E.: Encyclopedia of case study research. Sage Publications, California (2010)

Organizational Impact on Software Development of eServices Techniques

Maurizio Cavallari[1,2(✉)], Francesco Tornieri[1], and Marco De Marco[2]

[1] Università Cattolica del Sacro Cuore, Largo Gemelli 1, 20123 Milan, Italy
{maurizio.cavallari,francesco.tornieri}@unicatt.it
[2] Università Telematica Internazionale UniNettuno,
Corso Vittorio Emanuele, 39, Rome, Italy
marco.demarco@uninettunouniversity.net

Abstract. New Software Development techniques with respect to cloud computing and eServices had modified IS architectures which were well established and consolidated in the past. The new methodologies of (software, micro/e) "Services" has pushed towards the adoption of software development organization independent from traditional tiered-architecture with the result of reducing both scale-up and down times as well as interruption times due to migration to different platforms. The eServices development organization relays on MicroServices architecture so that it is decomposing the legacy architecture in micro-components, each one with an independent life-cycle but interconnected and correlated, i.e. eServices. Each eService is hosted within a single container which has a proper software lifecycle and with minimal set of executable operating system libraries. The analysis goes into details about the structure and the development of eServices with MicroServices architecture. The paper discusses the new technological tendencies under the lens of an Organizational approach.

Keywords: Organization · Software development · eServices · Information system · Architectures · Software life-cycle · MicroServices · Scalability

1 Introduction

In the IT world in recent years, there has been a strong push in terms of innovation aimed at creating technology that would simplify, where possible, and speed up the deployment of new, or the maintenance of existing, software.

Two trends have emerged:

- Micro Servicing
- DevOps.

These trends enable the development of the application in a distributed way and they also make parts of the business functionality services available to developers [1].

They help to overcome many of the development limits making it easier to work collaboratively and to write/correct code independently [2].

S. Za et al. (Eds.): IESS 2017, LNBIP 279, pp. 64–75, 2017.
DOI: 10.1007/978-3-319-56925-3_6

Emerging field of DevOps borrows practices from software engineering to tackle complexity. The need for automation to improve scalability and testability while simultaneously reducing the operators' work, has always present in the software development research field.

The knowledge/skills/abilities needed for both software development and operator, encounter perfectly the four perspectives of DevOps, i.e. collaboration culture, automation, measurement, and sharing.

Consequently, when we think about software developmental we can say that decomposing the applications business logic into independent functional components has many advantages [3]:

- allows a development team to be small and highly focused
- can choose whichever technology best suits their specific use
- deliveries are composed by a set of small individual services modules
- could then be deployable on different servers and infrastructure.

2 Ops, Dev and DevOps

The concept of micro-services within the IT landscape and, consequently, within corporate IT departments, introduces the problem of the interrelationship between and management of new professional figures, such as DevOps, and system engineers (operations) and developers.

Traditional application development has several limits. One of them is that the application developers, have to build the internal logic as a single unit with all the rules for handling requests running in a single process [2].

Executives now face the task of recompiling the company organisation chart (where tasks and duties had, hitherto, been well defined) to take into account the roles of these experts, as outlined below:

- Operational/systems engineer: the duties of this professional figure are related to the configuration of bare metal systems or traditional virtualisation (also called heavy virtualisation) involving the use of batch to perform complex operations. Mainly related to legacy architectures that are scarcely or poorly integrated with cloud systems.
- Ops for developers: figures mainly involved in the commissioning of services or operational frameworks in the cloud in order to enable the developer to focus solely on the development and deployment of applications.
- Developer: a traditional figure focused only on software development with little or no knowledge of the architectural part (exclusively managed by operational/systems engineers). Each new software project required liaison with ops in order to have an environment/employment framework.
- DevOp: a professional figure with expertise in both dev and ops. The figure originates from the need to simplify the deployment of new applications, taking advantage of new technologies (e.g. Cloud and MicroServices), and to limit confusion typical in legacy environments. The role includes, albeit in embryonic form, QA functions (Quality Assurance) using CI (Continuous Integration) and CD (Continuous Delivery) software.

Figure 1 illustrates the workflow in legacy environments (where there are only systems engineers and developers). Previous research findings showed that self-adaptive software systems are capable of adjusting their behavior at runtime to achieve certain functional or quality-of-service goals [2, 3]. This profoundly affects the organization of software development [1–5].

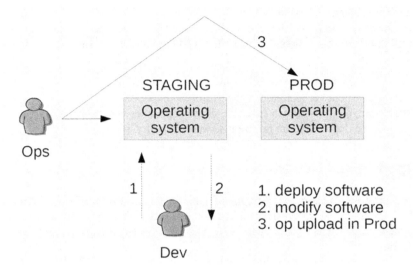

Fig. 1. Workflow in legacy environment

3   Literature Review

The new development approach for self-adaptive software relies on two innovations discussed by Malek et al. in 2010 [6]:

- an approach for representing engineers' knowledge of adaptation choices that are deemed practical, and
- an online learning-based approach for assessing and reasoning about adaptation decisions that does not require an explicit representation of the internal structure of the managed software system [6, 7].

Empirical evidences confirm that using a real-world self-adaptive software system could be of great help in eServices as it has shown for MicroServices ability to accurately learn the changing dynamics of the system while achieving efficient analysis and adaptation.

Scalability is achieved by replicating the application on different servers and through load balancing requests: this architectural structure is beneficial and is used successfully in data centers. The reasoning is that it divides the application into classes and functions under one deployable application [1].

The correlation between service and process mean it can be difficult to manage software changes and in case of scaling it requires to replicating the whole application.

Software development team organization is then greatly affected by this new approach [3, 5, 6].

A stream line of research has focused on Web applications as Services as the reference paradigm for next generation computing. The idea is of assembling application services into a network of services that can be loosely coupled to create flexible, dynamic business processes and agile software systems, independent from the organizations and from computing environments [6]. Developing such applications needs different approaching, from modeling and analysis issues in the early stages of their development, to issues related to implementation and on-the-run operational.

The eServices idea permits to host the application on different server so the resources can be more effectively attributes to those parts of the application that needs them [1, 2, 6].

However, the separation between the business logic and services there will need to be a communication system that permits the communications between these parts requires additional integration overhead [3].

These are other benefits of developing eServices utilizing MicroServices:

- Due to the small nature of each service it is easier for a developer develop and deploy it [1].
- Each service can be deployed separately from each other [2].
- Development and deployment becomes more efficient as small teams are required to work on individual service's components and they work independently using their own tools to completing the assignment without external limits [2, 3].
- Improved maintenance and troubleshooting, this is due to the fact that it becomes easier to find faults and isolate the problems in one service instead of including the others [1, 3, 5].
- It will be easier to introduce new and emerging technologies even in a monolithic environment as MicroServices can be built to proxy functionality and work alongside the monolithic application [1–3, 5].

The approach of dissecting a large and monolithic application needs to include the identification of areas that need to be either replaced or upgraded making sure that others services are not suffering [5, 6].

It is extremely important to remember to drive modularity through the requirement of change: this is to say that whatever needs to be changed needs to be changed at the same time and in the same module.

Notably services that change frequently needs to be kept separated from those who gets updated frequently in their modules [3, 6].

This also implicated that services who change with a frequent recurrence should be kept in the same module.

By comparing the use-case examples for functionality, the developer can divide the system using inbuilt functionality, which is beneficial, as the original developers designed them to be modular and independent [1, 3, 5].

When MicroServices cover only one well-defined business logic function, which has well defined inputs and outputs, the designer will invariably find the process of decomposing a massive enterprise application a lot easier [7]. The same results are shown from other research into applications with respect to supply-chain application domain [8].

Virtual Enterprises, mobile application security and electronic health record are two major examples of application domains where MicroService architecture and eServices solutions are seen as appropriate by independent research [9–11].

With respect to the organization of software development process, research has demonstrated that when the developer has chosen the best option to proceed with the decomposition of an existing monolithic application or to design and application, he will face the issue of integration [7].

There is no way around it, MicroServices makes integration tasks far more numerous and for those reason there is the need of a well-defined integration policy for protocols and technologies [1, 3, 5].

Other studies concentrate on services deliver to enterprise networks independently from the network shape and enabling efficiency and effectiveness of the business activities. Findings indicate that the deployment of eServices increases organizational productivity and effectiveness [12, 13].

4   The Organization of Legacy Team v/s DevOps

Figure 2 shows the workflow involving DevOps (dev and DevOps in this case may coincide). It shows how dev/DevOps carry out activities independently, in a completely different way from the past, creating the environment/framework as the need arises. This is enabled by the use of technologies such as containers, e.g. Dockers [1, 3, 6].

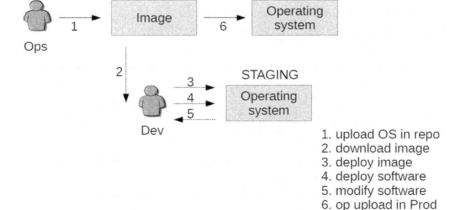

Fig. 2.   Dev/DevOps, carrying out of activities

It is necessary to highlight the differences between the professional figures in order to define the distinguishing characteristics of the DevOps and therefore the benefits.

Table 1 compares the legacy and DevOps approaches.

Table 1. Legacy and DevOps comparison.

Type	Ops/Dev (legacy)	DevOps
Project	Project	Small
Team structure	Group	Container
Deploy	Centric	Distributed
Information	Centric	Distributed
Service	Static	In deploy
Tecnology	Consolidated	New tech

Professionals using a traditional approach are involved in large projects (situations caused by maintenance/upgrade necessities of legacy services and the use of waterfall development methodologies), which necessarily involve an adequate workforce (ops and dev) to meet the deadline set in the design phase (usually the duration of this type of project is medium/long). In a context like this, there are few working groups (normally one per project) and information is kept within the working group (an approach known as silo mentality). By their nature, these projects exploit established technologies, run on internal or "private" cloud infrastructure and, as functions undergo few changes, service is static. The company will tend to make the best use of the existing infrastructure and legacy services and those in production [3–5].

The DevOps approach divides large projects into multiple small projects to reduce the complexity of implementation and the number of people involved on each the team (this is possible through an initial analysis and thanks to the use of agile development methodologies, for example, which are leaner than waterfall). This will result in a dramatic reduction in development time. In a situation like this, information must necessarily be shared between teams in order to stay aligned on the status of the project and, in order to facilitate members of the different teams, deployment espouses cloud logic ("public" or "hybrid"), thus permitting the use of micro-service technologies (which is also facilitated by the allocation of tasks imposed by division into micro groups). This approach encourages more dynamism in the teams and at the same time facilitates the introduction of new technologies/solutions typical of strongly cloud-based systems such as the CD/CI, as well as drastically reducing deployment time between updates.

On the basis of data collected through surveys conducted on medium/large companies, three different organizational models could be defined.

Figure 3 shows the organisation of a work team (Devs and Ops) in a traditional environment.

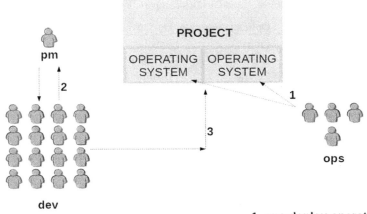

Fig. 3. Work team Devs and Ops in traditional environment

Often a representation that reflects the internal structure of the managed system is used to reason about its characteristics and make the adaptation decisions. This is particularly true in DevOps. As shown in Figs. 4 and 5, below.

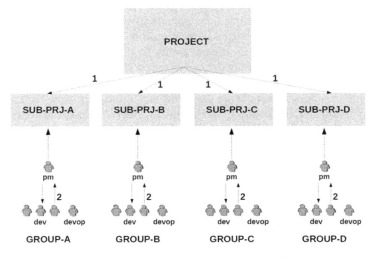

Fig. 4. DevOps working team structure with hierarchical subdivisions and PM under each

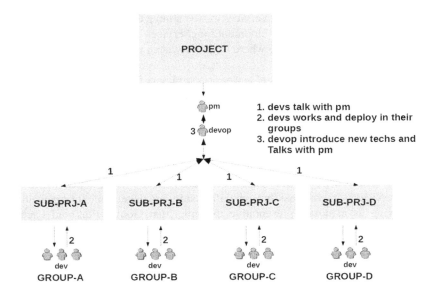

Fig. 5. DevOps working team structure with PM over all subdivisions

The companies applying the structure shown in Fig. 5, trigger problems in communication between the different groups and the PM and DevOps, thereby preventing adequate circulation of information both within the team and between the team and the PM/DevOps.

5   The eServices Development Protocols

The structure of eServices based on MicroServices development, show a communication with each other using messaging services protocols. The first protocol is synchronous and is used for communication which require an immediate action. The second are asynchronous and are used for subscribe/publish commands on line [7, 8].

MicroServices developments need to use a method of shared process communication between separate services.

The well-known method for inter-process communication is the Synchronous HTTP and is also used in eServices developments [7–9].

These types of services may present some performance problems in a congestion environment [1, 6, 8, 13].

Applications in a monolithic architecture are scaled by taking one instance of the application and replicating it how many times as is necessary to provide redundancy, performance and capacity so they need to be hosted on a different server and they are connected via load balancer devices, which will proxy the incoming client connections and forward them to a particular instance [14].

Load balancers maintain the state of the sessions (typically trough a session cookie) and assure that the client is connected to the correct instance of the application.

Creating an application using eServices can be a huge effort and take some time to have it run and ready. Software development for eServices only shows benefit when the applications reach a maturity stage where scalability is an issue as only then are all the benefits realized [10, 11].

Instead, it looks like development teams or contractors for startups use a monolithic architecture for fast and cheaper development to build applications ready to deploy and then they will address big issues of scalability at a later stage of business maturity.

6   Conclusions and Practical Implications

As discussed within the presented paper, MicroService and in particular the methodology of eServices is a real breakthrough within software development organization.

Considering as a most representative example, that Google-Plus and Facebook themselves are both based on MicroServices architecture, the most utilized social media platforms are developing highly scalable software utilizing eServices [14]. We can conclude that companies with extensive and substantial information systems, e.g. big banks or international corporations, could greatly benefit from research and practice of eServices software development. Continuous Integration – CI is a practice which has emerged to eliminate discontinuities between development and deployment. Similarly, the recent emphasis on DevOps and eServices shows that the integration between software development, Continuous Development – CD, and its operational deployment needs to be a continuous one [15].

The first conceptual finding is that eServices as part of DevOps and MicroService architecture can help industry to embrace changes in business and consequent software development [16]. New means of communications and understanding are emerging as the new software development team organizational forms, where appropriate and innovative language patterns in conjunction with a methodology to implement DevOps play as an efficient means for collaboration and automation purposes. Efficient collaboration and automation possible with eServices are proven to be the key enablers to implement continuous delivery and thus to react to changing business requirements timely.

The second research finding demonstrates that eServices sum up all concepts and strategies of new organizational forms for software development methodologies as indicated by scientific literature. Scholarly works have had constantly pointing to Platform-as-a-Service conceptual and architectural solution for large-scale and data intensive applications. This goal is achieved with eServices providing a customized and specific approach for PaaS platforms and integrating emerging paradigms such as DevOps for automate deployments [17].

As DevOps/eServices is a software development organization that permits software development team within enterprises, to rapidly deliver software product features through process automation, it favors greater collaboration and increased efficiency [18]. For this reasons two different adoption of DevOps eServices implementation within two different enterprises, would not be equal as each enterprise has unique characteristics and requirements and DevOps/eServices would adapt and reflect to particular organizational environment.

The response to increased demand of rapid developing and deploying, has been the eServices development environment where development and operations work close together. The response is even more appropriate as all software is being developed and operated in shared platforms with no formal requirement processes, and no analysis of the overall enterprise capabilities and architecture [19].

The eServices approach radically changes the enterprise internal software development methodology, CI and CD, so that an organizational change also occurs within software development teams. The corresponding compartmentalization die to granularity and independence of eServices, impacts the relationships among systems administrators, software development teams and business functions. This leads to conflicts between the mentioned three organizational functions, according to research findings [20]. Software development team support and cohesion, both channel the effects of relationship conflict and its management on team productivity, so eServices could act as a moderator in order to mitigate the mechanism of conflict that affects team performance.

Practical implications are related to the need of Business Continuity. From a (software) client point of view, all the 3 tier components of web based applications, are divided in a higher number of MicroServices. If in a frame – rss there would be a fault in the specific MicroService, the web application page would continue to function properly with respect to all service, excluding just the one in fault [21].

Other authors focus on a new organizational model to deliver transparent services and to improve effective collaboration. The identification of distinctive aspects of some services, gives support to present study outcomes [22]. Also Cloud Computing, heavily relying on eServices, and its significant impact on Corporate Responsability shall be considered with respect to general business strategy [23]. E-Government also is affected by development techniques of eServices [24] as highlighted in previous reserach.

This would be a real breakthrough for all enterprise and governmental applications where business continuity is a serious issue and even more when it is a legal requirements or compliance issue [25], e.g. in banks and large corporations.

7   Limitations and Scope for Future Research

The present research has been pursued both in theory and verified in practical environments, as test sessions in laboratory. A substantial empirical analysis is envisaged in order to confirm conceptual research findings and conclusions, in real organizational software development teams environment, living a huge scope for future investigation and research.

Acknowledgments. A preliminary version of the study was presented at the 13th conference of the Italian Chapter of AIS (Association for Information Systems) itAIS2016, Verona, I, October 7–8, 2016.

References

1. Newman, S.: Building Microservices: Designing Fine-Grained Systems. O'Reilly Media, Denver (2015)
2. Choudhary, B., Rakesh, S.K.: An approach using agile method for software development In: 1st International Conference on Innovation and Challenges in Cyber Security, ICICCS 2016, pp. 155–158 (2016). Article No. 7542304
3. Pokahr, A., Braubach, L.: Elastic component-based applications in PaaS clouds. Concurr. Comput. **28**(4), 1368–1384 (2016)
4. Cavallari, M.: Analysis of evidences about the relationship between organisational flexibility and information systems security. In: Information Systems: Crossroads for Organization, Management, Accounting and Engineering: ItAIS: The Italian Association for Information Systems, pp. 439–447 (2012)
5. Ciuffoletti, A.: Automated deployment of a microservice-based monitoring infrastructure. Procedia Comput. Sci. **68**, 163–172 (2015)
6. Autili, M., Tivoli, M., Goldman, A.: Thematic series on service composition for the future internet. J. Internet Serv. Appl. **7**(1) (2016). Article No. 3
7. Lwakatare, L.E., Karvonen, T., Sauvola, T., Kuvaja, P., Olsson, H.H., Bosch, J., Oivo, M.: Towards DevOps in the embedded systems domain: why is it so hard? In: Proceedings of the Annual Hawaii International Conference on System Sciences, pp. 5437–5446, March 2016. Article No. 7427859
8. Aiello, L., Dulskaia, I., Menshikova, M.: Supply chain management and the role of ICT: DART-SCM perspective. In: Ricciardi, F., Harfouche, A. (eds.) Information and Communication Technologies in Organizations and Society. LNISO, vol. 15, pp. 161–176. Springer, Cham (2016)
9. Spagnoletti, P., Za, S.: Securing virtual enterprises: requirements and architectural choices. Int. J. Electron. Commer. Stud. **4**(2), 327–336 (2013)
10. Cavallari, M., Adami, L., Tornieri, F.: Organisational aspects and anatomy of an attack on NFC/HCE mobile payment systems. In: Proceedings of the 17th International Conference on Enterprise Information Systems, ICEIS 2015, vol. 2, pp. 685–700 (2015)
11. Bonomi, S.: The electronic health record: a comparison of some European countries. In: Ricciardi, F., Harfouche, A. (eds.) Information and Communication Technologies in Organizations and Society. LNISO, vol. 15, pp. 33–50. Springer, Cham (2016)
12. Feitelson, D.G., Frachtenberg, E., Beck, K.L.: Development and deployment at Facebook. IEEE Internet Comput. **17**(4), 8–17 (2013). Article No. 6449236
13. Bellini, F., D'Ascenzo, F., Dulskaia, I., Savastano, M.: Digital service platform for networked enterprises collaboration: a case study of the NEMESYS project. In: Borangiu, T., Drăgoicea, M., Nóvoa, H. (eds.) IESS 2016. LNBIP, vol. 247, pp. 313–326. Springer, Cham (2016). doi: 10.1007/978-3-319-32689-4_24
14. Fitzgerald, B., Stol, K.: Continuous software engineering and beyond: trends and challenges. In: Proceedings of the 1st International Workshop on Rapid Continuous Software Engineering, RCoSE 2014, pp. 1–9 (2014)
15. Kuranuki, Y., Ushio, T., Yasui, T., Yamazaki, S.: A new business model of custom software development for agile software development. In: Proceedings of the International Workshop on Innovative Software Development Methodologies and Practices, InnoSWDev 2014, pp. 73–77 (2014)

16. Wettinger, J., Breitenbücher, U., Leymann, F.: DevOpSlang – bridging the gap between development and operations. In: Villari, M., Zimmermann, W., Lau, K.-K. (eds.) ESOCC 2014. LNCS, vol. 8745, pp. 108–122. Springer, Heidelberg (2014). doi:10.1007/978-3-662-44879-3_8
17. Tolosana-Calasanz, R., Bañares, J., Colom, J.-M.: On autonomic platform-as-a-service: characterisation and conceptual model. Smart Innov. Syst. Technol. **38**, 217–226 (2015)
18. Babar, Z., Lapouchnian, A., Yu, E.: Modeling DevOps deployment choices using process architecture design dimensions. In: Ralyté, J., España, S., Pastor, Ó. (eds.) PoEM 2015. LNBIP, vol. 235, pp. 322–337. Springer, Cham (2015). doi:10.1007/978-3-319-25897-3_21
19. Farroha, B.S., Farroha, D.L.: A framework for managing mission needs, compliance, and trust in the DevOps environment. In: Proceedings of the IEEE Military Communications Conference, MILCOM, pp. 288–293. IESS Press (2014). Article No. 6956773
20. Nesterkin, D., Porterfield, T.: Conflict management and performance of information technology development teams. Team Perform. Manage. **22**(5–6), 242–256 (2016)
21. Pokahr, A., Braubach, L.: Towards elastic component-based cloud applications. In: Camacho, D., Braubach, L., Venticinque, S., Badica, C. (eds.) Intelligent Distributed Computing VIII. SCI, vol. 570, pp. 161–171. Springer, Cham (2015)
22. Casalino, N., Buonocore, F., Rossignoli, C., Ricciardi, F.: Transparency, openness and knowledge sharing for rebuilding and strengthening government institutions. In: WBE 2013 Conference, pp. 866–871. IASTED-ACTA Press, Zurich, Innsbruck (2013). SSRN, https://ssrn.com/abstract=2553189
23. Patrignani, N., De Marco, M., Fakhoury, R., Cavallari, M.: Cloud computing: risks and opportunities for corporate social responsibility. In: Ricciardi, F., Harfouche, A. (eds.) Information and Communication Technologies in Organizations and Society, vol. 15, pp. 23–32. Springer, Cham (2016)
24. Casalino, N., Cavallari, M., De Marco, M., Gatti, M., Taranto, G.: Defining a model for effective e-government services and an inter-organizational cooperation in public sector. In: Proceedings of the 16th International Conference on Enterprise Information Systems, ICEIS 2014, vol. 2, pp. 400–408 (2014)
25. Cavallari, M.: A conceptual analysis about the organizational impact of compliance on information security policy. In: Snene, M. (ed.) IESS 2012. LNBIP, vol. 103, pp. 101–114. Springer, Heidelberg (2012). doi:10.1007/978-3-642-28227-0_8

Healthcare and the Co-creation of Value: Qualifying the Service Roles of Informal Caregivers

Maddalena Sorrentino[1(✉)], Nabil Georges Badr[2(✉)], and Marco De Marco[3]

[1] Department of Economics, Management and Quantitative Methods,
Università degli Studi di Milano, Milan, Italy
maddalena.sorrentino@unimi.it
[2] Grenoble Graduate School of Business, Grenoble, France
nabil.badr@alumni.grenoble-em.com
[3] Università Telematica Internazionale Uninettuno, Rome, Italy
marco.demarco@uninettunouniversity.net

Abstract. The study advances the debate on the co-creation of value in healthcare by treating the informal caregivers as a key organizational resource for the providers. Using the Dialogue, Access, Risk, and Transparency (DART) model developed by Prahalad and Ramaswamy as an interpretative key, this qualitative paper frames the role of the informal caregivers within the multiple experiences of value co-creation in which they are engaged. The central argument is that the informal caregiver performs three intersecting key roles: patient's advocate, system navigator and coordinator of care.

Keywords: Value co-creation · Co-production · Informal caregivers · Healthcare · Public sector

1 Introduction

There is not only growing acceptance of the need to include patients in the delivery of health services and increasing recognition of involving patients in the shared decision making of their treatment [1], but also increasing acknowledgement that healthcare customers can co-create value by integrating resources from healthcare providers and from others, outside the traditional healthcare setting (such as complementary therapies) and with the customer's private sources (such as peers, family, and friends), and through self-activities [2, p. 165, 3].

In a seminal article, Prahalad and Ramaswamy [4: 8–9] say that the real challenge for the healthcare provider is "to create an experience environment within which individual patients (consumers) can create their own unique personalized experience". In turn, the quality of that experience depends "on the nature of the involvement the customer (patient) has had in *co-creating* it with doctors, counselors, and others" [ibidem: our emphasis]. The two scholars identify four "building blocks of interaction" between the firm and consumers that shape and facilitate co-creation experiences [4]: the *dialogue* and communication between stakeholders; the ability of stakeholders to

© Springer International Publishing AG 2017
S. Za et al. (Eds.): IESS 2017, LNBIP 279, pp. 76–86, 2017.
DOI: 10.1007/978-3-319-56925-3_7

access and share data; the ability to monitor *risk-benefits*; and, finally, the *transparency* among stakeholders eliminating information barriers.

The premise of this paper is that growth and value creation are essential themes for public managers in a services-based economy [5]. In the healthcare field this means acknowledging that informal caregivers are a crucial link in the value chain of better healthcare outcomes [6]. A clear example of such an ecosystem is the healthcare services delivered to chronically ill patients. Hence, to get the highest benefit from the co-producing relationship, and thus boost value-creation potential, the provider must make an effort to learn about the patient and their personal and collective situation.

The informal caregivers who support patients in home settings are believed to improve ongoing and patient-tailored care but the extant research tends to treat them as an 'optional extra' at zero cost to the provider ultimately responsible for ensuring continuity of care. In other words, academia has not yet clarified what informal care-givers actually do when they co-create value *with* the healthcare provider. The paper attempts to bridge this knowledge gap, making the central argument that it is essential to understand the joint creation of value by the public healthcare provider and the informal caregivers in order to design and facilitate appropriate service delivery strat-egies. Thus, the main objective of our qualitative paper is to provide a deeper appreci-ation of the co-creation value of the informal caregivers within the multiple experiences in which they are engaged. This better understanding reflects one of the key priorities of research in service innovation [3].

We pursue this objective by (1) reviewing studies that examine the contribution of informal caregivers in care delivery from different perspectives; and (2) providing summaries of the outcomes and critical issues that, according to the academic research, are linked to the healthcare provider-informal caregiver interaction. The paper is a preliminary and, to the best of our knowledge, a first step in the effort to conceptualize the role of informal caregivers drawing on the four building blocks of interaction iden-tified by Prahalad and Ramaswamy in their DART (Dialogue, Access, Risk, and Trans-parency) framework.

2   Research Approach

A qualitative review of the studies on the informal caregivers' contribution to care delivery frames their role in the multiple co-creation of value experiences. Methodo-logically, the paper is articulated as a 'scoping study' [7, 8] with the aim of mapping research findings and identifying areas worth further attention.

The search for English-language articles published since 2000 that investigate informal caregiving from diverse academic perspectives generated a sample basket of 42 peer-reviewed papers (including empirical and conceptual contributions and several systematic reviews).

The thematic analysis of the selected papers proceeded in three steps. The first revealed the three key roles of informal caregivers as *patient's advocates, system navi-gators* or *gatekeepers,* and *coordinators of care.* The second, guided by the DART model developed by Prahalad and Ramaswamy, extrapolated the main topics of interaction between the informal caregivers and the public health provider. The third integrated the descriptive dimension (first step) with the value co-creation processes to better qualify the three roles in context.

The use of the model developed for the conceptualization of the relationship between firms and consumers might seem inappropriate here but serves our purpose well because the original DART framework identifies the 'blocks of interactions' between "the firm" and "the consumer". The different spatial and temporal settings in which the informal caregivers operate on behalf of the latter prompted the authors of this paper to apply the framework to the interactions between the health service provider(s) as a constituent of the "firm" and the informal caregivers acting *on behalf* of the patient as a representation of the "consumer".

Each author acted independently then discussed and compared the various viewpoints, addressing any interpretive variations and selecting the main themes for successive analysis.

3   Informal Caregivers

The content analysis of the 42 selected papers evidenced that the involvement of informal caregivers is a central theme in health systems: family and friends are the most important source of care for people with long-term conditions needs in OECD countries [9: 202]. The difference between an informal caregiver (or family carer or caregiver, informal carer or care provider or caregiver) and a healthcare professional (e.g. clinical staff, nurse, and physician) is that they are usually a family member, a close member of the patient's societal context, or a non-clinical social worker.

Most studies hold the view that informal caregiving is essential to better outcomes, more responsive care and lower costs [9, 10]. The literature acknowledges that the informal caregivers provide not only a range of emotional and instrumental support [11] and assistance with daily activities (e.g., transportation, meal preparation, shopping, locomotion and personal hygiene) [10], but also operate as *patient's advocates, system navigators* or *gatekeepers* of support and services and *coordinators of care.*

As *patient advocates* they provide a source of continuity of care for their care recipient during their transitions through care settings [11]. By assuming the role of spokespersons and intermediaries, they seek proper outcomes of decision-making on behalf of the care recipient [12]. They often act as "buffers from stresses stemming from responsibilities, anxieties, and discomforts" [13: 102]. Informal caregivers are *system navigators* who locate, evaluate and integrate relevant knowledge and information on behalf of the care recipient [11].

As *gatekeepers* of support and services, they assist the care recipient in the navigation of the often complex healthcare system [11]. Meanwhile *patient advocacy* helps to ensure continuous monitoring of patient risks [14], "and mitigate the consequences of inadequate levels of care provided by the formal caregivers" [12: 340].

Finally, as *coordinators of care*, informal caregivers facilitate and supplement the work of the medical staff [13], assisting with the coordination of care activities [15] such as paperwork management, solicitation of insurance approvals [10], scheduling medical appointments and coordinating care services [11], and negotiate care procedures by seeking information and establishing dialogue with formal care providers [12].

The literature review clarifies the multiple roles assumed by the informal caregivers as a result mainly of the implementation of co-productive models of care [14, 16, 17]. Another noticeable fact is that the focus of the mainstream co-creation and co-production research is mostly on the *dyadic* relation between the informal caregiver and the care recipient, i.e., the medical and wellbeing outcomes, and that, conversely, the studies that make the informal caregivers the main unit of analysis focus primarily on the psychological, economical and physical implications of the role on the individual carers. This highlights a discrepancy between the thrust of the research and what happens in the everyday experience of healthcare settings, a core feature of which is the *triadic* relation between the informal caregivers, the service recipient and the healthcare provider.

4   Applying the DART Framework

Applying the lens of the DART framework to the three interrelated roles of the informal caregivers identified in the previous section enables us to locate and track what actually happens on the ground.

4.1   Dialogue Between Formal and Informal Caregivers

Current research underscores the importance of constant communication and information exchange between formal and informal caregivers [12: 333]. The literature emphasizes that "the understanding of caregivers' needs, their varied experiences and the complex interactions between caregivers, healthcare professionals and patients is important if effective care is provided" [18: 154]. The prevailing tendency among providers is to focus on the patients and dismiss the needs of informal caregivers [15: 147]. Healthcare professionals recognize the informal caregiver's continuity-of-care role in the co-creation of value process [19] but often show them hostility and neglect [17], suggesting a divergence in the way formal and informal caregivers interpret and understand each other's needs. The general perception is that caregivers and care recipients are at the mercy of the healthcare services personnel [12]. This perceived injustice affects the wellbeing of the family caregivers, eroding their value potential instead of raising the professionals' awareness and appreciation of the caregiving tasks undertaken [20].

Patients and caregivers assign major importance to receiving both verbal and written information [24]. Extant literature suggests the consistent incorporation of novel ideas for patients and caregivers to manage their own conditions and to foster communication among the circle of care [21]. Informal caregivers place critical value on their relationship with professional care providers to receive recommendations, guidance and endorsement to sources of caregiving information [22]. This relationship is dynamic and ever changing [15]. A degradation of relational continuity between family caregivers and health professionals often leads to a loss of trust and the former's reluctance to receive guidance from the health professional [11: 542].

A relational coordination between formal providers and caregivers is supposed to improve caregiver's readiness to provide quality care [16] which is, in turn, positively associated with caregiving value (i.e., patients' freedom from pain, functional status, and mental health) [15]. This coordination would occur through frequent, high-quality communication that is supported by relationships of shared goals, shared knowledge, and mutual respect [23].

4.2   Access to Data for Informal Caregivers

The research [24] shows that ten years ago, when access to health records was restricted to solely the professionals, leaving the informal caregivers out of the loop, families were quicker to request information, guidance and support from the medical team and that the family carers valued the informal emotional support network that enabled them to share their experiences.

The internet then led to the proliferation of websites offering comprehensive information on the quality of care [25: 42], which, however, tended to become swiftly outdated or was too generic to be of specific use to the caregiver, often responsible for a home-bound care recipient [26]. To offset this, some websites offered a telephone helpline dedicated to informal carers requiring information on care and support services as well as related legal and financial advice (ibidem).

Today, both the patients and their caregivers can use a remote-processing device to access and upload data directly, reaping the benefits of the latest technological advances, such as assisted-living, telecare, telehealth, telemedicine, and the social media [27: 8]. "Sensors and context-awareness features allow for individualization and real-time information submission delivery" [28: 2]. Data can then be obtained in the form of gathering and analyses, monitoring and alerting (e.g., breathing monitors), diagnosis and treatment at distances (e.g. telepsychiatry), or communication [29]. Extending the caregiving assistance circle to the telecare monitoring-center workers would enable each call or alert to be assessed, also in terms of personnel requirements [30]. However, the as-yet unexplored ethics of remote assessment and treatment technology [27: 8], which could be perceived as a threat to personal autonomy and independence, makes integration into the care protocols a significant challenge [31: 7]. For example, the patient might deem the personal data too private to send over an open network and block its transmission [32]. Moreover, equipment checking/maintenance and other contingent care issues might add further pressure to the caregiver's responsibilities, triggering patient-carer tensions [31: 9].

Nevertheless, the positives of the ICT-based systems seem to far outweigh the negatives: secure messaging for communication among the patient care team [33]; platforms for scheduling medical appointments and coordinating treatment and services [11]; and even tailored educational materials for patient and informal caregiver [33]. Hence, in coordination with the formal caregivers [12: 337], the caregivers navigate the latest structured healthcare databases and systems that aggregate the information on patients, care professionals and care-related goals [33].

4.3 Risk Mitigation

Informal caregiving promotes the patient's psychological wellbeing [34]. Patients undergoing treatment in a familiar home setting tend to suffer less from depression than those who undergo disruptive hospital visits, especially in cases of chronic disease. However, the outcome for the caregiver - often described as a "stressful experience that may erode psychological well-being and physical health" [35] – is a longstanding concern. In fact, the psychological burden carried by the informal carers often translates into anxiety and depression, above all, if caring for the chronically ill [36] when the act of delivering care can seriously compromise the caregiver's quality of life [37].

To provide the proper measure of care, the caregiver must be inherently wired [35], i.e., they must possess physiological [20], psychological and mental readiness, as well as emotional [38], financial [37], and social preparedness [35, 39, 40]. For instance, the family caregivers must be able to lift and carry the patient without incurring personal injury. Hence, when modifying and adapting the home setting, it is doubly important to assess not only the needs of the patient and their clinical treatment, but also the carer's physical condition. The informal caregiver's depressed mental state often worsens if the patient is hospitalized. In fact, despite the support of the clinical structures and the delivery of an extended cycle of care by professional staff, the post-caregiving psychological impact can endure for weeks [41]. A distinction needs to be made between patient *information* and patient *education*. Informing the patient and the caregiver about the disease and its treatment improves their knowledge of the condition but does little to alleviate the caregiver's mood [42].

Other studies note that the lack of formal education made the informal caregivers' feelings of role overload worse [43] with the possibility, often underestimated, that they could be unwittingly exposed to a communicable disease. Such instances can lead to legal ramifications but, whilst the risk of wrongdoing in the professional healthcare sector is covered by accredited insurance, the activities of the informal care providers are not. Training and enhanced caregiver training programs were shown to increase the informal carer's self-efficacy for patient's symptom management [41]. The professional caregiver has the onus to build the competence of the informal caregiver through the facilitation of culturally-specific workshops, support and skills-training for caregivers [10]. This could be reinforced by adequate education and support from physicians and nurses during the activities of care [44]. However, some studies have shown that there were no reported changes in the levels of depression, anxiety, and burden [45].

4.4 Transparency of Information Flow Between Formal and Informal Caregiver

There is scant academic work to understand the different preferences of caregivers for receiving information [25]. In general, patients and caregivers lament the scarcity of the information issued by healthcare staff [46]. Studies emphasize that "healthcare professionals could resist giving differential information, citing causes of tensions and interpersonal conflicts" [25: 42]. This lack of transparency of information between care providers prevents the creation of a virtuous circle of care and suggests the need to organize seamless care among care providers [47].

Patient engagement and family caregiver interventions have been shown to improve self-regulated disease management [48]. Informal and family carers are motivated to obtain appropriate information, securing access to services and acting as informal advocates in negotiations with service providers [49]. Nevertheless, the cascade of care-work and responsibilities from professional caregivers to family carers changes the models of data quality governance and proliferation [30: 352]. Moreover, fear expressed by both patients and professionals regarding security and privacy of electronic data files impedes the transparency of information flow [47].

The intersection between the roles of the informal caregivers and the four building blocks of interaction shown in Table 1 suggests that there are multiple spaces and temporal settings of potential co-creation experiences. Further, the table indicates (in tune with Sharma and colleagues [5]) that co-created value is multifaceted, and that all three roles cast the informal caregiver as a potential "operant" resource who co-creates value with the "firm" being used for the benefit of the service recipient [50: 7]. Table 1 provides an initial response to the question asked by the public managers interested in working with the informal caregivers to build a more co-creative environment of "What are the critical attributes and behaviors to consider?"

Table 1. The Dialogue, Access, Risk, and Transparency (DART) map and the intersecting roles of informal caregivers (authors' elaboration)

	Patient's advocate	System navigator & Gatekeeper	Coordinator of care
D	Communication and information exchange between caregivers [11, 12, 15, 21, 22, 26]	Understanding of needs between formal and informal caregivers [12, 17, 18, 20]	Coordination supported by relationships of shared goals/shared knowledge [15, 16, 23]
A	Data access issues can impede the effectiveness of informal caregivers [27, 30–32]	Access to multiple sources enables the exchange of information on patient condition [11, 24–26, 29]	Technological access to useful data optimizes the coordination of care [12, 27, 28, 33]
R	Problem-solving approach to reduce patient health risks [34, 41]	Possible induced risk to the informal caregiver [36, 37, 39, 42, 43, 45]	Informal caregiver preparedness [10, 20, 35, 37–41, 44]
T	Care-recipient spokespersons and intermediaries with service providers [49]	Transparency of information flow impeded by fears over security and loss of privacy [30, 47]	Improved disease management through patient engagement and family caregiver assistance [48]

5   Final Remarks and Conclusions

As one of the first studies to explore the value co-creation role of informal caregivers, the authors build on the DART framework of Prahalad and Ramaswamy to evidence how the three roles of patient's advocate, system navigator and coordinator of care performed by the informal caregivers intersect with the four interaction blocks of dialogue, access, risk, and transparency.

The paper makes several contributions. First, it offers a rich body of evidence to reflect on the practice of value co-creation in healthcare settings. Second, it illuminates those areas in which the often hidden contribution of the informal caregivers occurs across the blocks of interaction with health service providers. Third, it confirms the essential role of the public health provider tasked with the overall management of the care service experience in both virtual and physical space. The study also identifies several hurdles to co-creation, such as the communication barriers cited in Sect. 4. This critical aspect suggests that the key condition of integrating the knowledge and experiential expertise of the laypeople in the organizational routines of the healthcare providers persists in its absence, resulting in the failure to channel these resources across boundaries and the waste of crucial assets.

The study's limitations certainly include the fact the DART approach focuses exclusively on the value creation inputs of the informal caregivers without going into their potential neutral or negative impacts on the value-creation ecosystem, such as when the caregiver incurs personal/social, physical, emotional, psychological and economic costs, as evidenced by quite a few studies [9, 11, 31, 51]. A no less important limitation is the multiple contexts (e.g., childcare, palliative medicine, long-term care), countries, patient populations, clinical conditions/stages and study designs addressed by the reviewed literature. Moreover, the inclusion and classification criteria are the personal choices of the authors of this paper.

Finally, the preliminary findings need to be elaborated and corroborated via a future systematic review of the studies that explore the value of informal caregiving in specific or particular circumstances.

References

1. Nambisan, P., Nambisan, S.: Models of consumer value cocreation in health care. Health Care Manag. Rev. **34**(4), 344–354 (2009)
2. McColl-Kennedy, J.R.: Health care in service science. Australas. Mark. J. **22**(3), 165–167 (2014)
3. McColl-Kennedy, J.R., Vargo, S.L., Dagger, T.S., Sweeney, J.C., van Kasteren, Y.: health care customer value cocreation practice styles. J. Serv. Res. **15**(4), 370–389 (2012)
4. Prahalad, C.K., Ramaswamy, V.: Co-creation experiences: the next practice in value creation. J. Interact. Mark. **18**(3), 5–14 (2004)
5. Sharma, S., Conduit, J.: Cocreation culture in health care organizations. J. Serv. Res. **19**(4), 438–457 (2016)
6. Gittell, J.H., Beswick, J., Goldmann, D., Wallack, S.S.: Teamwork methods for accountable care: relational coordination and TeamSTEPPS. Health Care Manag. Rev. **40**(2), 116–125 (2015)
7. Arksey, H., O'Malley, L.: Scoping studies: towards a methodological framework. Int. J. Soc. Res. Methodol. **8**(1), 19–32 (2015)
8. Anderson, S., Allen, P., Peckham, S., Goodwin, N.: Asking the right questions: scoping studies in the commissioning of research on the organisation and delivery of health services. Health Res. Policy Syst. **6**(1), 1–17 (2008)
9. OECD: Health at a Glance. OECD, Paris (2015)
10. Lum, J.: Informal caregiving. In: Focus Backgrounder, pp. 1–11 (2011)

11. Ghazzawi, A., Kuziemsky, C., O'Sullivan, T.: Using a complex adaptive system lens to understand family caregiving experiences navigating the stroke rehabilitation system. BMC Health Serv. Res. **16**(538), 1–10 (2016)
12. Bragstad, L.K., Kirkevold, M., Foss, C.: The indispensable intermediaries: a qualitative study of informal caregivers struggle to achieve influence at and after hospital discharge. BMC Health Serv. Res. **14**(1), 331–342 (2014)
13. Casey, V., Crooks, V.A., Snyder, J., Turner, L.: Knowledge brokers, companions, and navigators: a qualitative examination of informal caregivers' roles in medical tourism. Int. J. Equity Health **12**(1), 1–10 (2013)
14. Sorrentino, M., Guglielmetti, C., Gilardi, S., Marsilio, M.: Health care services and the coproduction puzzle: filling in the blanks. Adm. Soc. **9**, 1–26 (2015)
15. Weinberg, D.B., Lusenhop, W., Gittell, J.H., Kautz, C.M.: Coordination between formal providers and informal caregivers. Health Care Manag. Rev. **32**(2), 140–149 (2007)
16. Cramm, J.M., Nieboer, A.P.: The changing nature of chronic care and coproduction of care between primary care professionals and patients with COPD and their informal caregivers. Int. J. COPD **11**, 175–182 (2016)
17. Palumbo, R.: Contextualizing co-production of health care. Syst. Lit. Rev. Int. J. Public Sect. Manag. **29**(1), 72–90 (2016)
18. Docherty, A., Owens, A., Asadi-Lari, M., Petchey, R., Williams, J., Carter, Y.: Knowledge and information needs of informal caregivers in palliative care: a qualitative systematic review. Palliat. Med. **22**(2), 153–171 (2008)
19. Sorrentino, M., Marco, M., Rossignoli, C.: Health care co-production: co-creation of value in flexible boundary spheres. In: Borangiu, T., Drăgoicea, M., Nóvoa, H. (eds.) IESS 2016. LNBIP, vol. 247, pp. 649–659. Springer, Cham (2016). doi:10.1007/978-3-319-32689-4_49
20. Mohammadi, S., de Boer, M.J., Sanderman, R., Hagedoorn, M.: Caregiving demands and caregivers' psychological outcomes: the mediating role of perceived injustice. Clin. Rehab. 1–11 (2016)
21. Matthew-Maich, N., Harris, L., Ploeg, J., Markle-Reid, M., Valaitis, R., Ibrahim, S., Gafni, A., Isaacs, S.: Designing, implementing, and evaluating mobile health technologies for managing chronic conditions in older adults: a scoping review. JMIR mHealth uHealth **4**(2), 1–18 (2016)
22. Peterson, K., Hahn, H., Lee, A.J., Madison, C.A., Atri, A.: In the Information age, do dementia caregivers get the information they need? Semi-structured interviews to determine informal caregivers' education needs, barriers, and preferences. BMC Geriatr. **16**(164), 1–13 (2016)
23. Gittell, J.H.: Coordinating mechanisms in care provider groups: relational coordination as a mediator and input uncertainty as a moderator of performance effects. Manag. Sci. **48**(11), 1408–1426 (2002)
24. Magnusson, L., Hanson, E., Brito, L., Berthold, H., Chambers, M., Daly, T.: Supporting family carers through the use of information and communication technology. The EU project ACTION. Int. J. Nurs. Stud. **39**(4), 369–381 (2002)
25. Washington, K.T., Meadows, S.E., Elliott, S.G., Koopman, R.J.: Information needs of informal caregivers of older adults with chronic health conditions. Patient Educ. Couns. **83**(1), 37–44 (2011)
26. Lauriks, S., Reinersmann, A., Van der Roest, H.G., Meiland, F., Davies, R.J., Moelaert, F., Mulvenna, M.D., Nugent, C.D., Dröes, R.-M.: Review of ICT-based services for identified unmet needs in people with dementia. Ageing Res. Rev. **6**(3), 223–246 (2007)
27. Greenhalgh, T., Procter, R., Wherton, J., Sugarhood, P., Shaw, S.: The organising vision for telehealth and telecare: discourse analysis. BMJ Open **2**(4), 1–13 (2012)

28. Hamine, S., Gerth-Guyette, E., Faulx, D., Green, B.B., Ginsburg, A.S.: Impact of mHealth chronic disease management on treatment adherence and patient outcomes: a systematic review. J. Med. Internet Res. **17**(2), 1–15 (2015)

29. Gentles, S.J., Lokker, C., McKibbon, K.A.: Health information technology to facilitate communication involving health care providers, caregivers, and pediatric patients: a scoping review. J. Med. Internet Res. **12**(2), 1–17 (2010)

30. Milligan, C., Roberts, C., Mort, M.: Telecare and older people: who cares where? Soc. Sci. Med. **72**(3), 347–354 (2011)

31. Eccles, A.: The complexities of technology-based care: telecare as perceived by care practitioners'. Issues Soc. Sci. **1**(1), 1–20 (2013)

32. Stowe, S., Harding, S.: Telecare, telehealth and telemedicine. Eur. Geriatr. Med. **1**(3), 193–197 (2010)

33. Robben, S.H., Perry, M., Huisjes, M., van Nieuwenhuijzen, L., Schers, H.J., van Weel, C., Rikkert, M.G.O., van Achterberg, T., Heinen, M.M., Melis, R.J.: Implementation of an innovative web-based conference table for community-dwelling frail older people, their informal caregivers and professionals: a process evaluation. BMC Health Serv. Res. **12**(1), 251–263 (2012)

34. Folkman, S.: Positive psychological states and coping with severe stress. Soc. Sci. Med. **45**(8), 1207–1221 (1997)

35. Pinquart, M., Sorensen, S.: Spouses, adult children, and children-in-law as caregivers of older adults: a meta-analytic comparison. Psychol. Aging **26**(1), 1–14 (2011)

36. McCullagh, E., Brigstocke, G., Donaldson, N., Kalra, L.: Determinants of caregiving burden and quality of life in caregivers of stroke patients. Stroke **36**(10), 2181–2186 (2005)

37. Giordano, A., Cimino, V., Campanella, A., Morone, G., Fusco, A., Farinotti, M., Palmisano, L., Confalonieri, P., Lugaresi, A., Grasso, M.G.: Low quality of life and psychological wellbeing contrast with moderate perceived burden in carers of people with severe multiple sclerosis. J. Neurol. Sci. **366**, 139–145 (2016)

38. Hoerger, M., Coletta, M., Sörensen, S., Chapman, B.P., Kaukeinen, K., Tu, X., Duberstein, P.R.: Personality and perceived health in spousal caregivers of patients with lung cancer: the roles of neuroticism and extraversion. J. Aging Res. 1–7 (2016)

39. Haley, W.E., Roth, D.L., Coleton, M.I., Ford, G.R., West, C.A., Colllins, R.P., Isobe, T.L.: Appraisal, coping, and social support as mediators of well-being in black and white family caregivers of patients with Alzheimer's disease. J. Consult. Clin. Psychol. **64**(1), 121–129 (1996)

40. Alvira, M.C., Risco, E., Cabrera, E., Farré, M., Rahm Hallberg, I., Bleijlevens, M.H., Meyer, G., Koskenniemi, J., Soto, M.E., Zabalegui, A.: The association between positive–negative reactions of informal caregivers of people with dementia and health outcomes in eight European countries: a cross-sectional study. J. Adv. Nurs. **71**(6), 1417–1434 (2015)

41. Hendrix, C.C., Bailey Jr., D.E., Steinhauser, K.E., Olsen, M.K., Stechuchak, K.M., Lowman, S.G., Schwartz, A.J., Riedel, R.F., Keefe, F.J., Porter, L.S.: Effects of enhanced caregiver training program on cancer caregiver's self-efficacy, preparedness, and psychological wellbeing. Support. Care Cancer **24**(1), 327–336 (2016)

42. Hafsteinsdóttir, T.B., Vergunst, M., Lindeman, E., Schuurmans, M.: Educational needs of patients with a stroke and their caregivers: a systematic review of the literature. Patient Educ. Couns. **85**(1), 14–25 (2011)

43. Yates, M.E., Tennstedt, S., Chang, B.-H.: Contributors to and mediators of psychological well-being for informal caregivers. J. Gerontol. Ser. B Psychol. Sci. Soc. Sci. **54**(1), 12–22 (1999)

44. Lehoux, P.: Patients' perspectives on high-tech home care: a qualitative inquiry into the user-friendliness of four technologies. BMC Health Serv. Res. **4**(1), 1–9 (2004)

45. O'Toole, M.S., Zachariae, R., Renna, M.E., Mennin, D.S., Applebaum, A.: Cognitive behavioral therapies for informal caregivers of patients with cancer and cancer survivors: a systematic review and meta-analysis. Psycho-Oncol. 1–10 (2016)

46. Wachters-Kaufmann, C., Schuling, J., The, H., Meyboom-de Jong, B.: Actual and desired information provision after a stroke. Patient Educ. Couns. **56**(2), 211–217 (2005)

47. Van Durme, T., Macq, J., Anthierens, S., Symons, L., Schmitz, O., Paulus, D., Van den Heede, K., Remmen, R.: Stakeholders' perception on the organization of chronic care: a SWOT analysis to draft avenues for health care reforms. BMC Health Serv. Res. **14**(179), 1–9 (2014)

48. Wallerstein, N.: What is the Evidence on Effectiveness of Empowerment to Improve Health?. World Health Organization, Copenhagen (2006)

49. Glendinning, C.: Increasing choice and control for older and disabled people: a critical review of new developments in England. Soc. Policy Adm. **42**(5), 451–469 (2008)

50. Lusch, R.F., Vargo, S.L., O'Brien, M.: Competing through service: Insights from service-dominant logic. J. Retail. **83**(1), 5–18 (2007)

51. Sharma, S., Conduit, J., Rao Hill, S.: Organisational capabilities for customer participation in health care service innovation. Australas. Mark. J. **22**(3), 179–188 (2014)

Exploring the Relationship Between Corruption and Health Care Services, Education Services and Standard of Living

Paulo Morais, V.L. Miguéis$^{(\boxtimes)}$, and Ana Camanho

Faculdade de Engenharia da Universidade do Porto, Porto, Portugal
pauloteixeirademorais@gmail.com, {vera.migueis,acamanho}@fe.up.pt

Abstract. Understanding the impact of corruption in modern societies, namely in standard of living, health and education services, is an issue that has attracted increased attention in recent years. This paper examines the relationship between the Corruption Perception Index (CPI) provided by Transparency International and the Human Development Index (HDI) of the United Nations Development Program and its components. The analysis is done for clusters of countries with similar levels of development. For the countries with high levels of development, it was found a negative relationship between corruption and human development. Moreover, for these countries, higher corruption levels are related to poor health care services, poor education services and low standard of living. For the other clusters of countries, these relationships were not statistically significant. The results obtained reinforce the importance of efforts by international politicians and organizations in fighting corruption, particularly in highly developed countries, to promote development.

Keywords: Health care services · Education services · Corruption · Human development

1 Introduction

The Human Development Index (HDI), provided by the United Nations Development Program, has been extensively used in research on human development, being widely recognized as a robust and valid measure of the access to basic social services, especially health and education, and economic well-being of populations. Consequently, the HDI is based on the measurement of people's ability to live a long and healthy life, the ability to acquire knowledge and the ability to achieve a decent standard of living [1]. The components of the HDI are health (measured by the life expectancy at birth), education (measured by the adult literacy rate, the combined primary, secondary, and tertiary education enrollment rates) and income (measured by per capita gross domestic product). The index is a composite measure of these three indicators and varies from 0 (the worst) to 1.0 (the best).

© Springer International Publishing AG 2017
S. Za et al. (Eds.): IESS 2017, LNBIP 279, pp. 87–100, 2017.
DOI: 10.1007/978-3-319-56925-3_8

The HDI was proposed with the intention to overcome the limitations associated with the assessment of human development based only on economic issues [2]. Nevertheless, the index has its own limitations. Some authors noted that the index cannot account for the intrinsic variability of the basic indicators, as distributional issues are not reflected in the HDI measure. Harttgen and Klasen (2012) [3] proposed an evaluation of the distribution of human development within a country, developing a method to assess human development at the household level. This procedure was applied to 15 developing countries. Permanyer (2013) [4] explored the spatial distribution of HDI, using census data from Mexico municipalities. Variability was found within these spatial levels in each of the studies, but it was explained by different factors, while in Mexico inequality was associated to the wealth component [4], in the 15 countries analysed by Harttgen and Klasen (2012) [3], inequality was associated to the educational and income components of HDI.

Other authors discussed the limitations of the HDI [5] and argued that this measure is too reductionist [6] invoking the need for an extended human development measure. Furthermore, Herrero et al. (2012) [7] criticized the type of normalization adopted and the approached used to evaluate the educational achievements. Other authors [8–10] contested the equal weighting given to the three components of the index. The United Nations were not indifferent to these claims, which resulted in changes to the computational procedure used to estimate HDI over time. Until 2009 the HDI was calculated using an equally weighted sum of the three components. The most significant change occurred in 2010, when this index started to be calculated using a multiplicative approach to aggregate the three dimensions [11].

Despite the unsettled controversy regarding the specification of the HDI, it remains widely used in the literature [12–14]. It is considered to be a reflexive image of the environment in which people can fully develop their potential to a better life.

Corruption is defined by the Transparency International (TI), a international non-governmental organization founded in 1993 and based in Berlin, as an "abuse of public office for private gain". Corruption deprives citizens of access to basic goods and services they need to live, such as health care and education. The mission of TI is to stop corruption and promote transparency, accountability and integrity at all levels and across all sectors of society [15]. One of its major contributions is the presentation of an annual ranking of countries, based on a corruption perception index (CPI). This index is largely used as a proxy for corruption in academic research [12,14,16,17]. As stated by Grae and Mehlkop (2003) [18], "the measurement of perceived corruption includes (...) assessments of the extent of illegal behaviour in a country in general".

The CPI measures the degree to which corruption is perceived to exist in the public sector. The CPI is a composite index that draws on several polls and surveys carried out by independent institutions. Some of these institutions are the African Development Bank, Bertelsmann Foundation, Economist Intelligence Unit, Freedom House, World Bank and World Economic Forum. Following a rigorous review process, in 2012 some important changes were made to the methodology used to obtain the CPI. The method used to aggregate the different data sources was simplified, and now only one year of data is used from each data source, to enable the assessment of the evolution of the index over time [19].

Ugur and Dasgupta (2011) [20] addressed the problems of dealing with perceptions-based data in scientific studies, as these may not be reliable. The literature has noted that the perception of corruption may suffer the effect of reverse causality, meaning that perception may be influenced by the economic performance. Ugur and Dasgupta (2011) [20] demonstrated that this effect does not hold when long-run economic growth is controlled for, and that the alleged bias of the surveys, over-representing business representatives, is less serious than suspected. So, despite the inherent subjectivity that results from the fact of attempting to measure, the CPI is considered a valid indicator, with strong correlations with other international indices, such as the Black Market Activity Index and Excess Regulation Index [21].

The aim of this paper is to explore the relationship between corruption (measured by the CPI) and countries' standard of living, health care and education services. These are the three main drivers of human development, which can be measured by the Human Development index (HDI). Although the link between human development and corruption has been largely explored, a robust quantitative assessment of the impact of corruption on the HDI and its primary indicators, e.g. health and education services, has been neglected. This paper proposes to use a cluster analysis in order to separate countries in a few groups with homogeneous development levels, meaning they are "neighbours in development".

The remainder of this paper is organized as follows. Section 2 provides an overview of the literature on the relationship between human development and corruption. Section 3 introduces the methodology and data used in this study. Section 4 discusses the results, analyzing the relationship between the HDI and the CPI within the clusters identified. The paper finishes with the conclusion and discussion of some policy implications.

2   Factors Influencing Living Standards and Social Services

It is commonly assumed that economic growth is associated with enhancements to living standards and social services, defined as human development. Ranis et al. (2000) [22] provided evidence of a significant mutual relationship between human development and economic growth. In particular, both dimensions are perceived as being interconnected, by a two-way chain. The chain that links economic growth to human development is particularly affected by public expenditures on health and education services. Human development, on the other hand, can lead to economic growth by changing investment rate and income distribution. The authors suggest that economic reforms must focus on human development from the beginning, as "economic growth itself will not be sustained unless preceded or accompanied by improvements in HD" [22]. Suri et al. (2011) [23] provided further evidences of this strong connection, arguing that "human development levels are important for determining growth trajectories and that policies to improve human development must precede or at least complement growth-oriented policies if growth is to be accelerated and sustainable". Baldacci et al. (2008) [24] analyzed the individual factors of the HDI and demonstrated that public spending in health and education services raises education and health capital, which has a positive and significant indirect effect on growth.

Several researchers conducted analysis on other factors with a positive effect on countries development, namely economic freedom, globalization and foreign direct investment (FDI).

The literature provides evidence of the positive link between economic freedom and growth, although the strength of the connection between these variables varies among countries [25, 26].

The influence of globalization on development is not so clear. Regarding globalization, Lalountas (2011) [16] concluded that globalization has no impact on human development in low income countries. In this context GDP appears to be the most important driver of human development. Singh (2011) [27] suggested that globalization influences micro and macro-economic factors that influence human development.

Foreign direct investment (FDI), an important source of capital for developing countries, has also been found to be linked to human development as the inflow of capital should promote economic growth. However, the impact of FDI is variable, depending on factors such as national policies and levels of corruption. The influence of FDI inflow on human development is greater when countries adopt policies to guide the investment to strategic areas (not leaving the decision up to the investors), and when the levels of corruption are low [14].

The identification of the factors with a negative impact on human development has also been analyzed in several studies, with corruption standing prominently. Corruption and development have been described as "decisively linked" [28] because corruption hinders economic growth and underdevelopment creates environments that favor unethical practices. Corruption has a very distinctive implantation worldwide, shaped by, along other factors, the existence of long-lasting institutions that go back to colonial times [29], protestant traditions or exposure to democracy [30] and trade openness [31]. Corruption is also a problem in many developing countries which are resources rich [32]. Educational achievements of the population are important in tackling corruption, as citizens, provided with a more transparent government can hold their members accountable [32] and change social norms of tolerance towards corruption [28].

The findings of studies on the impact of corruption on development are not consensual and further research is necessary to clarify this topic. Previous studies showed that corruption may be a promoting factor of economic growth by "greasing the wheels" of growth [33–35]. This perspective was contradicted by empirical evidences provided by Mauro (1995) [36], Mo (2001) [37] and

Meon (2005) [38], that suggested that corruption directly prevents growth. Gupta (2002) et al. [39] reported that corruption is associated to income inequality and poverty. Mo (2001) [37] found that a 1% increase in the corruption level may lead to a 0.72% reduction of growth rate. Political instability is the most important channel affecting economic growth, while corruption also reduces the level of human capital and the share of private investment. Mauro (1998) [40] also found that corruption affects the composition of government expenditure, namely that education services are found to be adversely affected by corrupt practices. Other studies confirm the impact of corruption on public spending, enlarging the identification of the components affected. Delavallade (2006) [41] proved that public corruption reduced the portion dedicated to social expenditure, such as education, health and social protection, which are key instruments of human development.

However, other studies found no association between corruption and human development. This is the case of Ayadi (2004) [13] that concluded that standard economic performance by itself does not prevent pervasive corruption, nor is economic success an infallible sign of innocence. Using Spearman and Hotelling-Pabst tests, these authors found no association between CPI and HDI, which indicates that corruption is not an exclusive disease of underdeveloped countries. Also Frechette (2006) [42] studied the corruption determinants and found disturbing evidence that an increase in education attainment or in income may increase corruption, considering that corruption has a procyclical nature.

These somehow contradictory results show the pertinence of investigating the relationship between corruption and human development. While economic conditions seem to play an important role in human development, the relationship between corruption and human development is not clear. On the other hand, some studies provided evidence that analysis with aggregate data can be misleading, as countries or world regional areas, with their individual context, can disrupt linear correlations [14,16,21,43]. For instance, Hysa (2011) [44] conducted research on the corruption and Human Development in Western Balkan Countries, and found that the relationship between corruption and human development is strong in Former Yugoslav Republic of Macedonia, Serbia, Montenegro and Albania, while in other cases this relationship is weak (e.g. Croatia) or meaningless (e.g. Bosnia and Herzegovina). In another study, Hysa (2011) [45] found a much stronger relationship between corruption and human development in Albania than in EU countries.

Although previous studies have used different methodologies and focused on different time periods and geographical contexts, it is clear that the relationship between corruption and human development is worth exploring with more detail. It is of utmost importance to gain insights about how the force of the relationship between corruption and human development varies for different levels of the dimensions underlying human development. Furthermore, it is also important to analyze the strength of the relationships between corruption and the different components of HDI.

3   Data and Methodology

International organizations such as the United Nations, the World Bank and the International Monetary Fund have been carrying out an ongoing debate and battle against corruption, considering that it hampers human development and economic growth. In particular, the fight against corruption has been considered of utmost importance to achieve the millennium development goals and to promote human development. The objective of this paper is to explore the relationship between corruption and human development in natural clusters of countries based on the components of human development, called development neighbours, using recent data. Furthermore, it intends to assess the impact of corruption levels on specific components of human development, namely health, education and economic aspects.

The CPI data was collected from the repository provided by Transparency International. This data corresponds to the corruption perception index of 2013, which is defined for 177 countries. The HDI data and its components were collected from the United Nations Development Programme repository and refer to 2013. The data is available for 187 countries.

The study is focused on 171 countries, which are common to both the CPI and HDI rankings. This set of countries includes 31 European Union countries, 31 American countries, 46 African countries, 26 Asian-Pacific countries, 18 East-European countries and 19 Middle East countries. Table 1 presents the summary statistics of the indices used.

Table 1. CPI, HDI and subindexes of human development.

Geographical area	CPI		HDI		Education		Health		Income	
	Mean	St Dev	Mean	St Dev	Mean	St Dev	Mean	St Dev	Mean	St Dev
European Union	65.19	15.82	86.27	4.07	82.01	4.97	90.65	4.96	86.54	5.37
America	44.32	18.67	72.82	8.68	64.57	10.63	83.76	6.28	71.81	10.38
Africa	33.50	11.28	49.36	10.57	43.37	12.11	59.35	10.09	48.46	15.23
Asian-Pacific	41.88	23.37	69.12	14.80	59.50	17.69	81.16	10.41	69.30	16.92
East Europe	32.72	10.45	72.25	5.22	70.19	6.45	78.88	5.98	68.69	9.44
Middle East	37.00	17.63	72.98	11.29	60.95	13.50	82.70	7.72	77.95	14.35
Total	42.79	19.83	68.34	16.07	61.45	17.57	77.42	14.04	68.17	18.39

As mentioned before, the health component of HDI is measured by the life expectancy at birth; education is measured by the adult literacy rate, the combined primary, secondary, and tertiary education enrolment rates; and income is measured by per capita gross domestic product.

It is important to note that CPI ranges between 0 to 100, where 0 represents the highest level of perceived corruption, and 100 the lowest level of perceived corruption (meaning higher transparency). The HDI and its components range between 0 and 1 (0 represents the lowest level of development and 1 the highest level of development). For our analysis we converted the HDI values to a scale between 0 and 100, such that both indices were measured in the same scale and in both cases a higher level of the indicator is desirable.

A close observation of the data displayed in Table 1 allows us to note that Africa presents the lowest levels of human development, both as evaluated by the HDI or by each of its individual components. The lowest level of transparency is registered for East European countries. On the other side, the European Union presents the lowest levels of corruption and the highest scores of all the components of the HDI. These remarks, however, do not entitle us to draw conclusions on the association between HDI and CPI.

The methodology proposed involves grouping countries according to the information provided by the human development components, using a cluster analysis. Despite its popularity in other contexts, the development of cluster analysis in order to find natural groups of countries considering their levels of health, education and standard of living in the literature is very incipient. Yang and Hu (2008) [46] adopts a cluster analysis to explore China's HDI data to classify China's provinces into four groups based on the HDI components and Rende and Donduran (2013) [47] adopts a cluster analysis based on HDI components as a means of encouraging multi-country cooperation in learning about the policies that promote better lives for the citizens. To the best of our knowledge this is the first study that develops a cluster analysis on the HDI components to support the specification of the relationship between human development and corruption at different development levels. This relationship is measured by means of a correlation analysis between CPI and HDI and between CPI and each HDI component (related to health, education and economic conditions).

Clustering aims to map each data object into one of several categorical classes, based on similarity metrics (see Jain (1999) [48] for a review). Most clustering algorithms can be categorized in partitional, hierarchical or model-based. Partitional clustering algorithms divide data objects into nonoverlapping groups, such that each object belongs to exactly one group. These algorithms optimize a criterion (e.g. the square-error) related to similarity of objects within clusters or dissimilarity of objects among clusters. These algorithms require the specification of the number of clusters in which the data objects should be grouped. Hierarchical algorithms can be classified as agglomerative or divisive. The agglomerative algorithms specify a cluster for each data object and subsequently merge them until all items are in the same cluster. In each iteration the two most similar clusters are merged. The divisive algorithms start with one cluster with all data objects and iteratively divides it into smaller clusters. Model-based methods attempt to optimize the fit between the dataset and a specific mathematical model. Such methods are often based on the assumption that the data are generated by a mixture of probability distributions, each representing a different cluster.

In this paper, to identify natural clusters of countries in terms of HDI componentes, we used a TwoStep Cluster Analysis [49,50], which is based on hierarchical clustering. The algorithm involves two stages: preclustering and hierarchical clustering. In the precluster stage, the data objects are grouped into several small clusters. The hierarchical clustering stage uses these clusters as input and groups them into larger clusters. This procedure can automatically select the optimal number of clusters, based on wellknown statistics, such as the Akaike Information Criterion (AIC). In this paper we used this criterion to determine the appropriate number of clusters. Having identified the clusters of countries, a correlation analysis between CPI and HDI (and its components) within clusters is also performed.

4   Findings and Discussion

In order to get insights about the relationship between CPI and HDI in groups of countries defined by their similarity in terms of health services, education services and standard of living, we conducted a data mining cluster analysis. These three components of HDI variables were included in the clustering analysis in order to capture the key specificities of the HDI. Applying the TwoStep Cluster Analysis, it was concluded that the most appropriate number of clusters was four. Figure 2 illustrates the countries included in each cluster. The proportion and characterization of the of countries in each cluster is presented in Table 2 and Fig. 2.

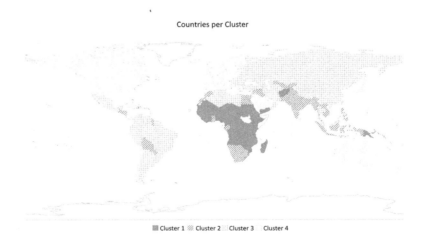

Countries per Cluster

■ Cluster 1 ⧄ Cluster 2 ⬚ Cluster 3   Cluster 4

Fig. 1. Distribution of the countries by cluster.

Figure 2 shows that most countries included in Cluster 1 are African countries. Cluster 2 is mainly composed by Asian-Pacific countries but also integrates a significant number of American and African countries. Cluster 3 is mainly composed by American countries and European-Eastern countries. Cluster 4 mainly includes European countries. As it could be anticipated, this analysis seems to suggest that there is a relationship between the level of human development and the geographical zone where the country is located (Fig. 1).

Based on the analysis of Table 2 and Fig. 2 it is possible to observe that the clustering analysis resulted in grouping together countries with similar human development values. Cluster 1 corresponds to countries with the lowest human development indicies, while Cluster 4 corresponds to countries having the highest values on these indices. On average, Cluster 3 encompasses countries with higher human development than Cluster 2. Regarding CPI, Cluster 1 includes countries

Table 2. CPI, HDI and subindexes of human development per cluster.

Cluster	#Countries	CPI		HDI		Education		Health		Income	
		Mean	St Dev	Mean	St Dev	Mean	St Dev	Mean	St Dev	Mean	St Dev
1	42	28.14	9.10	45.66	6.21	39.12	9.19	57.41	7.85	44.06	10.98
2	34	32.97	12.01	62.20	4.62	53.31	8.69	75.01	6.35	61.19	7.88
3	52	41.37	13.64	75.10	3.24	68.69	6.47	83.08	4.76	74.66	5.54
4	43	66.58	17.25	87.19	3.51	80.94	7.07	92.00	4.05	89.40	5.62

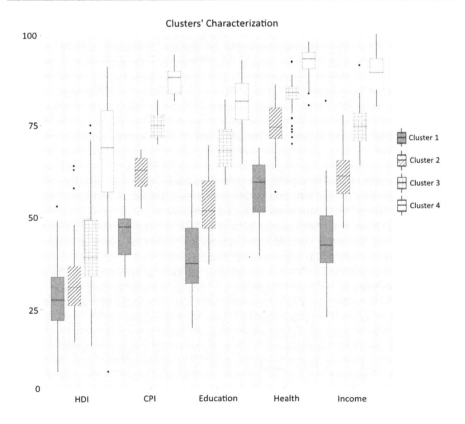

Fig. 2. CPI, HDI and subindexes of human development per cluster.

Table 3. Correlation between CPI and HDI within clusters.

		Cluster 1	Cluster 2	Cluster 3	Cluster 4
HDI	R	0.08	0.08	0.23	0.58***
	p-value	(0.6053)	(0.6385)	(0.1070)	(0.0000)
Education	R	0.25	0.02	0.03	0.34*
	p-value	(0.1071)	(0.9266)	(0.8244)	(0.0267)
Health	R	0.00	−0.27	0.35*	0.37*
	p-value	(0.9860)	(0.1237)	(0.0118)	(0.0135)
Income	R	−0.10	0.31	0.07	0.37*
	p-value	(0.5341)	(0.0783)	(0.6202)	(0.0145)

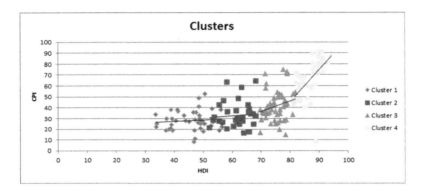

Fig. 3. Correlation between CPI and HDI within clusters.

having the highest corruption perception indices and Cluster 4 includes countries having the lowest corruption perception indices. Cluster 2 includes countries presenting higher level of corruption than Cluster 3.

The strength of the relationship between CPI and HDI, and CPI and each HDI component, for each cluster is shown in Table 3. Complementarily, Fig. 3 depicts the same correlation between CPI and HDI.

The results reported in Table 3 reveal that there is a positive and significant relationship between human development and transparency in the countries from Cluster 4. This relationship is observed for all components of the HDI, as there is a significant positive relationship between transparency and social services (health and education) and standard of living (income). This demonstrates that an increase in the transparency levels enhances the quality of public services, which are essential to the promotion of better quality of life and life expectancy of citizens.

Regarding the groups of countries that present lower levels of development, the relationship between the levels of corruption and the levels of human development is not significant. The only exception is the positive and significant relationship between transparency and health services level observed in cluster 3. This general picture suggests that the link between corruption and human development observed in European countries, is not equaly strong in developing countries in other regions. In fact, in countries from sub-Saharan Africa and from Asia, where the levels of health and education services are low, we found no evidence that the development of public services is prevented by corruption.

Figure 3 highlights that developed countries are not able to achieve the highest levels of human development without winning the battle against corruption.

5   Final Considerations

This paper evaluated the relationship between corruption and human development within the groups of countries presenting similar human development components. This analysis involved a cluster analysis that grouped together countries presenting similar HDI components, called development neighbours. Furthermore, this paper analyzed the impact of corruption on the countries' living standards and social services.

The data presented highlights the different impacts of corruption on human development and its components, within different clusters of countries. This more complex picture allows the identification of the world regions where corruption impacts most the components of HDI, thus opening doors to more directed policy strategies.

The results presented offer a sound justification for the efforts of international organizations, such the International Monetary Fund, the World Bank and the United Nations, in fighting corruption in European countries, in order to promote improvements in human development. In fact, in the countries with the highest levels of human development indices, the relationship between CPI and HDI is essential to explain the success or failure of policies aiming to provide better living conditions. Furthermore, transparency and development share an evolution along the same track, which is also observable for the primary components of HDI, including health and education services. This link between corruption and human development is less significant for the countries with lower levels of HDI.

To conclude, it is widely accepted that wealthy societies that invest in better health and education services, are rewarded with direct improvements in human development. The results obtained in our study suggest that in order to achieve the highest levels of human development, it is also crucial to avoid corruption, as it was demonstrated that high levels of development are only accomplished when there is transparency.

References

1. UNDP: Human Development Report 2015. Oxford University Press, New York (2015)
2. Stiglitz, J.E., Sen, A., Fitoussi, J.P.: Report by the commission on the measurement of economic performance and social progress. Technical report, Commission on the Measurement of Economic Performance and Social Progress (2009)

3. Harttgen, K., Klasen, S.: A household-based human development index. World Dev. **40**(5), 878–899 (2012)
4. Permanyer, I.: Using census data to explore the spatial distribution of human development. World Dev. **46**, 1–13 (2013)
5. Kulkarni, A.P., Doke, P.P.: Exploring computation of modified human development index by using proxy indicators. J. Public Health Res. Dev. **6**(3), 139–144 (2015)
6. Ranis, G., Stewart, F., Samman, E.: Human development: beyond the human development index. J. Hum. Dev. Capabilities **7**(3), 323–358 (2006)
7. Herrero, C., Martnez, R., Villar, A.: A newer human development index. J. Hum. Dev. Capabilities **13**(2), 247–268 (2012)
8. Despotis, D.K.: A reassessment of the human development index via data envelopment analysis. J. Oper. Res. Soc. **56**(8), 969–980 (2005)
9. Sagar, A.D., Najam, A.: The human development index: a critical review. Ecol. Econ. **25**(3), 249–264 (1998)
10. Chowdhury, S., Squire, L.: Setting weights for aggregate indices: an application to the commitment to development index and human development index. J. Dev. Stud. **42**(5), 761–771 (2006)
11. Tofallis, C.: An automatic-democratic approach to weight setting for the new human development index. J. Population Econ. **26**(4), 1325–1345 (2012)
12. Akhter, S.H.: Is globalization what it's cracked up to be? Economic freedom, corruption, and human development. J. World Bus. **39**(3), 283–295 (2004)
13. Ayadi, O.F., Chatterjee, A., Ademiluyi, A.: The o-ring theory, geographical distribution of misery and corruption. Int. Bus. Econ. Res. J. **3**(4), 61–66 (2004)
14. Reiter, S., Steensma, H.K.: Human development and foreign direct investment in developing countries: the influence of FDI policy and corruption. World Dev. **38**(12), 1678–1691 (2010)
15. Transparency International: Mission, vision and values (2016). http://www.transparency.org/whoweare/organisation/mission_vision_and_values
16. Lalountas, D.A., Manolas, G.A., Vavouras, I.S.: Corruption, globalization and development: How are these three phenomena related? J. Policy Model. **33**(4), 636–648 (2011)
17. Swamy, A.V., Knack, S., Lee, Y., Azfar, O.: Gender and corruption. J. Dev. Econ. **64**(1), 25–55 (2001)
18. Graeff, P., Mehlkop, G.: The impact of economic freedom on corruption: different patterns for rich and poor countries. Eur. J. Polit. Econ. **19**(3), 605–620 (2003)
19. Transparency International: Corruption perceptions index 2012: Technical methodology note (2012). http://transparency.ch/de/PDF_files/CPI/CPI2012_Technical_Methodology_Note.pdf
20. Ugur, M., Dasgupta, N.: Corruption and economic growth: a meta-analysis of the evidence on low-income countries and beyond. MPRA Paper 31226, University Library of Munich, Germany (2011)
21. Wilhelm, P.: International validation of the corruption perceptions index: implications for business ethics and entrepreneurship education. J. Bus. Ethics **35**(3), 177–189 (2002)
22. Ranis, G., Stewart, F., Ramirez, A.: Economic growth and human development. World Dev. **28**(2), 197–219 (2000)
23. Suri, T., Boozer, M.A., Ranis, G., Stewart, F.: Paths to success: the relationship between human development and economic growth. World Dev. **39**(4), 506–522 (2011)
24. Baldacci, E., Clements, B., Gupta, S., Cui, Q.: Social spending, human capital, and growth in developing countries. World Dev. **36**(8), 1317–1341 (2008)

25. Berggren, N., Jordahl, H.: Does free trade really reduce growth? Further testing using the economic freedom index. SSRN Scholarly Paper ID 933981, Social Science Research Network, Rochester, NY (2006)

26. Berggren, N., Elinder, M., Jordahl, H.: Trust and growth: a shaky relationship. Empirical Econ. **35**(2), 251–274 (2008)

27. Singh, A.: Financial globalization and human development. J. Hum. Dev. Capabilities **13**(1), 135–151 (2012)

28. Truex, R.: Corruption, attitudes, and education: survey evidence from nepal. World Dev. **39**(7), 1133–1142 (2011)

29. Acemoglu, D., Johnson, S., Robinson, J.A.: Reversal of fortune: geography and institutions in the making of the modern world income distribution. Working Paper 8460, National Bureau of Economic Research (2001)

30. Treisman, D.: The causes of corruption: a cross-national study. J. Public Econ. **76**(3), 399–457 (2000)

31. Tella, R.D., Ades, A.: Rents, competition, and corruption. Am. Econ. Rev. **89**(4), 982–993 (1999)

32. Kolstad, I., Wiig, A.: Is transparency the key to reducing corruption in resource-rich countries? World Dev. **37**(3), 521–532 (2009)

33. Bardhan, P.: Corruption and development: a review of issues. J. Econ. Lit. **35**(3), 1320–1346 (1997)

34. Beck, P.J., Maher, M.W.: A comparison of bribery and bidding in thin markets. Econ. Lett. **20**(1), 1–5 (1986)

35. Lien, D.H.D.: A note on competitive bribery games. Econ. Lett. **22**(4), 337–341 (1986)

36. Mauro, P.: Corruption and growth. Q. J. Econ. **110**(3), 681–712 (1995)

37. Mo, P.H.: Corruption and economic growth. J. Comp. Econ. **29**, 66–79 (2001)

38. Meon, P.G., Sekkat, K.: Does corruption grease or sand the wheels of growth? Public Choice **122**(1–2), 69–97 (2005)

39. Gupta, S., Davoodi, H., Alonso-Terme, R.: Does corruption affect income inequality and poverty? Econ. Governance **3**(1), 23–45 (2002)

40. Mauro, P.: Corruption and the composition of government expenditure. J. Public Econ. **69**(2), 263–279 (1998)

41. Delavallade, C.: Corruption and distribution of public spending in developing countries. J. Econ. Finance **30**(2), 222–239 (2006)

42. Frchette, G.R.: Panel data analysis of the time-varying determinants of corruption. CIRANO Working Paper 2006s–28, CIRANO (2006)

43. Smith, R.B.: Global human development: accounting for its regional disparities. Qual. Quant. **43**(1), 1–34 (2009)

44. Hysa, E.: Corruption and human development correlation in western balkan countries. EuroEconomica **30**(4), 148–157 (2011)

45. Hysa, E.: Corruption and human development: Albania and EU-27. In: 6th Annual International Conference of Albanian Institute of Sociology, Tirana/Albania (2011)

46. Yang, Y., Hu, A.: Investigating regional disparities of chinas human development with cluster analysis: a historical perspective. Soc. Indic. Res. **86**(3), 417–432 (2008)

47. Rende, S., Donduran, M.: Neighborhoods in development: human development index and self-organizing maps. Soc. Indic. Res. **110**(2), 721–734 (2013)

48. Jain, A.K., Murty, M.N., Flynn, P.J.: Data clustering: a review. ACM Comput. Surv. **31**(3), 264–323 (1999)
49. Chiu, T., Fang, D., Chen, J., Wang, Y., Jeris, C.: A robust and scalable clustering algorithm for mixed type attributes in large database environment. In: Proceedings of the Seventh ACM SIGKDD International Conference on Knowledge Discovery and Data Mining, KDD 2001, pp. 263–268. ACM, New York (2001)
50. Zhang, T., Ramakrishnan, R., Livny, M.: BIRCH: an efficient data clustering method for very large databases. In: Proceedings of the 1996 ACM SIGMOD International Conference on Management of Data, SIGMOD 1996, pp. 103–114. ACM, New York (1996)

Examining the Impact of Social Networking Sites on Performance of Service Firms: Evidence from Romania

Gheorghe Militaru[✉], Dana-Corina Deselnicu, and Alexandra Ioanid

Departement of Management, Faculty of Entrepreneurship, Business Engineering
and Management, University Politehnica of Bucharest, 313 Splaiul Indepdendenţei,
006042 Bucharest, Romania
gheorghe.militaru@upb.ro

Abstract. The purpose of this study is to extend research on social networking sites and the role of these tools on business performance of service firms. To date, there has been little investigation of the contribution of these tools in order to identify the opportunities for improving business performance. Using cross-section data from a sample in Romania, we empirically investigated the mediating effects of service innovation on the relationship between social networking sites and business performance. Confirmatory Factor Analysis (CFA) was used to perform reliability and validity checks, and structural equation modelling (SEM) was used to test the research model. The findings fill the gap in the literature by demonstrating social networking sites influence the collaboration, ideas and knowledge sharing among employees, customers, and business partners, leading to growth the business performance. The results also indicate that innovation capability played a key role on business performance but its mediating role between social networking sites and business performance of service firms was not confirmed. Finally, the theoretical and practical implications are discussed.

Keywords: Social networking sites · Innovation · Service firms · Business performance

1 Introduction

In today's extremely competitive market create huge challenges for service firms because of shortening product life cycles and vertical and horizontal specialization. The extent to which service firms use their resources to introduce new services to remain competitive could determine their success or failure. Service value is a better form of service experience and a higher service innovation results in better performance because the service sector is undergoing rapid changes. In this regard, service firms need to engage with their customers in greater depth [1].

Social networking sites (SNSs) are informal communication channels which facilitates receiving and sharing information and knowledge. They are based more on informal than formal sources of information in order to understand customers' tastes and preferences. The growing use of social networking sites offers new ways of combining external knowledge mainly from the virtual platforms and internal base of knowledge exploiting market

© Springer International Publishing AG 2017
S. Za et al. (Eds.): IESS 2017, LNBIP 279, pp. 101–112, 2017.
DOI: 10.1007/978-3-319-56925-3_9

opportunities to capture the complexity of markets [2]. For example, sunk costs bring additional layers of complexity and managers sometimes exaggerate their technological achievement and simplify obstacle or challenges.

In this regard, there is a gap between innovation and market, good innovation is not enough for business success. Social networking sites such as Facebook, Twitter, LinkedIn, and YouTube represent a huge opportunity for service firms because they are more dynamic in exploit opportunities using intensively knowledge [3]. SNSs usage should be viewed as a complementary tool for achieving optimal effectiveness or sustainable competitive advantages through converting onetime customers into repeat buyers, identifying loyal and profitable customers and targeted advertising.

Although, in the literature there are many studies focusing on the innovation performance, a large part of these refer to the relationship between service innovation and firms performance or between innovation performance and various facilitators [4–7]. Surprisingly, with all these, there is little empirical research investigating whether social networking sites affects innovation performance of service firms and their business performance or improving business processes. The literature remains largely fragmented. Consequently, a study in this direction could provide a theoretical model and could show the advantages that service firms could benefit from social networking sites.

The purpose of this research is to examine the role of service innovation to mediate the relationship between social networking sites and business performance of service firms. This paper contributes to the literature in two important ways. First, by proposing and testing the mediating role of service innovation between SNSs usage and business performance of service firms, emphasizing the influence of SNSs on business performance is achieved by a conceptual model. The model represents a novel approach to understanding co-creation customers and service innovation. By incorporating SNSs into the research model, our findings should contribute to improve innovation capabilities of service firms. Second, a contribution to the literature by presented the main advantages of using SNSs in service innovation process.

The remainder of this paper is structured as follows. In Sect. 2 of this paper we provide a state of the art by summarizing of the relevant literature on social networks, and service innovation processes. Next, the conceptual model and hypothesis will be presented. The development of research measurement items and data collection methods will then presented, following by an analysis of the data. Finally, we present key findings and theoretical and managerial implication will be highlighted, followed by a discussion of our conclusion, the limitations of our study, and some directions for future research.

2   Theoretical Foundations and Hypotheses Development

From a theoretical point of view academic researchers and business managers recognize the importance of service innovation as key driver of organic growth because new offerings yield opportunities to increase business performance and can be a source of sustainable competitive advantage [8]. A service innovation can be considered as an offering not previously available to the firm's customers that require modifications in the sets of competences applied by service provider and customers. Furthermore, failure is part of the innovation process. Previous studies have shown that personal interaction among employees, customers, and other value-network partners facilitates innovation by the exchange of new ideas and knowledge in order to add value and complementary capabilities [9].

Fitzsimmons et al. (2008) argue that service innovation is the process of creating new services, improving current services, processes, policies, and procedures that will meet some specific needs to achieve a strategic positioning in the market [10]. In this regard, process innovation is related to the firm's competitive capabilities. In the competitive worldwide market, the capability for continuous innovation is the critical driver to attaining competitive advantage for service firms because a new or creative idea lead to a successful new service on the market is easily imitated by competitors [11].

The role of front-line employees exerted by internal capabilities to extract new ideas and knowledge from customers contribute to boost service firm's innovation performance to offer innovative and high value added services. In this regard, Campbell stated that the performance of service innovation and business depends on the alignment between service innovation and exploiting possible complementarities such as social networking sites [12].

Different online platforms such as Facebook, Twitter, Research Gate, and LinkedIn can substantially affect the interaction of employees with customers because this process depends of interpersonal trust and feedback [13]. Service firms tend to develop innovative ideas through a dynamic and collaboration approach to generate new knowledge and services by integration resources into attractive value proposition and minimizing their investment risk [14]. The innovation capability has significant impact on business performance. A strategic alignment between service innovation and business activities is the key to improve business performance in service sector and to exploit innovation to meet customers' needs [15].

As indicated by prior research, the role of social networking sites in relationship to innovation and business performance can be capitalized in a few benefits: SNSs contribute to reduce the time in which a new service is marketed, speed up the adoption of new services and reduce the new services development cost for service firms by exchanging ideas with customers or other business partners [16]. These digital platforms facilitate employees' interactions with customer, co-created through collaboration in innovation process, service customization, and increase brand loyalty and reputation [17]. Indeed, Facebook facilitates marketing communication, Twitter as microblogging, LinkedIn as professional networking, and You Tube as video sharing. These digital platforms can be used to identify market trend and to generate customer and business partners' insights. Social networking sites facilitate to bring new innovation ideas and knowledge from customers and reduce costs of convert this ideas into an innovation [18].

The profitability of innovation comes from its adoption and even its commercialization, not only from its creation, limiting service firms' capability to remain competitive in dynamic business environments. Therefore, firm use various systems to capture, store, analyse the date to extract value and generate market for new services. Therefore, social networking sites activities can attract and retain high-quality employees, improve business processes and minimizing their investment risk [18]. On the basis of this theoretical reasoning, we propose the following hypothesis:

Hypothesis 1: *The effects of social networking sites are positively associated with business performance of service firms*

In previous studies is shown that information technology has become a strategic tool for service innovation [18]. Facebook, LinkedIn, Twitter integrate digital communication and enable to build connections among employees, customers and business partners. They enable frontline employees to share information, ideas, and information. Customers bring new ideas and resources that need to be integrate with firm capabilities. Social networking sites might serve as a facilitator the promotion of new services to potential customers and together with their customers service firms co-create innovative new service by convert their valuable ideas [18]. Creativity is the key of the co-creation process through generate, selecting new and innovative ideas, develop new service concept, and launching this new offerings into crowded markets.

A virtual network structure is required in order to support the innovation process in all its stages and extract greater business value [19]. The service innovation focus on the value co-creation based on a collaborative process in which costumers and other business partners play a key role to create value. Furthermore, the innovation process needs to apply appropriately SNSs in a timely and congruent way with business processes, market requirements, business goals, organizational culture and business capabilities.

Broadly speaking, the strategic use of SNSs corresponds to the extent to which service firms use these digital platforms to improve their innovation capabilities, especially in highly changeable and dynamic market requires consistent and congruent efforts. SNSs as a resource for innovation play a key role by collecting multiple information and knowledge from various sources. It is interesting to note that service firms will achieve better performance in terms of sales growth and increased market share than these do not use these tools. Being able to differentiate by offering unique service experience the service firms can use SNSs to target useful information about its offerings to better segment and target customers. To examine whether prior findings on customer collaboration using digital platforms about judgments on innovation performance are true and influence the degree of interaction with customers and service customization, we advance the following hypothesis:

Hypothesis 2: *The social networking sites is positively associated with the innovation performance of service firms*

Operating and improving processes are essential activities for improving the business performance in service industry. Innovation process involves resource-intensive used to create new service and to find the best ways to commercialize them on the market. Radical innovation involves new services and existing services combination that are offerings on the market. While incremental service innovation involve adaptation, refinement, improving of existing services, processes, procedures, policies based on the existing base of knowledge to produce insight and business value. The positive effect of using SNSs is that new service is more innovative, better suited to the market, and more attractive to customers.

Previous studies have found that innovation performance has a positive effect on business performance in service industry [19]. Clarifying the role of SNSs in service innovation can lead to more effective use of resources and capabilities for new service innovation. SNSs are expected to complement rather than substitute effect and have a positive impact on service innovation. We tested this possibility by examining whether SNSs has a positive influence on the innovation performance and lead to better business and service innovation performance. Therefore, we propose that there is theoretical support for the following hypothesis:

Hypothesis 3: *The service innovation performance is positively associated with the business performance of service firms*

Our theoretical arguments and hypotheses are summarized in the conceptual research model shown in Fig. 1.

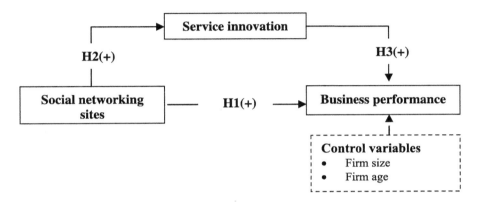

Fig. 1. The conceptual research model

3   Research Methodology

A questionnaire survey was designed to test the theoretical model. The data for the quantitative analysis has been drawn from a sample of engineering students were recruited from University Politehnica of Bucharest, Romania, during December 2016. This university is the largest and the oldest technical university in Romania. All respondents were working in service firms. Those who had no experience with service firms were excluded from sample. In order to ensure the validity of the study, all the variables included in the questionnaire were obtained from empirical observations and theoretical reviews.

3.1   Construct Operationalization

Scales were taken from the existent literature and previous research and adapted to our requirements. All items were measured using seven-point Likert scales ranging from strongly disagree to strongly agree. We are concerned with the degree to which employees of a service firm use social networking sites to improve service innovation capabilities and business performance. Many researchers analyse organizations' innovation with reliable valid measurement scales [20, 21]. When possible, construct measures were created based on previously validated survey instrument. In addition, individual measures were averaged to obtain a simple value for each construct. *Service innovation* was measured using 3 items: "Do you consider that lately the firm you work for improved its processes, procedures or the rules used when providing their services (SI1)", "Do you consider that lately the firm you work for improved its existing services (SI2)" and "Do you consider that lately the firm you work for developed new services (SI3)". *Social network sites* were measured using 3 items: "In the workplace, do you consider that firms' employees use social networks in order to collect valuable ideas from clients (SNS1)", "In the workplace, do you consider that firms' front-line employees use social networks in order to improve the existing services as an effect of social interaction (SNS2)" and "In the workplace, do you consider that firms' front-line employees use social networks in order to develop new services (SNS3)". *Business performance* was measured using 3 items: "Do you consider that lately the innovation efforts of the firm you work for leaded to increasing the sales (BP1)", "Do you consider that lately the innovation efforts of the firm you work for leaded to raising the share market (BP2)" and "Do you consider that lately the innovation efforts of the firm you work for leaded to raising the profitability (BP3)".

Control variables. Following previous studies, we controlled for firm age and size that were not of direct interest for testing our hypotheses but could be theoretically related to the business performance and might provide reasonable alternative explanations for our findings [22]. We expect that firm size has a greater impact on innovation and business performance in large firms because they have the potential to attract talent and capital. Firm size has a positive effect on performance and also on using SNSs in service innovation because the biggest firms usually have more resources to invest in training, analytical tools and digital platforms. In this regard, large organizations have more specialized, professional and skilled workers, who can benefit from SNSs. We controlled firm size by taking into account the natural log of the number of full-time employees of individual firms in our sample. The natural log was taken in order to normalize the data.

Firm age was determined by the number of years since service firm was established. Indeed, experience, reputation, technical skills, individual capabilities, and organizational competencies are formed through time, and these characteristics help service firms to develop their operations more efficiently, including the innovation processes and business performance. Firm age was operationalized as the logarithm of the number of years since the firms was founded in order to normalize the data.

3.2 Data and Sample

We used a cross-sectional survey for data collection. A multi-item method was used to form the questionnaires. The methodology applied was a structured questionnaire and a procedure stratified sample with proportional allocation. We recruited students from master programs because they have at least some knowledge about service sector and IT skills. Thus, by this solution we reduced the bias of our collected data. The respondents were asked to indicate the extent to which they use the social networking sites in current activity. The respondents hold various job titles and have more 2 years of experience with service firms. This survey was completed from a respondent of each service firms. The response rate was 90% and resulted in 68 valid questionnaires (N = 68). Among the valid respondents, 65% of the respondents were males and 35% of the respondents were women. Most of the service firms have ranged from 1 to 15 years. The average age was 4.87 years. The data were assessed for the extent of missing values. The missing values were completely at random and the means substitution method was used the replace them. Because the survey was to be conducted in a Romanian context, a good translation was performed to ensure that the respondents understand the questions and that no less of information was present in the Romanian version. The composition of the sample is presented in Table 1.

Table 1. Composition of sample (N = 68)

Variable	Frequency	Percentage
Firm size (number of employees)		
Less than 20 employees	45	66
Between 20 and 100 employees	12	18
More than 100 employees	11	16
Firm age (number of years since foundation)		
Less than 3 years	17	25
Between 3 and 8 years	36	53
More than 8 years	15	22
Gender (respondents)		
Female	24	35
Male	44	65

3.3 Data Analysis and Results

To verify the reliability and validity of our research model, we used SPSS version 20 and a structural equation model with the LISREL techniques. The descriptive statistics such as means, standard deviations (SD), and matrix of correlations among variables are presented in Table 2. The summary statistics and correlations among the study variables correlation values among all variables are generally low to moderate. The highest correlation 0.74, it is an acceptable level. The general rule is that correlation values should not exceed 0.75 [23]. Our results indicate that there are no serious multicollinearity problems.

Table 2. Descriptive statistics and correlations

Variable	Mean	SD	1	2	3	4	5	6	7	8	9	10	12
1. SNS1	4.84	1.5	1										
2. SNS2	4.81	1.5	.74	1									
3. SNS3	4.81	1.6	.47	.63	1								
4. SI1	5.69	1.0	.18	.16	.1	1							
5. SI2	5.58	1.2	.08	.2	.31	.24	1						
6. SI3	5.63	1.1	.22	.29	.2	.23	.39	1					
7. BP1	6.56	0.8	.08	.14	.23	.1	.12	.07	1				
8. BP2	6.53	0.8	.1	.23	.1	.1	.19	.1	.5	1			
9. BP3	6.38	0.9	.17	.12	.1	.02	.05	.04	.46	.44	1		
10. FS	17.25	6.7	.02	−.1	−.1	−.2	.04	.02	−.3	−.1	−.1	1	
11. FA	4.87	3.2	−.1	−.0	−.1	−.2	.03	−.1	−.1	.02	.1	.65	1
12. GR	0.34	0.5	−.2	−.2	−.1	.00	−.1	.08	.06	.04	−.31	−.33	−.34

Note: FS = Firm Size; FA = Firm Age: GR = Gender; *p < 0.01; N = 68

As can be observed in Table 3, the value of Cronbach's α for all latent variables indicates adequate internal consistency of scales because it is greater than the threshold value of 0.7 [24]. All item loadings are significant at p < 0.05 level. The composite reliability (CR) of each latent variable is also greater than the recommended value of 0.6 [25]. The evaluation of the convergent validity of constructs was carried out by analysing the significance of the factor loadings (λ) and the average extracted variance (AVE). The average variance extracted for all of the value is greater than a value of 0.5, indicating a reasonable degree of convergent validity.

Table 3. Scale proprieties for the measurement model

Construct, indicators	Factor loadings	t-value	Cronbach's α	Composite reliability (CR)	Average variance extracted (AVE)
BP					
BP1	0.908	–	0.902	0.927	0.8083
BP2	0.904	10.23			
BP3	0.885	8.22			
SI					
SI1	0.943	–	0.931	0.9498	0.863
SI2	0.908	10.66			
SI3	0.936	12.11			
SNS					
SNS1	0.971	–	0.976	0.979	0.9396
SNS2	0.963	20.83			
SNS3	0.974	22.17			

Note: BP = Business Performance; SI = Service Innovation; SNS = Social Network Sites

4   Hypothesis Testing

The hypothesized relationships among variables were examined by means of structural equation modelling analyses (SEM) using LISREL 8.80 program. A path-analytic model was developed and tested. Path analysis allows researchers to test direct and indirect effects of multiple independent variables on multiple dependent variables, as illustrated in Fig. 2. To test the mediation effect of service innovation, we follow the methodology proposed by Baron and Kenny (1986). First, we need to show that there is a significant relationship between SNSs and BP. Second, we need to identify a significant relationship between SNSs and SI. Third, we investigate that there is a significant relationship between the mediation variable (SI) and BP. Finally, we must show that the effect of SNSs on BP is less when the mediator variable is included in the model [22].

The final path model is presented in Fig. 2. One mediating variable (service innovation) and one dependent variable (business performance) were included. The fit indices for this model were as follows: $\chi^2 = 25.05$, df $= 24$, p-value < 0.01 (χ^2/df < 5 is preferred), the root mean square error of approximation (RMSEA) is 0.026 (good model < 0.06), the comparative fit index (CFI) is 0.95 (>0.9 is preferred) [25]. With respect to the fit statistics this indices indicate a good fit for the covariance observed. Consequently, this model illustrates the best results of the hypothesis testing.

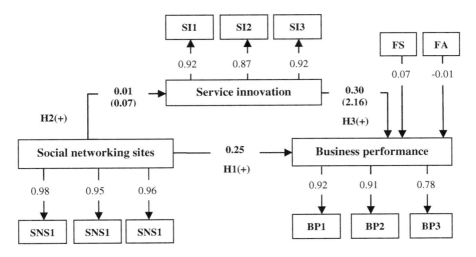

Fig. 2. Estimated path model

Hypothesis 1 states that social networking sites have a significant impact on business performance. The results provide support for this hypothesis ($\beta = 0.25$, $p < 0.05$). Therefore, Hypothesis 1 was supported. Hypothesis 2 states that social networking sites are positively associate with the innovation performance of service firms. We note in Fig. 2 that this hypothesis is not significant for innovation services to capture value from SNSs and it was not supported. In contrast, Hypothesis 3 states that service innovation is positively associate with the business performance of service firms ($\beta = 0.3$, $p < 0.001$). In line with previous studies [18], this suggests that innovation has a positive impact on the business performance. Therefore, Hypothesis 3 was supported. However, the mediating effect of service innovation on the relationship between social networking sites and business performance was not confirmed because the Hypothesis 2 was not supported. This suggests that statistically social networking sites are not used significantly to increase the innovation capabilities of service firms. Firm size and firm age, which reflect different businesses' stages over time, are not significant for innovation performance ($\beta = 0.07$, $p > 0.1$ and $\beta = -0.01$, $p > 0.1$, respectively).

5   Discussion and Conclusion

Our exploratory study is a response to the demand for improvement the innovation capabilities of service firms and extends the stream of research on SNSs. Service firms must proactively develop their innovation capabilities using IT if the customers is actively involved in the value co-creation, with an explicit focus on the SNSs solutions and their role to improve the business performance. Today, service firms use various systems to capture, store, analyse the data to extract value and generate market for new services. The findings confirm that service innovation provide strong support for improving the business performance of service firms.

The key managerial insight provided by this study is that it provides managers some guidelines about how they can use the business performance using SNSs in all stages of innovation process. Our findings indicate that SNSs can be used to improve the business performance because they contribute to faster time to market or faster to service adoption on the market. Observed in more detail, SNSs usage contributes to bring ideas into de market and identifying solutions to minimize risk while seeking to maximize opportunities for growth. SNSs generate interactions that result in innovation and operational effectiveness. However, service firms should have an innovative culture to enhance innovation and business performance.

Although a large body of research has investigate the contribution of SNSs to improve the innovation capabilities of service firms and the mediating effect of this variable on the relationship between SNSs and business performance, our findings do not support this hypothesis because the utilization of digital platforms in process is very low. There are several explanations, for example, employees do not have the appropriate digital' skills or managers do not encourage to use these digital platforms. One unanticipated finding was that there is mistrust about the information generated by these platforms. The trust is the key factor to support the collaboration between employees and customers or business partners. This study has augmented that the existence of bad data might hinds opportunities for new service development. Bad data would have little relevance in extracting business value. Therefore, verification is necessary to generate relevant data.

This study has several limitations, which can be addressed in future research. First, for each firms included in sample was a single respondent as the source of information. We suggest that the future research should use multiple sources of survey data. Second, our research is based on cross-sectional data. Furthermore, SNSs adoption can be regarded as a dynamic process. Consequently, a longitudinal study may extend our research on mediating the effects of the relationship between SNSs and business performance. Thus, future research will be needed to investigate changes in patterns and factors. Third, while we have investigated the influence of SNSs on business performance and the mediating effect of service innovation, further work by included other factors may offer useful insights. For example, future research can refine our research model by investigation of mediating effect of operating efficiency. SNSs could impact on the operating efficiency or quality provided the firms from service industry. We believe that this is a promising research area and our model offers an important point of departure for this.

Acknowledgement. This work has been funded by University Politehnica of Bucharest, through the "Excellence Research Grants" Program, UPB – GEX. Identifier: UPB–EXCELENTA–2016; research project title: *Improving the performance of small and medium – size enterprises in Romania by implementing the integrated risk management* (Acronym: PERFORM), contract no. 55/2016.

References

1. Wang, K.: Determinants of mobile value-added service continuance: the mediating role of service experience. Inf. Manage. J. **52**, 261–274 (2015)
2. McKelvey, M., Lassen, A.H.: Knowledge Intensive Entrepreneurship: Engaging, Learning and Evaluating Venture Creation. Edward Elgar Publishing, Cheltenham (2013)
3. Malerba, F.: Knowledge Intensive Entrepreneurship and Innovation Systems: Evidence from Europe. Routledge, Abingdon (2010)
4. Melton, H.L., Hartline, D.M.: Customer and frontline employee influence on new service development performance. J. Serv. Res. **13**(4), 411–425 (2010)
5. Demirel, P., Mazzucato, M.: Innovation and firm growth: is R&D worth it? Ind. Innov. **19**(1), 45–62 (2012)
6. Gunday, G., Ulusoy, G., Kilic, K., Alpkan, L.: Effects of innovation type on firm performance. Int. J. Prod. Econ. **133**(2), 662–676 (2011)
7. Nordman, E.R., Tolstoy, D.: Technology innovation in internationalising SMEs. Ind. Innov. **18**(7), 669–684 (2011)
8. Campbell, B.: Alignment: resolving ambiguity within bounded choices. In: Proceedings of the PACIS, Bangkok, Thailand, pp. 1–14 (2005)
9. Ghoshal, S., Korine, H., Szulanschi, G.: Interunit communication in multinational corporations. Manage. Sci. **40**(1), 96–110 (1994)
10. Hang, H.L.: Performance effects of aligning service innovation and the strategic use of information technology. Serv. Bus. **8**, 171–195 (2014)
11. Fitzsimmons, J.A., Fitzsimmons, M.J.: Service Management: Operations, Strategy, Information Technology, 6th edn. McGraw-Hill, New York (2008)
12. Leyden, D.P., Link, A.N.: Toward a theory of the entrepreneurial process. Small Bus. Econo. **44**(3), 475–484 (2015)
13. Stelzner, M.A.: Social media marketing industry report: How marketers are using social media to grow their businesses, [online] social media examiner (2013). http://www.socialmediaexaminer.com/SocialMediaMarketingIndustryReport2013.pdf. Accessed 19 June 2016
14. Hanna, R., Rohm, A., Crittenden, V.L.: We're all connected: the power of the social media ecosystem. Bus. Horiz. **54**, 265–273 (2011)
15. Tussyadiah, I., Zach, F.: Social media strategy and capacity for consumer co-creation among destination marketing organizations. In: Cantoni, L., Xiang, Z. (eds.) Information and Communication Technologies in Tourism 2013. Springer, Heidelberg (2013)
16. Arts, S., Veugelers, R.: Technology familiarity, recombinant novelty and breakthrough invention. Ind. Corp. Change **24**(6), 1215–1246 (2015)
17. Del Giudice, M., Maggioni, V.: Managerial practices and operative directions of knowledge management within inter-firm networks: a global view. J. Knowl. Manage. **18**(5), 841–846 (2014)
18. Scuotto, V., Ferraris, A., Bresciani, S.: Internet of things: applications and challenges in smart cities: a case study of IBM smart city projects. Bus. Process Manage. J. **22**(2), 357–367 (2016)
19. Ordanini, A., Rubera, G.: How does the application of an IT service innovation affect innovation and firm performance? A theoretical framework and empirical analysis on e-commerce. Inf. Manage. J. **47**(1), 60–67 (2010)
20. Hurley, R.F., Hult, G.T.M.: Innovation, market orientation, and organizational learning: an integration and empirical examination. J. Mark. **62**(3), 42–54 (1998)
21. Hansen, M.T., Birkinshaw, J.: The innovation value chain. Harv. Bus. Rev. **85**(6), 121–130 (2007)
22. Baron, R.M., Kenny, D.A.: The moderator-mediator variable distinction in social psychological research: conceptual, strategic, and statistical considerations. J. Pers. Soc. Psychol. **51**, 1173 (1986)

23. Tsui, A.S., Ashford, S.J., Clair, L.S., Xin, K.R.: Dealing with discrepant expectations: response strategies and managerial effectiveness. Acad. Manage. J. **38**(6), 1515–1543 (1995)
24. Fornell, C., Larcker, D.F.: Evaluating structural equation models with unobserved variables and measurement error. J. Mark. Res. **18**(1), 39–50 (1981)
25. Hair Jr., J.F., Money, A.H., Samouel, P., Page, M.: Research Methods for Business. Wiley, Chichester (2007)

Service Systems Analysis and Design

A Conceptual Modelling of the Key Components and Relations of Service Systems

Bertrand Verlaine[(⊠)]

Department of Business Administration, PReCISE Research Center,
University of Namur, Rempart de la Vierge 8, 5000 Namur, Belgium
`bertrand.verlaine@unamur.be`

Abstract. A conceptual model helps to know, understand, communicate and make simulations on the kind of objects that it represents. Although the main concepts of service science have been deeply discussed by many authors, their conceptual modelling is lacking. Therefore, in this paper, we progressively model the concept of *service system*, which is incontestably the fundamental element of service science. We define and depict the generic components of service systems as well as the relations between them. We also apply the recursion principle in order to model more complex service systems, i.e., service systems composed of other services systems. The key contribution of this work, i.e., the conceptual model of a generic service system, will help researchers and practitioners to argue on service systems as well as on their key components and relations.

Keywords: Service science · Service system · Conceptual model · Service system modelling · Service-dominant logic · Recursion principle

1 Introduction

The service-dominant orientation of our current economy, which is handled in the service science literature, is indubitably a raising research domain. It is built around the concepts of *service* and *value co-creation*. These concepts and their relations have been discussed and defined by several authors (e.g., [1–4]), although their conceptual modelling is still missing.

A conceptual model is an abstract and generic representation of an object, which is a *service system* in the scope of this work. *Conceptual* means that the model is created based on a conceptualization or, in other words, a formal generalization and abstraction of the reality [5]. Thanks to a conceptualization, the model helps the readers to know, understand, communicate and make simulations of the kind of objects that it represents. The elements in a conceptual model, which often results in a graphical representation [6], depict the concepts of the modelled domain as well as their relations [7]. This approach is followed for a long time in, i.a., mathematics and computer science. In these fields, conceptual models mainly help to conduct some reasoning on systems. Seeing that

© Springer International Publishing AG 2017
S. Za et al. (Eds.): IESS 2017, LNBIP 279, pp. 115–130, 2017.
DOI: 10.1007/978-3-319-56925-3_10

the key element of service science is the *service system*, its conceptual modelling improves, and will help to improve, the knowledge about the service science research domain.

The contribution of this paper is twofold. Firstly, based on the key concepts and relations of service systems, we introduce and define the notion of internal interface between the components of a service system, as well as the notion of external interface linking two service modules. These interfaces are used to decompose service systems, which enables to hide some complexity of these service systems, to ease their understanding and to reason upon them. The clear identification, differentiation and modelling of those interfaces will, i.a., facilitate the analysis of service networks as well as the underlying value chains [8, Chap. 10].

As explained, the key concepts and relations of service systems are depicted in an abstract model. On this basis, the second contribution is the discussion of the *recursion* principle, which is actually present in many service systems, and its inclusion in the conceptual model proposed.

The paper proceeds as follows. After this introduction, the fundamentals of the research domain related to this work are presented in Sect. 2. In Sect. 3, the main related works are identified and positioned in comparison with the contributions detailed in the rest of this paper. In Sect. 4, we discuss the expressions *Service Science* and *Service-dominant logic* (SDL), which are the starting point of the modelling work proposed. Then, in Sect. 5, we develop our reasoning in order to progressively build the conceptual model of service systems. In Sect. 6, we extend this conceptual model thanks to the *recursion* principle and we illustrate the contributions proposed through an example. Lastly, Sect. 7 contains the conclusion and future work.

2   Foundations: The Shift Towards a Service-Dominant Logic

In the economic and management literature, the notion of service has been discussed several times. One of the most cited definitions is the one proposed by Ted Hill in 1977. He said that a service is "a change in the condition of a person, or of a good belonging to some economic unit, which is brought about as the result of the activity of some other economic unit, with the prior agreement of the former person or economic unit" [9]. In the scope of the service science research domain, one of the first and most outstanding definitions is the one proposed by John Rathmell. In [10], he said that "a service is a deed, a performance, an effort", whereas a good is "an object, an article, a device or a material". He clearly argued that the service has to be considered as a set of activities instead of being the result of those activities. Several other authors concur with this fundamental idea. For instance, a service is "an activity or series of activities of more or less intangible nature that normally, but not necessarily, take place in interactions between the customer and service employees and/or physical resources or goods and/or systems of the service provider, which are provided as solutions to customer problems" [11]; a service is "an act or performance offered by one party to another. Although the process may be tied to a physical product, the performance is essentially intangible and does not normally result in ownership of any of the factors of production" [12].

All these definitions of the service notion underline the fact that a *service* is actually understood as a transformation process and not any more as the results of a transformation process. For researchers and practitioners in the service field, the notion of *service* corresponds to the orchestration of resources in order to co-create value in a win-win situation [13]. Indeed, "customers do not buy goods or services: they buy offerings which render services which created value. [...] It is not a matter of redefining services and seeing them from a customer perspectives; activities render services, things render services" [14]. The consequence of this viewpoint, which is the first and the main argument of *service science*, is a shift from a good-dominant logic towards an SDL of our economy. This means that we moved from a focus on the *value-in-exchange* to a focus on the *value-in-context* [2]: the "value is being co-created with and determined by customers" [3]. A good-dominant logic is characterized by (i) a focus on the product, which is the basic unit for economic exchanges, (ii) a possibility to create a demand in order to sell products, (iii) the value provided to customers is integrated into the products during the manufacturing process, (iv) a distinction between the production time, during which the value is created, and the consumption time, during which the value is destroyed, and (v) a product standardization for higher efficiency [15,16]. SDL focusses on the service notion, which becomes the basic exchange mean [16]. This shift is the foundation of service science, which is defined by Definition 1.

Definition 1. *Service science is "an interdisciplinary field that combines organizations and human understanding with business and technological understanding to categorize and explain the many types of service systems that exist as well as how service systems interact and evolve to co-create value"* [4].

3   Related Work

The syntax *service* is used in many disciplines such as, e.g., marketing, operations and computing science. However, its understanding varies a lot depending on the discipline. In these disciplines, it exists many solutions to model *specific* and *existing* services. As examples of these, we can cite WSDL [17] or SoaML [18] used to describe, respectively, services in the service-oriented computing, SML [19] for defining services from a cloud computing perspective (cf. the SaaS layer), or USDL [20] aiming at unifying the technical and the business perspectives of a given service. However, the *abstract* representation of the service science concepts and relations has been little tackled. In the scope of this research, we focus on the abstract and conceptual modelling of the elements of service systems, which should later be applicable to any service system.

Some authors focused a part of their research work on the modelling of service systems. In [21], the authors propose an ontological foundation of services science by representing a general model of *service* based on the UML class diagram. This general service model is related to DOLCE, a core ontology. In our work, we focus on the concept of service system. We also introduce the notions of internal and external interfaces, and of service modules thanks to the recursion principle. This enables to reason on less complex service systems given their decomposition.

Tukker presents eight different types of Product-Service Systems (PSS) [22]. His research objective is to categorize these PSS types. His work partially represents the architecture of each of the eight PSS types. In our paper, we argue for a single conceptual representation of the service system concept. Indeed, the basic resources of service systems and their relations are identical.

Maglio and his colleagues propose an abstract and normative model, that they called ISPAR [23]. This model aims at representing the interactions between service systems. They identify ten possible interaction outcomes. Although different, this work could be related to the research presented in our paper (see the future work, in Sect. 7.1, for more information).

In [1], Alter discusses some fundamental elements of service systems and proposes three frameworks. His starting point is the *work system framework*. He separates the customers from the service system which, in fact, does not respect the SDL principles.

Some other related works discuss tools for representing business models based on service systems or on product-service systems (e.g., [24,25]). Other tools can be used to model existing services. The e3 value model is probably one of the most known modelling solutions for representing the value chain network of specific service systems [26]. These kind of tools and modelling solutions are used to model specific and/or existing services and service systems. As a reminder, the objective of this paper is to propose a conceptual and abstract modelling of the service system concept based on the existing concepts and relations of service science. It is independent of any existing and specific service, although it could be used to instantiate existing service systems.

Lastly, it is also important to mention the work of Grönroos and Voima [27]. They use several conceptual models to discuss the notions of value creation and co-creation from an SDL viewpoint. In our work, we focus on the service system concept.

4   Discussion and Schematic Description of the Key Concepts of Service Science

4.1   An Analysis of the Terms Contained in the Expression *Service Science*

The expression *Service science* is actually the short version of the expression *Service science, management, engineering and innovation* (SSMEI). The first term of this expression, *service*, corresponds to the mobilization and the integration of resources provided by the service provider and consumer in order to co-create value (see Definition 2 defining the concept of *service*). The key point to underline in Definition 2 is that a service is not a result, but it includes all the actions conducted in order to reach an expected result.

Definition 2. *A* ***service*** *is the application of competencies to mobilize and integrate the resources of the service provider and consumer in order to co-create value for their benefit* [4, 15, 28].

Concerning the other terms of the expression SSMEI, *science* refers to the ways to create knowledge – "what service systems are and how to understand their evolution" [28]–, *management* covers the ways to structure, organize and present the knowledge related to the service concept – "how to invest to improve service systems" [28]–, *engineering* is the ways to apply this knowledge – "how to invent new technologies that improve the scaling of service systems" [28]– and *innovation* refers to the new and improved ways for co-creating value in service systems [29].

This means that, service scientists have to study the engineering and the management of service systems, the generic innovation possibilities in service systems and solve fundamental issues, such as the definition of the quality of service or the modularity in service systems [30]. Making use of a conceptual model of a generic service system should help them in achieving this work.

4.2   The Service-Dominant Logic: Modelling of the Transformation Process from an Outside Perspective

The SDL key characteristics are:

– an inclusion of customers in the transformation process, which therefore become co-creators of the value generated following the execution of the service system,
– a description of the value propositions made by service providers, called offerings, which comprises a combination of goods and services (see Definition 6 for an explanation about the differences between the notion of service, in the singular, and of services, in the plural form),
– a significant focus on the customers and on the relations between the co-creators of value, and
– the assessment of the actual value obtained by the service beneficiaries.

These SDL characteristics enable us to model the transformation process from an SDL viewpoint, along with its inputs and outputs. Figure 1 is an illustration of this transformation process. The latter is actually the service system from an external perspective. Broadly speaking, the service system is the way to associate several resource types in order to co-create value –we discuss and model the service system in Sect. 5. This is why the inputs of the transformation process belong to the service co-creators, that is to say the service provider and the service customer. The possible types of inputs are people, technology, information and organizations. These resources, defined in Sect. 5, have to be coordinated in

Fig. 1. The service transformation process, along with its inputs and outputs, from an SDL viewpoint

order to adequately transform these inputs into outputs, which are called the *offerings*. An offering comprises one or several goods and/or services. Services, in the plural form, are possible results of the transformation process or, in other words, of the service system.

In the next two sections, we study in detail the generic components and relations of the inside of this service transformation process, which consists in the conceptual modelling of service systems.

5   Modelling of the Key Components and Relations of Service Systems

5.1   The Service System and Its Four Types of Resources

The basic theoretical element of service science is the concept of *service system* [23]. It represents the transformation process leading to the co-creation of value. When we mention the concept of *value co-creation*, we refer to the mobilization of competencies and the integration of resources of different entities –which are people, groups of people and/or organizations–, by bringing them together in order to mutually and reciprocally create a new condition for these entities and/or for their belongings, which is estimated by these entities as better than before [31].

A service system is thus a specific type of system. This notion comes from the systemics. This is the research domain focussing on the holistic study of sets of given components and of their relations in order to understand the possible behavior(s) of these combined components. Indeed, the outputs of a system depend on the constituting components, but also on the relationships between them. Definitions 3 and 4 respectively define the concepts of *system* and *service system*.

Definition 3. *A **system** is a configuration of interacting and/or interdependent components forming a complex whole in which the properties and behavior of the configuration is more than the properties and behavior of the individual components* [23,32].

Definition 4. *A **service system** is "a dynamic value co-creation configuration of resources, including people, organizations, shared information (language, laws, measures, methods), and technology, all connected internally and externally to other service systems by value propositions"* [23].

As underlined by Definition 4, the types of components of service systems are thus people, organizations, information and technology. They are the four generic components of service system, which can be combined in many ways (see Sect. 5.2). They are defined as follows [15,23,28,33]:

1. People: they are human beings, acting as stakeholders in service systems, i.e., the representatives of providers, of customers, of the authorities or of competitors, as well as the individuals acting in these organizations.
2. Technology: it encompasses material supplies, transformation assets, processes and management solutions.
3. Information (also called *shared information*): this resource type could be a language, a law or a measure.
4. Organizations: this resource type consists of structured groups of people, sharing similar high level objectives, which affect and are affected by their environment (e.g., a corporation, a non-profit association, a city, a school, a country, and so on).

Each of these four resource types can vary in terms of intensity, what is often visible in the value propositions of service systems. If one of the resource types is substantially represented, this gives one of the following particular kinds of service systems:

People-intensive service system such as a session with a physiotherapist or legal advice given by a lawyer.

Technology-intensive service system such as the purchase of a car coming from an automated production line or the use of software.

Information-intensive service system such as the modelling of the business processes of a company or the determination of some KPI in a production line.

Organizations-intensive service system such as the creation of a new law at the parliament of the European Union or the creation of a joint venture.

From an SDL viewpoint, the smallest possible service system is a human being; at the opposite, the largest service system comprises the global economic world. A prerequisite for service science is the division of the work. According to Durkheim, the division of labour is "the higher law of human societies and the condition of their progress" [34, p. 1]. Of course, this division of labour requires a reintegration and a coordination of all its components. The value co-created by the customer and the provider during the service system execution is actually determined by the integration of its components [2]. Service providers can make value propositions, which are called *offerings*.

Definition 5. *An **offering** represents the value proposition of a service system describing a change or a set of related changes that should be preferred by customers* [35].

As a reminder, the value proposition could be different in comparison with the effective value co-created after the execution of the service system. Offerings are goods or services (in the plural, see Definition 6), or an association of both of them as depicted by Fig. 1. What is important for the service co-creators is that the offerings represent a win-win situation [33]. This means that, in each value proposition, both the service customer and the service provider should derive a benefit from the future executions of the service system. Of course, other kinds of people and organizations could be impacted by the service execution or present in the environment of the service execution. Apart from service providers and customers, these are principally the authorities and the competitors. Maglio et al. [23] created the ISPAR model (ISPAR stands for Interact-Service-Propose-Agree-Realize) in which they described ten possible outcomes for the four possible types of stakeholders. The latter are the service customer, the service provider, the authorities and the competitors. In the scope of this discussion, we only consider the service customer and provider. An extension of the models proposed in order to include the authorities and competitors is left for future work (see Sect. 7.1).

Definition 6. *Services –in the plural– are possible instances of service systems and are parts of offerings.*

As a reminder, the term "service" –in the singular– refers to the application of competencies in the scope of this work (see Definition 2).

5.2   The Relations Between the Resources of Service Systems

In service systems, resources may be operant (see Definition 7) or operand (see Definition 8). The former means that the resource is able to act upon other resources in order to produce an effect, while the latter is defined as a resource which need to be acted upon [3]. This means that, in a service system, operand resources become valuable thanks to operant resources. Among the four types of resources, people, organizations and technologies could be both operant and operand resources. Information is always an operand resource –this fact is significant for the building of the conceptual model of a generic service system depicted by Fig. 3 and described in Sect. 5.3. This is explained by the fact that the information resource cannot act upon other resources. It always needs to be associated with an operant resource to be valuable.

Definition 7. *An **operant resource** is a resource which acts upon another resource to produce an effect.*

Definition 8. *An **operand resource** is a resource which is acted upon by another resource to produce an effect.*

Each relation between an operant and an operand resource results in an *internal interface*, concept which is defined by Definition 9. An internal interface can also relate an operant and an operand resource of the same type (except for resources being of the type *information* seeing that this can only be an operand resource). For instance, in a service system, a person is related by an internal interface with another person if the first person massages the second one.

Definition 9. *An **internal interface** is a relation between an operant resource and an operand resource resulting in an effect and constituting (partially) a service system.*

Figure 2 abstractly depicts an internal interface between two resources. The arrow indicates that the resource on the left, which is the operant, is applied on the resource on the right, which is the operand, in order to produce an effect. These operant and operand resources associated by the internal interface constitute (a part of) a service system.

Note that the use of these modelling artefacts is illustrated in Sect. 6.2.

Fig. 2. Generic representation of an internal interface connecting two resource types

5.3   Conceptual Model of the Generic Service System

Fig. 3 depicts the conceptual model of service system, which is made up of the four types of resources. The line between the resource types indicate that these resources can be related together via internal interfaces. As a reminder, the resource of the type *information* is only an operand resource. This explains why there is no possible internal interface between two resources of this type. This is depicted in Fig. 3 through the absence of line from and to the resource type *information*.

6   Discussion and Modelling of the Recursion Principle

6.1   Applying the Recursion Principle to Service Systems

The principle of the **recursion** is expressly introduced in the definition of a service system –as a reminder, service systems are connected *internally* and externally to other service systems (cf. Definition 4). The recursion principle applied to an object means that this object can be defined in terms of one –in the case of a single recursion– or several –in the case of multiple recursions– other objects of the same type. In the scope of this paper, the application of this principle means that service systems could be defined in terms of other service systems, what is essential in service science. Indeed, many service systems are quite complex in the current economic world. By considering that they can

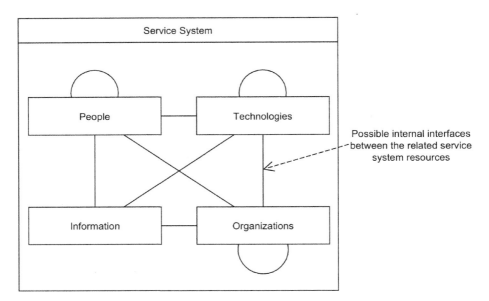

Fig. 3. Conceptual model of the generic service system constituted of the four main resource types related by all the possible internal interfaces

be decomposed into several other and, obviously, less complex service systems which, in turn, can also be decomposed into other service systems, and so on, we can analyse each of these (smaller) service systems individually. Therefore, it is possible to adopt different levels of abstraction regarding a particular (composed) service system when analysing it or reasoning on it.

In order to take into account the recursion principle in the conceptual modelling work proposed, we introduce the notion of *service module*, which is defined hereafter.

Definition 10. *A **service module** is a service system which composes at least one other service system.*

Service modules, associated through value propositions, have to be considered as a new and composite service system, which is a potential source of value cocreation also described thanks to a value proposition. This new service system can be executed on its own. It could also be a component of a service module. In this case, this service system becomes a service module when it is associated with other service modules in order to compose a bigger and, presumably, more complex service system. In turn, the latter can also be a service module in the case where it is associated with other service modules to form an even more complex service system, and so on.

Note that a given service system can be executed on its own and, in other circumstances, it could also be executed as a service module which composes one or several larger service systems. For example, let's consider a hairdressing service. It can be executed on its own in the case where you go to the hairdresser and ask for a new haircut. This same hairdressing service could also be considered as a service module if it is one of the service modules composing, e.g., a makeover service. This example is further developed in Sect. 6.2.

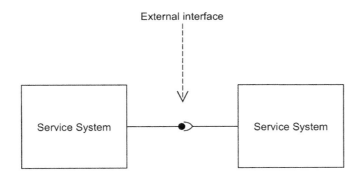

Fig. 4. External interface between two service systems

As we consider that a service system is composed of service modules, interfaces between them are required. These interfaces are called *external interfaces*. This concept is depicted by Fig. 4 and defined hereafter.

Definition 11. *An **external interface** is a value proposition, which combines two service modules, describing an expected co-creation of value-in-context.*

6.2   Model of a Generic Service System

Figure 5 sums up this discussion. It illustrates a generic service system, which is composed of three service modules. These service modules are associated via external interfaces. One of these service modules is also composed of three other service modules. Of course, the two other service modules as well as the service modules composing the service module (C) could be, in turn, composed of other service modules.

Let's go back to the example of the makeover based on Fig. 5. We consider that the main service system is thus the makeover service. Its three service modules are as follows:

– Service module (A): Determining a new style
– Service module (B): Advice and support in the purchase of new clothes
– Service module (C): Modification of the facial appearance
  Service module C is, in turn, composed of three other service modules. The service modules related with an external interface are *Hair dyeing* and *Haircut*; the third service module is an *Eyebrows care*.

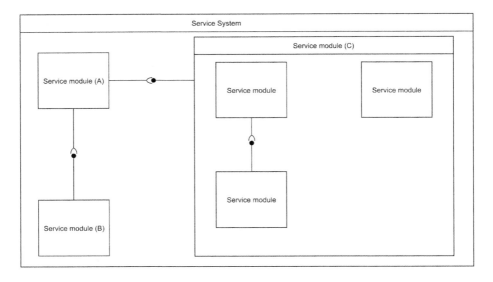

Fig. 5. Illustration of a generic service system composed of service modules

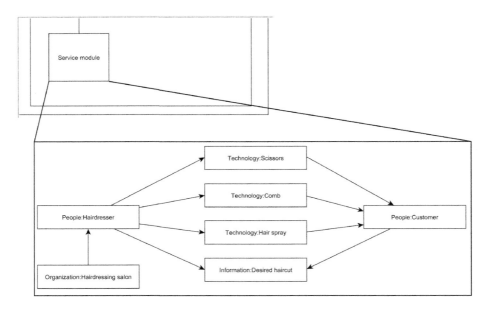

Fig. 6. Illustration of the resources associated to build the service module *Haircut*

Figure 6 illustrates the resources associated to build the service module *Haircut* (cf. Fig. 5). The different resource types –i.e., people, information, technology and organizations– are associated by internal interfaces. Of course, the description of the resources associated to form a service module can be carried out for each of the service modules illustrated in Fig. 5.

7   Conclusion

By creating a conceptual model of a service system, which is the fundamental element in service science, we first highlight the need for a common understanding of the main concepts of service science as well as of their relations. Similarly to other researchers, e.g., [21], we argue that a conceptual model is a very interesting solution for understanding, communicating and reasoning on a specific and defined part of the reality, which is a generic service system in the scope of this paper. The proposed conceptual model depicts the four types of resources associated through internal interfaces, which form service systems. The proposed modelling also takes into account the recursive principle, which enables us to introduce the notion of service modules. The latter are a specialization of service systems, which compose other service systems, while being able to be executed on their own too. As the starting point of our discussion, we also model the generic transformation process from an SDL viewpoint. Other scientific fields –e.g., organization theory, operations, marketing, economics and so on– have studied parts of this transformation process and/or some of their relations. However, they often focus on a very specific part of service systems. SDL provides a shared perspective and vocabulary, but it also provides the same hypothesis for the engineering and the management of service systems as well as for innovating through them [2, 13, 15, 23].

One important remark concerns the internal and external interfaces. Indeed, their use enables to constitute networks of value co-creation by combining resources and service systems together [3, 36]. These value networks enable to co-produce service offerings through the exchange of service offerings of lower levels in the service system in order to co-create value within a whole supply chain [37]. The proposed generic model can be used to model the components of these value networks and to reason upon them.

A last point to underline is that, until now, we do not mention the notion of product-service systems (PSS) –i.e., "product(s) and service(s) combined in a system to deliver required user functionality in a way that reduces the impact on the environment" [38]. PSS is a particular case of servitization [8, Chap. 1]. PSS are thus service systems. This means that researchers and practitioners could use the conceptual modelling proposed in this paper in order to conduct some reasoning on PSS.

7.1   Future Work

One limit of this paper is the lack of effective use of the proposed conceptual model, although we illustrated the concept introduced through a makeover service system and one of its service modules –i.e., the haircut service. One future work is the use of these contributions in different environments in order to prove the robustness of the conceptual model introduced here.

In order to pursue the research work described in this paper, we identify two possible directions. The first one comes from the following observation: we summarize the possible interactions between service systems thanks to the notion of *external interface*. However, there are many possible interaction types such as underlined in the ISPAR framework [23]. Therefore, a promising research direction is the investigation of the possible association of the ISPAR model with the conceptual model of the service systems proposed in this paper.

The second future work is the analysis of the service system life cycle development in the light of the conceptual modelling proposed in this paper. This should enable to better understand and control it.

References

1. Alter, S.: Service system fundamentals: work system, value chain, and life cycle. IBM Syst. J. **47**(1), 71–85 (2008)
2. Vargo, S.L., Maglio, P.P., Akaka, M.A.: On value and value co-creation: a service systems and service logic perspective. Eur. Manag. J. **26**(3), 145–152 (2008)
3. Vargo, S.L., Lusch, R.F., Akaka, M.A.: Advancing service science with service-dominant logic: clarifications and conceptual development. In: Maglio, P.P., Kieliszewski, C.A., Spohrer, J.C. (eds.) Handbook of Service Science, pp. 133–156. Springer, Heidelberg (2010)
4. Maglio, P.P., Spohrer, J.: Fundamentals of service science. J. Acad. Mark. Sci. **36**(1), 18–20 (2008)
5. Thalheim, B.: Towards a theory of conceptual modelling. J. Univ. Comput. Sci. **16**(20), 3102–3137 (2010)
6. Larkin, J.H., Simon, H.A.: Why a diagram is (sometimes) worth ten thousand words. Cogn. Sci. **11**(1), 65–100 (1987)
7. Wieringa, R.: Real-world semantics of conceptual models. In: Kaschek, R., Delcambre, L. (eds.) The Evolution of Conceptual Modeling. LNCS, vol. 6520, pp. 1–20. Springer, Heidelberg (2011). doi:10.1007/978-3-642-17505-3_1
8. Cardoso, J., Fromm, H., Nickel, S., Satzger, G., Studer, R., Weinhardt, C. (eds.): Fundamentals of Service Systems. Springer, Cham (2015)
9. Hill, T.P.: On goods and services. Rev. Income Wealth **23**(4), 315–352 (1977)
10. Rathmell, J.M.: What is meant by services? J. Mark. **30**(4), 32–36 (1966)
11. Grönroos, C.: Service management and marketing: managing the moments of truth in service competition. Lexington Books, Issues in Organization and Management Series (1990)
12. Lovelock, C., Vandermerwe, S., Lewis, B.: Services Marketing. Prentice Hall Europe, London (1996)
13. Vargo, S.L., Lusch, R.F.: Service-dominant logic: what it is, what it is not, what it might be? In: Lusch, R.F., Vargo, S.L. (eds.) The Service-Dominant Logic of Marketing: Dialog, Debate, and Directions. ME Sharpe, Armonk (2006)
14. Gummesson, E.: Relationship marketing: its role in the service economy. In: Glynn, W.J., Barnes, J.G. (eds.) Understanding Services Management. Wiley, New York (1995). Chap. 9

15. Vargo, S.L., Lusch, R.F.: Evolving to a new dominant logic for marketing. J. Mark. **68**(1), 1–17 (2004)
16. Vargo, S.L., Lusch, R.F.: Service-dominant logic: continuing the evolution. J. Acad. Mark. Sci. **36**(1), 1–10 (2008)
17. Chinnici, R., Moreau, J.J., Ryman, A., Weerawarana, S.: Web Services Description Language (WSDL) version 2.0 part 1: core language. W3C Recommendation, World Wide Web Consortium (W3C), June 2007. http://www.w3.org/TR/wsdl20/
18. Editorial group of the object management group: Service oriented architecture Modeling Language (SoaML) Specification. Normative document, The Object Management Group, May 2012. http://www.omg.org/spec/SoaML/1.0.1/
19. World Wide Web Consortium: Service modeling language, version 1.1. W3C Recommendation. The Object Management Group (2009). http://www.w3.org/TR/sml
20. Barros, A., Kylau, U., Oberle, D.: Unified Service Description Language 3.0 (USDL) overview. Technical report, SAP Research (2011)
21. Ferrario, R., Guarino, N., Janiesch, C., Kiemes, T., Oberle, D., Probst, F.: Towards an ontological foundation of services science: the general service model. In: 10th Internationale Tagung Wirtschaftsinformatik, pp. 47–56 (2011)
22. Tukker, A.: Eight types of product-service system: eight ways to sustainability? Experiences SusProNet. Bus. Strateg. Environ. **13**(4), 246–260 (2004)
23. Maglio, P.P., Vargo, S.L., Caswell, N., Spohrer, J.: The service system is the basic abstraction of service science. Inf. Syst. e-Bus. Manag. **7**(4), 395–406 (2009)
24. Kindström, D.: Towards a service-based business model - key aspects for future competitive advantage. Eur. Manag. J. **28**(6), 479–490 (2010)
25. Zolnowski, A., Semmann, M., Böhmann, T.: Metamodels for representing service business models. In: Proceedings of SIGSVC Workshop, paper 477 (2011)
26. Gordijn, J., Akkermans, H., Van Vliet, J.: Designing and evaluating e-business models. IEEE Intell. Syst. **16**(4), 11–17 (2001)
27. Grönroos, C., Voima, P.: Critical service logic: making sense of value creation and co-creation. J. Acad. Mark. Sci. **41**(2), 133–150 (2013)
28. Spohrer, J.C., Maglio, P.P., Bailey, J.H., Gruhl, D.: Steps toward a science of service systems. IEEE Comp. Soc. **40**(1), 71–77 (2007)
29. Spohrer, J., Maglio, P.P.: The emergence of service science: toward systematic service innovations to accelerate co-creation of value. Prod. Oper. Manag. **17**(3), 238–246 (2008)
30. Maglio, P.P., Srinivasan, S., Kreulen, J.T., Spohrer, J.: Service systems, service scientists, SSME, and innovation. Commun. ACM **49**(7), 81–85 (2006)
31. Prahalad, C.K., Ramaswamy, V.: Co-creation experiences: the next practice in value creation. J. Interact. Mark. **18**(3), 5–14 (2004)
32. Backlund, A.: The definition of system. Kybernetes **29**(4), 444–451 (2000)
33. Spohrer, J., Maglio, P.P.: Service science: toward a smarter planet. In: Salvendy, G., Karwowski, W. (eds.) Introduction to Service Engineering, pp. 3–30. Wiley, New York (2010). Chap. 1
34. Durkheim, E.: The Division of Labor in Society. The Free Press, New York (1984). Translation of "De la division du travail social" published in 1933
35. Edvardsson, B., Gustafsson, A., Roos, I.: Service portraits in service research: a critical review. Int. J. Serv. Ind. Manag. **16**(1), 107–121 (2005)
36. Spohrer, J.C., Maglio, P.P.: Toward a science of service systems: value and symbols. In: Maglio, P.P., Kieliszewski, C.A., Spohrer, J.C. (eds.) Handbook of Service Science, pp. 157–194. Springer, Heidelberg (2010)

37. Lusch, R.F., Vargo, S.L., Tanniru, M.: Service, value networks and learning. J. Acad. Mark. Sci. **38**(1), 19–31 (2010)
38. Baines, T.S., Lightfoot, H.W., Evans, S., Neely, A., Greenough, R., Peppard, J., Roy, R., Shehab, E., Braganza, A., Tiwari, A., et al.: State-of-the-art in product-service systems. In: Proceedings of the Institution of Mechanical Engineers, pp. 1543–1552. Sage Publications (2007)

Social Media Marketing and Value Co-creation: A Dynamic Performance Management Perspective

Francesca Costanza[✉]

University of Palermo, Via Montepellegrino 163, 90142 Palermo, Italy
f.costanza82@gmail.com

Abstract. The diffusion of web-based technologies has transformed the today consumers, shifting them from mere users to content generators, able to influence each other's opinions and choices. Such trend discloses important consequences for firms, products and brands. In order to support strategic and operational decision-making, this study frames the phenomenon of social media marketing into the value co-creation paradigm and shows how to design an analytical framework combining performance management and System Dynamics.

Keywords: Social media marketing · Value co-creation · System dynamics · Performance management

1 Introduction

Today consumers are taking an increasingly active role in co-creating everything from product design to promotional messages [1]. In addition, social media technologies, such as social networks and blogs, have witnessed explosive growth in recent years [2, 3]. They enable individuals to create, share, and recommend information, extending the spheres of marketing influence up to disclose a new business model [4, 5].

In spite of traditional advertising, social media permits individuals to share educational contents about products and brands, communicating and interacting in an ample community [6]. Thus, social digitization has transformed consumer behaviour [7, 8], with important consequences for firms, products, and brands [8–10].

Companies aiming at a competitive advantage [11] must learn how to use social media in coherence with their business plans [12], especially when product launches are involved, in consideration that firms typically allocate approximately half of their marketing budgets for new products [13].

However, many companies are reluctant to allocate resources to develop new media-based strategies [14]. Such attitude is due to a lack of understanding of how to use social media proficiently [7, 14]. There exist a general gap between the increasing complexity of markets and the companies' abilities to respond to new demands with innovative and complex thinking [8, 15]. It follows the need for further researches in this area, integrating the traditional analytical tools with new approaches to support strategic and operational decision-making.

S. Za et al. (Eds.): IESS 2017, LNBIP 279, pp. 131–143, 2017.
DOI: 10.1007/978-3-319-56925-3_11

This paper aims to make a step forward in the understanding of social media driven dynamics of value co-creation. To achieve this scope, it proposes a qualitative insight model embracing performance management and System Dynamics principles [16–19]. Such analytical framework, representing a complex hypothesis still under validation, consists of a stock-and-flow structure integrating socio-economic feedbacks. Key-variables and links come from critical literature selection and media observations, conducted in pace with a system-wide performance management analysis.

The paper's organization is as follows. After framing the phenomenon of social media marketing into the value co-creation theory, it follows the explanation of the main steps of model conceptualization and design. Finally, conclusion and call for future researches will take place.

2   Background

Contents and connections in an online community are user-generated, so that firms cannot directly control consumer-to-consumer messages surrounding their brands [12]. For the above reason, social media messages often have higher credibility and trust than traditional media [20].

The social media interactions can be read in light of the value co-creation paradigm and its amplification due to digital technologies' advancements [21–25]. For a today company, improving the performance management means not to focus to its limited boundaries, but it refers to the understanding of the more complex system of value co-creation. Traditionally, the attention devoted to consumers has been relatively scarce [26–28], in spite of the dominating paradigms of transaction costs [29], firm positioning [30], resource-based view [31], dynamic capabilities [32].

The definition of value is controversial [33]. Value has been conceived in strategic management from the supply side, exclusively created by producers [34], as reflected for example in the common term 'value added' [35]. At the contrary, value is multi-dimensional and its creation lays on complexity. In addition, consumers may customise values and meanings to achieve their lifegoals [36, 37].

In co-creation processes, both consumers and producers collaborate or participate in creating value. Such consumer's active role is often tackled to have transformed the economic logic, and shifted the power from producers to consumers [38], bringing a revolution even within the strategic management literature. The 'free consumer' [39], is now a threat to marketers [40], and the main challenge is to stimulate shared meaning and a common sense of purpose between customer and company [41].

Three main research streams have addressed the concept of value co-creation. Each of them focuses specific aspects, but they all have a common point: the consumer can co-create value with companies and with other consumers in order satisfy individual and social lifegoals.

The first theory is the Working Consumer, where the so-called 'prosumers' co-create value using their skills and knowledge, producing what they consume [42, 43], or providing immaterial labor to the companies [39, 40, 44].

According to a second theory, the Service-dominant Logic [25, 45], value comes from exchanges of services between different actors, each of them bringing and using resources, knowledge and skills, to benefit the others. In such mutuality, when customers experience goods and services, the co-creation of value takes place dynamically and 'in context' [46]. Then the concept of value is relational, whereby the knowledge and experience of customers, not simply passive buyers but also active agents, become crucial [47].

The third literature stream is Consumer Culture Theory [48], where value co-creation depends on how customers perceive, interpret and interact with market offerings [49]. Value does not depend just upon utility of goods or services, but also on the consumer's interpretation of consumption objects, such as products, brands, and services [50]. Since consumer culture and social resources are characters of the value co-creation process, companies should explore the symbolic meaning of consumptions in relation with consumers' life projects [51].

In this regard, the context of web 2.0 stimulates the sociality in value co-creation. Consumers use the available information to construct their meaning of life, socialization reinforce such mechanisms [52, 53]. Indeed, people sharing online their consuming practices and ways to face reality, form new entities based on the links between the practical activity and its representations [54, 55]. As a result, they build 'social consensus' [56] around products and brands.

3   The Methodological Approach

This section of the paper shows how to combine System Dynamics (SD) with basic elements of performance management (PM), in order to design a 'Dynamic Performance Management System' (DPMS) [16, 57].

System Dynamics is a methodology for policy analysis and design aided by computer simulations. It applies to dynamic problems/issues arising in systems characterized by interdependence, mutual interaction, information feedback, and circular causality.

In SD, the representation of the structure of a system is through feedback networks of stock-and-flow diagrams tracking processes of accumulations [19]. These diagrams are made of three elements, defined as follows.

Stocks reflect the level of accumulation of material, people, money and information (identified with units of an item at a certain time). Flows are rates able to affect the stocks' level (identified in units per time). Finally, auxiliary variables help in calculations or represent exogenous parameters or constants.

Stock accumulates their inflows less their outflows, so mathematically stock-and-flow structures are system of integrals and differential equations [19]. Figure 1 shows a simplified stock-and-flow structure: stock 1 has an inflow increasing its level and an outflow reducing it, whilst stock 2 only has an inflow. Each stock influences, directly or by mean of an auxiliary variable, a flow linked to the other stock, creating a feedback loop.

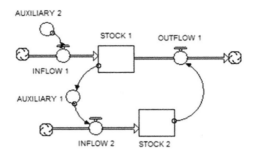

Fig. 1. Core elements of a system dynamics model

Among the SD models' advantages, there is the possibility to run simulations of key-variables' patterns over the time in response to alternative policies' activation. Using a simulation software means to specify the stock-and-flow structures with a list of equations representative of decision rules and behavioral relationships.

In this study, SD and PM work together since the author believes them complementary. On the one hand, traditional focus of PM is the financial dimension [58], a narrow perspective to face today's complexity, and then PM spectrum should embrace other perspectives related to programs' quality and outcomes [59].

In this respect, SD models can help to capture the multidimensional aspect of value-creation and support company decision-makers to better recognizing and then measuring key-performance indicators and the factors affecting them. At the same time, by adopting a Dynamic Performance Management approach [16, 57], it is possible to enclose in the SD modelling some guiding principles coming from PM. Figure 2 represents a general framework to conduct this kind of analysis.

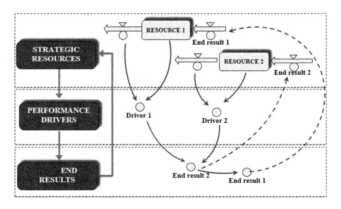

Fig. 2. The general Dynamic Performance Management framework (Bianchi, 2010)

To design a DPMS, it is necessary to adopt an instrumental view [16, 57], requiring the identification of three elements: first, the relevant 'strategic resources' that the organization is able to accumulate/deplete through external acquisition/disposal or internal building/depletion processes.

The strategic resources' representation is by the SD stock's notation and they are able to influence the acquisition of other resources and the capacity of the organization to pursue economic, competitive and social objectives.

The second category to identify is the 'performance drivers', modeled as auxiliary variables, referring to the critical success factors and outlining the capability of an organization, in relationship with the performance of external economies or counter-parts, to influence the end-results.

Finally, the 'end-results', modeled as inflows and outflows, can be expressed in economic, competitive or social objectives and are able to affect the processes of accumulation/depletion of the strategic resources over the time.

3.1   The Model Building

A three-step process from synthesis to analysis conceptualized the stock-and-flow insight model of value co-creation.

The first phase was based on PM and implied the mapping of value co-creation macro-processes.

The second phase was the enlargement of the initial framework into a more complex one based on the instrumental view, namely the formulation of a value co-creation dynamic performance management framework.

The last step was to make explicit the variables' links and to convey the modelling efforts into a stock-and-flow feedback model.

Since the output of each modelling step is just a more analytical version of the model iteratively built in previous phases, in this paper the terms 'model' and 'framework' can be used either to identify the final output and the intermediate ones.

In the model, company's perspective is the brand orientation, aimed at the creation of a brand identity for the target consumers, to achieve lasting competitive advantages [60]. Brand consists of a cluster of rational and emotional values enabling a promise of a unique and welcomed experience [61]. Here the terms 'brand' and 'company' are used interchangeably to highlight the brand's goal to communicate the uniqueness of a company and its products.

Consumers are considered in their social dimension as participants to online communities in light of Consumer Culture Theory, with the gradual development of shared meanings and practices in consumer and marketplace cultures [46, 49]. Then, the unit of analysis of the model is a cohort of individuals characterized by shared lifestyle interests, meanings and practices, held together through emotional links and members' commitment [54].

Inputs to the model building were the literature and empirical findings. In particular, twelve social media of fashion, beauty and lifestyle (five blogs, the integrated social networks pages, and two content communities) were observed along with one-year period, starting from January 2015 to January 2016. The author joined communities and social networks as a member and observed each media community's interactions (member-to-member, influencer-to-members, and influencer-to-influencer). In-depth interviews to one of the bloggers provided further inputs for the model building.

Blogs, content communities and social networks were preferred for the study due to their relevance in terms of number of users [14, 62]. In addition, their links and synergies were judged suitable for emphasizing the social level of consumption. Therefore, for the purpose of the model, the umbrella concept of 'online community' intends to capture the web-based social level of conversation and consumption, such as brand community, subculture of consumption, consumer tribe, co-consuming groups [63].

3.2   First Step: Co-creation Value Chain

The first modelling step was the mapping of value co-creation macro-processes. In particular, there were identified the products resulting from the co-creation process, and the clients, benefitting from the product.

In Fig. 3, the 'final' product consists of the creation of improved/new products, able to meet better the consumers' exigencies. Such selection is due to the consideration that the innovation capacity directly affects the brand quality, ultimately the main outcome of value co-creation processes, since it implies customer satisfaction and, by this way, enduring success for a company.

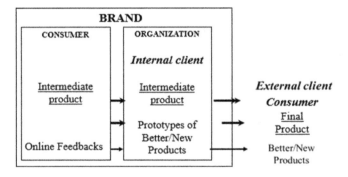

Fig. 3. Value co-creation macro-processes

Upward from the final product, it is identified a synthetic system of 'intermediate' products resulting from processes fulfilled by consumer and organization. The consumer plays the double-role of external client and internal actor, since it provides the intermediate product 'online feedbacks' and thereby gives inputs to the company's innovation processes to satisfy his/her needs.

Such representation, depicting both consumer and organization as 'internal clients' of the process is in line with the Working Consumer and the Service-Dominant theories, where value is created through interactions between resources and competencies of consumers and producers', whose roles are less distinct than traditional business models, that would otherwise limit the role of consumer just to external client.

3.3   Second Step: Dynamic Performance Management Framework

Using the scheme in Fig. 3 as a start, the author, moving backward from final to instru-mental products, progressively collected inputs from literature and media observations. This way it was possible to hypothesize the dynamic performance management cycle resulting from the interaction of strategic resources, performance drivers and end-results. The output of this phase is the framework in Fig. 4.

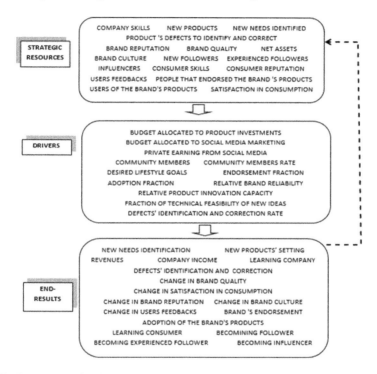

Fig. 4. The Dynamic Performance Management framework adapted to social media-driven value co-creation

The author started with identifying the strategic resources; both tangible (e.g. new products) and intangible (e.g. company skills). Their allocation is crucial since it may influence performance drivers and thereby feed back into the end-results.

Then it was possible to include the performance drivers, representing the system's capacity to pursue its objectives through critical success factors. The model contains financial drivers (e.g. budget allocated to products' investments, budget allocated to social media marketing, private earning from social media); technical drivers (fraction of technical feasibility of new ideas); 'social' drivers (community members rate over the potential community members). In addition, competitive drivers, such as the product innovation capacity or the brand reliability (capacity to identify and correct existing products' bugs), are modelled like the ratio of company new products or products' defects referred the competitors' ones.

Finally, it was possible to highlight end-results in terms of economic (e.g. revenues), competitive (e.g. adoption of the brand's products) and social objectives (e.g. change in satisfaction in consumption, change in users' feedbacks), pursued by the performance system in a certain range of time, and able to re-affect the processes of strategic resources' accumulation/depletion.

3.4   Third Step: Stock-and-Flow Structure

The above framework was translated into a SD model, where all the links between end-results, drivers and strategic resources are made explicit and the system's feedback structure is captured. This step is possible by using the SD notation for representing the Dynamic Performance Management framework. In particular, the end-results shift into inflows and outflows; strategic resources are modelled as stocks, performance drivers get into auxiliary variables.

Figure 5 depicts the resulting stock-and-flow structure. It considers Service-Dominant theory, since there are no clear boundaries separating companies and consumers. The feedback structure is made of seventeen stocks and captures the multidirectional exchanges of information flowing via social media: from customers to customers, from customers to brands, and from brands to customers.

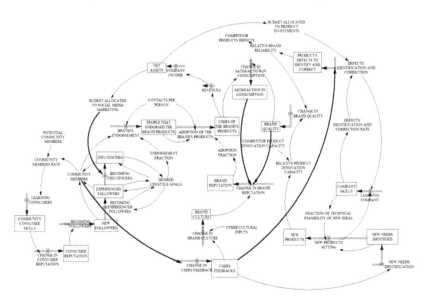

Fig. 5. The stock-and-flow structure of the model

On the left side, there is the space of social conversation between online community members. Going online, consumers try not only to fulfil information needs about goods, but they also enlarge their social life and by sharing habits they co-create self-identity within a group of belonging [36, 64]. Therefore, they develop *mental models*, or *moments of truth* [65] about each other's consuming behaviours and opinions, triggering mechanisms of shared understanding of the world [66, 67].

Such elements have been called generically 'Lifestyle Goals' and induce people to participate to a community. Due to the cumulated experiences in the social group, they flow from the stock 'New Followers' to 'Experienced Followers'. When they get an advanced experience that make them able to modify opinions, perceptions and purchasing choices, they become 'Influencers'.

Influencers' role is getting quite relevant because of their hybrid nature between consumers and workers. Since they receive financial or goods' benefits for conveying online conversations to certain products/topics, their motivation to participate in virtual communities depends not just on 'Lifestyle Goals', but also on incentives coming from 'Firms' Budget Devoted to 'Social Media Marketing'.

The cumulated experience of 'Followers', 'Experienced followers' and 'Influencers', affects the community members' 'Consumer Skills', and by this mean the 'Consumer Reputation' within the community, to be conceived like the social standing in a media setting, able to attract new interactions.

In the centre of Fig. 5, there is the adoption of the brand's products, not ascribable to traditional marketing levers only, but also influenced by social media word-of-mouth. The stock of 'People that endorsed the brand's products' captures the dynamics occurring whenever the continuous online conversation about products and services leads to a gradual perception, endorsement and validation by the group, even without a real purchasing.

On the right side, favourable loops of consumers' feedback create competitive advantages for the companies. The active consumers have a double effect on the product strategy: firstly, they stimulate the innovation capacity by increasing the company skills of new needs identification and new products' development. Second, they increase the company reliability by helping in the detection and correction of defects in existing products. Brand reliability and product innovation capacity relatively to the competitors are able to contribute to 'Brand's Quality' and then to 'Brand's Reputation' and firm's revenues [68].

The brand' reputation depends on 'Brand Quality', 'Brand Culture' and 'Satisfaction in Consumption', in turn affected by the quality of the products and on the number of users of the brand products.

The model captures learning and skill formation mechanisms on both the demand and the supply side by the stocks 'Brand Skills' and 'Consumer Skills', whose changes are attributed to the 'Learning Company' and 'Learning Consumer' flows. These figures fit the Service-Dominant logic, where the 'operant' resources (skills and knowledge, organizational routines, relational competences) are the main responsible for the competitive advantage and value-creation, since they are able to operate on the other kinds resources to solve problems and fulfil consumers' needs [25, 69].

4   Conclusion and Future Researches

This paper proposed a model hypothesis about how value is co-created in pace with company and consumer's skills formation, thereby disclosing the strategic potential of experimenting social media-based innovation.

In the study, value co-creation comes from consumers' feedbacks, and the challenge for companies is to enhance their skills to fulfil consumer needs, in turn more sophisticated due to products evaluation skills' development. This requires a mind-set shift from considering brand quality as a mere company's matter to a more complex process going beyond the organizational boundaries.

The latter would imply for companies to search for new connections with customers, for example by the creation of inventive communities intertwining organizational knowledge with external learning and innovative inputs [70–72].

Limitations of the model can be the discretional variables' selection and the possible bias within the empirical observations. Indeed, the model neglects the potential integration of social media strategy with the traditional marketing mix and does not capture the links between different social media. Future research steps will fill such gaps.

The model still lays at qualitative level: next avenues of this work will also be the formalization of a simulating SD model, implying the setting of a differential equation system. Data gathering may come from the data analytics of the social media already observed, surveys to online community members, as well as the involvement of companies' case study through management and personnel interviews and surveys. After statistical validation of the simulated behavior patterns, the model will effectively help in scenario analyses and decision-making.

References

1. Berthon, P.R., Pitt, L.F., McCarthy, I., Kates, S.: When customers get clever: managerial approaches to dealing with creative consumers. Bus. Horiz. **50**(1), 39–48 (2007)
2. Singaraju, S.P., Nguyen, Q.A., Niininen, O., Sullivan-Mort, G.: Social media and value co-creation in multi-stakeholder systems: a resource integration approach. Indus. Mark. Manag. **54**, 44–55 (2016)
3. Onishi, H., Manchanda, P.: Marketing activity, blogging and sales. Int. J. Res. Mark. **29**, 221–234 (2012)
4. Hanna, R., Rohm, A., Crittenden, V.L.: We're all connected: the power of the social media ecosystem. Bus. Horiz. **54**, 265–273 (2011)
5. Hansen, D., Shneiderman, B., Smith, M.A.: Analyzing Social Media Networks with NodeXL: Insights from a Connected World. Elsevier, Boston (2011)
6. Weinberg, B.D., Pehlivan, E.: Social spending: managing the social media mix. Bus. Horiz. **54**(3), 275–282 (2011)
7. Kaplan, A., Haenlein, M.: Users of the world, unite! The challenges and opportunities of social media. Bus. Horiz. **53**(1), 59–68 (2010)
8. Lipiäinen, H.: Digitization of the Communication and its Implications for Marketing. Jyväskylä University Printing House, Jyväskylä (2014)
9. Muntinga, D.G., Moorman, M., Smit, E.G.: Introducing COBRAs: exploring motivations for brand-related social media use. Int. J. Advertising **30**(1), 13–46 (2011)

10. Wymbs, C.: Digital marketing: the time for a new "academic major" has arrived. J. Mark. Educ. **33**(1), 93–106 (2011)

11. Paquette, H.: Social media as a marketing tool: a literature review. Unpublished master's paper, University of Rhode Island, Kinston, RI (2013)

12. Mangold, G.W., Faulds, D.J.: Social media: the new hybrid element of the promotion mix. Bus. Horiz. **52**, 357–365 (2009)

13. Urban, R., Hauser, J.: Design and Marketing of New Products. Prentice Hall, Upper Saddle River, NJ (1993)

14. Kietzmann, J.H., Hermkens, K., McCarthy, I.P., Silvestre, B.S.: Social media? Get serious! Understanding the functional building blocks of social media. Bus. Horiz. **54**, 241–251 (2011)

15. Day, G.S.: Closing the marketing capabilities gap. J. Mark. **75**(4), 183–195 (2011)

16. Bianchi, C.: Enhancing performance management and sustainable organizational growth through system dynamics modeling. In: Groesser, S.N., Zeier, R. (eds.) Systemic Management for Intelligent Organizations: Concepts, Model-Based Approaches, and Applications, pp. 143–161. Springer, Heidelberg (2012)

17. Forrester, J.W.: Industrial Dynamics. MIT Press, Cambridge (1961)

18. Meadows, D.H.: The Unavoidable A Priori. In: Randers, J. (ed.) Elements of the System Dynamics Method, pp. 23–57. MIT Press, Cambridge (1980)

19. Sterman, J.D.: Business Dynamics: System Thinking and Modeling for a Complex World. Mc Graw-Hill, Boston (2000)

20. Blackshaw, P., Nazzaro, M.: Consumer-Generated Media (CGM) Word-of-Mouth in the Age of the Web-Fortified Consumer. BuzzMetrics Inc., New York (2006)

21. Ballantyne, D., Varey, R.J.: Creating value-in-use through marketing interaction: the exchange logic of relating. Commun. Know. Mark. Theor. **6**(3), 335–348 (2006)

22. Grönroos, C.: Service logic revisited: who creates value? and who co-creates? Eur. Bus. Rev. **20**(4), 298–314 (2008)

23. Prahalad, C.K., Ramaswamy, V.: The Future of Competition: Co-creating Unique Value with Customers. Harvard Business School Press, Boston (2004)

24. Schau, J.H., Muñiz Jr., A.M., Arnould, E.J.: How brand community practices create value. J. Mark. **73**(5), 30–51 (2009)

25. Vargo, S.L., Lusch, R.F.: Evolving to a new dominant logic for marketing. J. Mark. **68**, 1–17 (2004)

26. Brief, A.P., Bazerman, M.: Editor's comments: bringing in consumers. Acad. Manag. Rev. **28**, 187–189 (2003)

27. Farjoun, M.: Towards an organic perspective on strategy. Strateg. Manag. J. **23**, 561–594 (2002)

28. Hunt, S.D., Lambe, C.J.: Marketing's contribution to business strategy: market orientation, relationship marketing and resource-advantage theory. Int. J. Manag. Rev. **2**, 17–43 (2000)

29. Williamson, O.E.: The vertical integration of production: market failure considerations. Am. Econ. Rev. **61**, 112–123 (1971)

30. Porter, M.E.: Competitive Strategy. Free Press, New York (1980)

31. Barney, J.B.: Firm resources and sustained competitive advantage. J. Manag. **17**, 99–120 (1991)

32. Eisenhardt, K.M., Martin, J.A.: Dynamic capabilities: what are they? Strateg. Manag. J. **21**, 1105–1121 (2000)

33. Sánchez-Fernández, R., Iniesta-Bonillo, M.Á.: The concept of perceived value: a systematic review of the research. Mark. Theor. **7**(4), 427–451 (2007)

34. Priem, R.L.: A consumer perspective on value creation. Acad. Manag. Rev. **32**(1), 219–235 (2007)

35. Woodruff, R.B., Flint, D.J.: Marketing's service-dominant logic and customer value. In: Lusch, R.F., Vargo, S.L. (eds.) The Service-Dominant Logic of Marketing: Dialog, Debate and Direction, pp. 183–195. M.E. Sharpe Inc., Armonk (2006)

36. Firat, A.F., Dholakia, N., Venkatesh, A.: Marketing in a postmodern world. Eur. J. Mark. **29**(1), 40–56 (1995)

37. Vargo, S.L., Lusch, R.F.: Service-dominant logic: continuing the evolution. J. Acad. Mark. Sci. **36**(1), 1–10 (2008)

38. Pongsakornrungsilp, S., Schroeder, J.E.: Understanding value co-creation in a co-consuming brand community. Mark. Theor. **11**(3), 303–324 (2001)

39. Zwick, D., Bonsu, S.K., Darmody, A.: Putting consumers to work: 'co-creation' and new marketing govern-mentality. J. Consum. Cult. **8**(2), 163–196 (2008)

40. Cova, B., Dalli, D.: Working consumers: the next step in marketing theory? Mark. Theor. **9**(3), 315–339 (2009)

41. Roser, T., Samson, A., Humpreys, P., Cruz-Valdivieso, E.: Co-creation: new pathways to value: an overview (2009). http://personal.lse.ac.uk/samsona/cocreation_report.pdf. Accessed 29 Apr 2016

42. Bagozzi, R.P., Warshaw, P.R.: Trying to consume. J. Consum. Res. **17**(2), 127–140 (1990)

43. Xie, C., Bagozzi, R.P., Troye, S.V.: Trying to prosume: toward a theory of consumers as co-creators of value. J. Acad. Mark. Sci. **36**(1), 109–122 (2008)

44. Arvidsson, A.: brands a critical perspective. J. Consum. Cult. **5**(2), 235–258 (2005)

45. Vargo, S.L., Lusch, R.F.: The nature and understanding of value: a service- dominant logic perspective. Rev. Mark. Res. **9**, 1–12 (2012)

46. Honkanen, M.: Expanding understanding of value co-creation: a cultural approach. Unpublished master's thesis, Aalto University School of Business, Helsingfors, FIN (2014)

47. Edvardsson, B., Tronvoll, B., Gruber, T.: Expanding understanding of service exchange and value co-creation: a social construction approach. J. Acad. Mark. Sci. **39**, 327–339 (2001)

48. Arnould, E.: Service-dominant logic and consumer culture theory: natural allies in an emerging paradigm. Res. Consum. Behav. **11**, 57–76 (2007)

49. Arnould, E., Thompson, C.: Consumer Culture Theory (CCT): twenty years of research. J. Consum. Res. **31**(4), 868–882 (2005)

50. Lawrence, T.B., Phillips, N.: Understanding cultural industries. J. Manag. Inquiry **11**(4), 430–441 (2002)

51. Arnould, E.J., Price, L.L.: Authenticating acts and authoritative performances: questing for self and community. In: Ratneshwar, S., Mick, D.G., Huffman, C. (eds.) The Why of Consumption: Contemporary Perspectives on Consumer Motives, Goals, and Desires, pp. 140–163. Routledge, New York (2000)

52. O'Guinn, T.C., Shrum, L.J.: The role of television in the construction of consumer reality. J. Consum. Res. **23**(4), 278–294 (1997)

53. Richins, M.L.: Valuing things: the public and private meanings of possessions. J. Consum. Res. **21**(3), 504–521 (1994)

54. Cova, B., Cova, V.: Tribal marketing: the tribalisation of society and its impact on the conduct of marketing. Eur. J. Mark. **36**(5), 595–620 (2002)

55. Reckwitz, A.: Toward a theory of social practices: a development in culturalist theorizing. Eur. J. Soc. Theor. **5**(2), 243–264 (2002)

56. Deighton, J., Grayson, K.: Marketing and seduction: building exchange relationships by managing social consensus. J. Consum. Res. **21**(4), 660–676 (1995)

57. Bianchi, C.: Improving performance and fostering accountability in the public sector through system dynamics modelling. From 'external' to an 'internal' perspective. Syst. Res. Behav. Sci. **27**(4), 361–384 (2010)

58. Fitzgerald, L.: Performance measurement. In: Hopper, T., Northcott, D., Scapens, R.W. (eds.) Issues in Management Accounting, pp. 223–241 (2007)
59. Chenhall, R.H., Langfield-Smith, K.: Multiple perspectives of performance measures. Eur. Manag. J. **25**(4), 266–282 (2007)
60. Urde, M.: Brand orientation: a mindset for building brands into strategic resources. J. Mark. Manag. **15**(1–3), 117–133 (1999)
61. De Chernatony, L.: Towards the holy grail of defining 'brand'. Mark. Theor. **9**(1), 101–105 (2009)
62. Berthon, P.R., Pitt, L.F., Plangger, K., Shapiro, D.: Marketing meets web 2.0, social media, and creative consumers: implications for international marketing strategy. Bus. Horiz. **55**, 261–271 (2012)
63. Arnould, E., Price, L., Moisio, R.: Making contexts matter. In: Belk, R.W. (ed.) Handbook of Qualitative Research Methods in Marketing, pp. 106–125. Edward Elgar Publishing, Cheltenham and Northampton (2006)
64. Taylor, D.G., Strutton, D., Thompson, K.: Self-enhancement as a motivation for sharing online advertising. J. Interact. Advertising **12**, 13–28 (2012)
65. Normann, R.: Service Management: Strategy and Leadership in Service Businesses. Wiley, New York (1984)
66. Berger, P.L., Luckmann, T.: The Social Construction of Reality: A Treatise in the Sociology of Knowledge. Penguin, London (1967)
67. Du Gay, P., Pryke, M.: Cultural economy: an introduction. In: Du Gay, P., Pryke, M. (eds.) Cultural Economy: Cultural Analysis and Commercial Life, pp. 1–20. Sage, London (2002)
68. Rishika, R., Kumar, A., Janakiraman, R., Bezawada, R.: The effect of customers' social media participation on customer visit frequency and profitability: an empirical investigation. Inf. Syst. Res. **24**(1), 108–127 (2013)
69. Hunt, S.D., Derozier, C.: The normative imperatives of business and marketing strategy: grounding strategy in resource-advantage theory. J. Bus. Indus. Mark. **19**(1), 5–22 (2004)
70. Brodie, R.J., Winklhofer, H., Coviello, N.E., Johnston, W.J.: Is e-marketing coming of age? An examination of the penetration of e-marketing and firm performance. J. Interact. Mark. **21**(1), 2–21 (2007)
71. Coviello, N.E., Milley, R., Marcolin, B.: Understanding IT-enabled interactivity in contemporary marketing. J. Interact. Mark. **15**(4), 18–33 (2001)
72. Kao, T.-Y., Yang, M.-H., Wu, J.-T.B., Cheng, Y.-Y.: Co-creating value with consumers through social media. J. Serv. Mark. **30**(2), 141–151 (2016)

Creative Industries and Big Data: A Business Model for Service Innovation

Giovanna Morelli[1](✉) and Francesca Spagnoli[2]

[1] Università di Teramo, Campus Universitario di Coste Sant'Agostino, 64100 Teramo, Italy
gmorelli@unite.it
[2] Istituto Nazionale di Fisica Nucleare, Laboratori Nazionali di Frascati,
00044 Frascati, Rome, Italy
francesca.spagnoli@lnf.infn.it

Abstract. Creative Industries have largely contributed to employment, GDP growth and social cohesion, even during recent economic crises. Despite their relevance, there is a lack for monitoring the impacts, especially for new technologies involved into their business. The paper aims to appraise it when specifically related to the use of Big Data. It evaluates the considerable economic benefit on creative business performance linked to exploiting vast new flows of information. A multi-criteria methodology for assessing these effects on Creative Industries, and a model for implementing business performance through collaborative and virtual value chains are presented. The model shows positive spillovers resulting from the collaboration among Digital Creative Industries usually in the fields of innovation, technology and intellectual property benefitting from Big Data applications, distinguishing a macro from a microeconomic level of effectiveness, since transforming data into captured value for the firms, despite their size and volume capacity, increases business performance.

Keywords: Creative Industries · Digital Creativity · Big Data · ICT · Impact Assessment (IA) · Organisational semiotics

1 Introduction

Creative Industries have arisen as a strategic topic in the political agenda all around the world, due to their increasing role in post-industrial knowledge-based economies. According to the first, original, UK definition, these industries are grounded on "individual skill and talent with the potential to create wealth and jobs through developing intellectual property" [1], transforming them into economic value. Usually, the classification framework groups Creative Industries under three broad headings (art and culture, design, and media), while they belong to a wider range of sectors (i.e., performing and visual arts, art crafts, advertising, architecture, design, fashion, audio-visual and multimedia, finance).

The common denominator is the term 'creativity', "the true wealth of nations in the 21st century", according to Newbigin and the United Nations [2, 3]. It indicates a specific organisation's capacity of innovation, flexibility and autonomy, all values that in some

© Springer International Publishing AG 2017
S. Za et al. (Eds.): IESS 2017, LNBIP 279, pp. 144–158, 2017.
DOI: 10.1007/978-3-319-56925-3_12

respect are considered to have replaced operational efficiency and strategic planning as primary sources of competitive advantages in business. What unifies these activities is that they all trade with creative assets in the form of intellectual property. Creative Industries were more resilient than others to the global economic crisis: until 2011, world exports of ideas and creativity-centered industries, from art to design, to new media and audiovisuals, continued to grow up to US\$ 624 billion, doubling their 2002 level, with an increase in the annual growth rate of 8.8%, on a decade basis, exceeding in volume its pre-crisis peak of 2008, despite a severe drop in global trade. Investment in creative sectors reveals to be a powerful development option: despite the 2007 ongoing recession in advanced countries, the Creative Economy - the socio-economic potential of activities that trade with creativity, knowledge and information - showed strong dynamism, growing more rapidly than other sectors, particularly in the global South [2–4]. In the last decade, its rise offset the fragility of the post-crisis recovery in developed countries, sick of abnormal public deficit, currency volatility, and high level of unemployment. From a semantic point of view, various definitions and models on Creative Economy exist, from WIPO model to Throsby's "concentric cycles", to several distinct national approaches, depending on whether a statistical, economic, sociological, or political approach is chosen [5–8].

As all fast-growing sectors accounting for a significant part of the world's economy, Creative Industries are moving away from a strict good-dominant logic to a 'Creative Economy' environment, where the service-dominant logic becomes prominent due to the increasingly use of data-intensive technologies. Since digital and creative sectors are merging due to technological change, many international public bodies refers to these two as a single economic group. In such a scheme, related services are the real central figure of interactions and transactions among agents. Creative Industries largely contribute to GDP, employment, design-driven innovation, and industrial development, positively affecting individual business performance, collective social cohesion and development. In real terms, in 2011, in the 27-EU countries, these activities generated 860 million of Euros in value-added, approximately equal to 6.8% of GDP, offering work for about 14 million people, a 6.5% of the total EU workforces [9, 10]. More than others, this sector is able to produce value from "ideas" and generate jobs with far less impact on the environment using digitisation. On the trail of Florida's "3T" model (talent, technology, and tolerance) [11], attributes to be present simultaneously in a place to stimulate and anchor economic growth, the "Europe 2020 Strategy" considers 'creativity' central to drive economic success. For countries with rich cultures and a pool of local creative talent, the Creative Economy opens a path to build fair economic value. Especially smaller countries, with poor capacity of exploiting economies of scale, can gain better opportunities for growth than many other sectors, thanks to the innovative capacity of these industries.

The paper aims to explain Creative Industries' advantages arising from ICT, the Big Data, providing some guidelines for benefitting from the use of Big Data. It combines multidisciplinary analyses from different fields (i.e., economics, social analysis, organisational semiotics, business informatics, cultural studies and innovation technology), and evaluates the impacts of Digital Creativity through a combined methodology. It considers the value of Big Data as a specific variable, and proposes a model to facilitate collaboration among Digital Creative Industries.

The paper is organised as follows: Sect. 2 summarises the institutional reports and academic literature on Creative Industries, Big Data, and High Tech Industries also focusing on current models' weaknesses. Section 3 outlines the research objectives and methodology, and defines our Impact Assessment Methodology. Section 4 explains the rationale of the methodology for Digital Creativity Services, by identifying a list of socio-economic and cultural impacts' indices able to capture the potential benefits deriving from the collaborative networking activity among agents. Section 5 illustrates the major contribution of this work: the collaborative value network business model. It has been developed for supporting Creative Industries' business worldwide to cooperate through a common digital platform and generate service innovation, by taking into account users' needs and current Creativity Industries' market features. Finally, in Sect. 6, the main conclusions are drawn with some suggestions for further research work.

2   Background Analysis

Literature on Creative Industries has generally questioned how much of the recent growth in services is due to a substitution effect of activities previously carried out within the firms and now outsourced. Nowadays, the most stimulating phenomenon is the transformation of the organisational models, which followed both at national and international level a growing fragmentation of production in the global value chains. The main challenge in developing a service system is to design a model that ensures sufficient flexibility and adaptability to the external and internal environment, crucial for its durability and viability, as Beer's "Viable System Model" predicts [12].

Entrepreneurs have to manage simultaneously structures, processes, results, experiences, and the past, present and future scenarios in real time. The use of ICT tools, and of Big Data, represent a big procyclical task, since the data sets now available are so large or complex that traditional data processing applications and business analytics are often inadequate to deal with, as well as most of the firms [13, 14]. The demand of economic agents to measure and manage more about their activities explains the recent explosion of digital data, "a management revolution", since we do not have better algorithms but just more data, directly translating information and knowledge into improved decision-making and performance [15]. In this view, 'creativity' produces a high degree of both expressive, and functional values. The use of ICT tools is crucial for the growth of Creative Economy, even with respect to spatial and locational aspects, and digitisation increases innovation impact by driving clusters' growth [16–18]. Indeed, Creative Industries are characterized by the innovative interpretation and application of knowledge, the readiness to adopt original technologies and new business models, and to use technology to interact closely with their customers. Through ICT tools, they are developing disruptive innovation *à la* Christensen [19, 20], by linking the informal economy to the formal structures of the economic activities, adding value to products through technologies, generating new business models, producing knowledge spillovers and developing valuable economic impacts. Hence, ICT and digitisation provide the capacity to transcend the traditional barriers of service, and of traditional manufacturing production.

Within the digital context, the Big Data technology and services market, due to its difference from traditional analytics in terms of volume, velocity, and variety [13–15], grew six times faster than the overall ICT market in the last decade and, by 2020, is expected to reach US$59 billion [21]. Although between 2015 and 2020, the revenue rate of growth sourcing from infrastructures, software, and professional and support services is estimated to be between 20.3% and 27.5%, year-over-year it is expected to reduce slightly as the market matures, and Big Data become a segment of broader business analytics market in its entirety. They improve creativity by providing less cost-effective solutions.

On the hypothesis that data-driven companies would be better performers, they contribute also to increase Creative Industries' efficiency by broadening competition in a market, reducing companies' market power, cutting entry barriers. Indeed, they can be a catalyst for innovation, particularly when new business models require development to incorporate new strategies elaborating inferential data for maximising the economic value-added. Big Data can assist industries and other stakeholders in making proactive knowledge-driven business decisions, and the "extraction of embedded intelligence and data insights" [22]. Improvements in efficiency might also support innovative business models by reducing entry barriers and, for a new entrant, making less risky to launch new products or services. Thus, the availability of strong value chain linkages are essential for Creative Industries to share infrastructure, skilled staff and services, by encouraging collaboration and capturing knowledge spillovers, but also to increase networking and knowledge sharing across business sectors, of fundamental relevance for private and government strategies [18]. Creative enterprises are more active than other sectors of the economy, showing a particular ability to synthesize various data sets varying in scales, visualizing and correlating their results to produce new insights, and explain linkages and relationships within the production process. The value-added of Big Data is that, sometimes, data required for large-scale analyses are both large and highly fragmented-housed in different database so that they require time to generate ad hoc results, no longer optimal at the time they are available [23]. Within this context, the distributed collaborative value network model uses digital technologies to enable multiple organisations to collaboratively design, move and manage products, thus increasing their value, and reduce costs. The model prevents companies from unleashing the power of technology to enhance their efficiency and drive their business goals [24].

Despite the broad interest on Creative Economy led to a growing academic literature, there is still a lack of monitoring for related industries: none approved consolidated methodologies are still available for cost-effectiveness analysis to capture simultaneously their economic, social and cultural impacts. Traditional innovation indicators (i.e., education, expenditures and personnel on R&D, patents applications, etc.) fail to highlight many pioneering activities, and their pivotal role. This is the case of intangible goods, from the impact on knowledge and intellectual capital, to the ability to better target users' needs [18]. Therefore, what is relevant in the analysis of Creative Industries is the production scheme as a whole, starting from resources and inputs, and taking into account also processes, outputs and final results. Until now, current research analysed the Creative Industries and ICT as two separate worlds, regardless of the presence of directional causal linkages relationships operating between them and generating reciprocal spillovers. Scholars have studied how to measure these positive effects especially at the macroeconomic level, analysing the determinants of the sector's contribution to the economy in terms of growth, employment, and value-added.

Hence, the role of managing innovation within creative organisations at enterprise level, giving evidence of differences among firms' size and performance thereof in terms of cost control, has been somewhat ignored [3, 11, 16, 17, 25]. Research in this field remains fragmented, with the least tendency to highlight the single nature of each subsector (i.e., performing and visual arts, design, communications, marketing, e-commerce) [26], instead of a more comprehensive approach. A general model for tracking the progress of ICT in the creative process by measuring e-readiness, the intensity of its usage, impact and outcome respectively on business organizations, the Creative Economy, and the economy as a whole, is still missing. Qualitative methods and indicators testing hypothesis for Creative Industries, benefitting from Big Data, are essential [27, 28]. Current methodologies for evaluating the impacts of Creative Industries, such as the United Nations' approach, take into account at most four cluster indicators, valid to capture distinctive features and advantages that yet confuse macro and micro levels: employment, time use, trade and value-added, copyright and Intellectual Property Rights. New combinations of Creative Industries with ICT cannot be properly captured and described via traditional industry taxonomy, otherwise they could be misleading causing a bias in the estimated results [29].

In our view, weaknesses are still present in conceptualising the effects of the multi-dimensional nature of creative innovation, and analysing the complex innovation process therein providing a comprehensive approach of the role played by innovation technologies in transforming 'creativity' and generating positive impacts. The expected result is to open a viable playing field for partnerships among Creative Industries to stimulate new businesses, boost exports, communicate their value and capabilities worldwide, and increase the access to finance for this new creative sector, enabling a company to be much more agile than its competitors [30]. The open challenge within Digital Creativity is to provide an analysis of the role played by Big Data and ICT vs. Creative Industries, in transforming them and generating positive impacts, facilitating connections between Creative Industries, investors and potential investees, and building clever concentrations of creative development activity in the established ecology.

3   Research Objectives and Methodology

The study is twofold. It provides a methodology for assessing the social, economic and cultural effects of ICT tools on these activities, and sketches a model to enhance collaboration among Digital Creative Industries, by performing business solutions for the global trade through better predictions and smarter decisions using real-time information offered by Big Data. The main research questions are: What are the socio-economic and cultural impacts of ICT on Creative Industries, and the potential indicators for assessing them? What the appropriate methodology for evaluating the efficiency and effectiveness of their Digital Creativity qualitative and quantitative contents? How can be structured the collaboration among Digital Creative Industries to increase value-chain linkages, and international partnerships, access to finance and product/services exports? How to capture and maximise the value-added of Big Data in a Digital Creativity process?

The analysis is grounded on a critical literature review, using market and SWOT analyses, emergent business models, and sustainability plans. We exploit organisational semiotics to understand and articulate the interplay between physical and digital worlds, and to develop the model for depicting a general "Digital Creativity ecosystem" for value co-creation. Assessing the impact of 'creativity' effects, relevant and very different spillovers are produced (i.e., knowledge, product, network, training and artistic). Taking into account the current models' weaknesses, and by considering the lack of tradition for assessing the impacts and results of ICT tools on Creative Industries, the methodology can be enriched with quantitative and qualitative analysis. According to the United Nations [3], the Value Chain Analysis is a good decision support tool, able to reveal competitive advantages exploring structure and functioning of these industries. It evaluates individual action and context in institutions and networks, and it also explicitly counts Creative Industries' governance structure for companies with a high level of complexity. By exploiting the concepts of organisational semiotics, we identified and presented a model for implementing collaborative and virtual value chains in this sector by using ICT tools, encompassing the entire industrial traditional approach, and transforming Big Data into "captured" value. For example, in retailing, since e-commerce started moving large volumes of trade, insights on design, individual/group customer experiences and preferences, and user relationship became basic information that can come from social networks, images, sensors, the web, or other unstructured sources, thus increasing evidence-based decision making.

4   Impact Assessment Methodology

The following methodology is based on a quali-quantitative approach to measure Big Data spillover effects on Creative Industries, and is built on Cost-Benefit [31], and Multi-Criteria analysis [32]. These methods are complementary, as they frame both impacts that can be represented in monetary, and non-monetary terms. Their combination enables to consider a wide spectrum of effects, and to combine variables measured in different ways. The model depicts an ecosystem conceptualising Creativity, Big Data and Innovation by measuring their social, economic and cultural effects. Since the lack of studies in the literature proving a methodology for analysing the impacts of Big Data on Creative Industries, we have proposed several indices, starting from the results of the literature review in terms of impacts of technologies on these industries, and by complementing previous analyses with the potential impacts of Big Data in terms of social, economic and cultural effects. It is based on three main steps: background analysis and literature review, definition of a list of impact indicators, and their validation by experts that will follow this preliminary study.

At the current stage of analysis, we present the indices selected; the aim is not to create new indicators, but to provide a first list of useful indices and sub-indices to analyse Creative Industries, following the background analysis and the literature review. Indeed, references from the critical review are next to each sub-index. The Impact

Table 1. Impact sub-indices

Social impact sub-indices	Economic impact sub-indices	Cultural impact sub-indices
	At macro level:	
Support the rise of living standards [2]	Increase employment, labour productivity, and gross value-added [2, 3, 16, 17, 25, 34–37]	Increase the effectiveness of cultural rights [3, 34]
Increase democratic processes [33]	Increase of international trades and exports [30, 35]	Improve cultural diversity practices [3, 34]
Improve freedom of expression [33]	Increase development and growth, especially via tourism [3]	Increase the capability to shape cultural experiences and identities [2, 3]
Support the creation of new sharing communities [2]	Improve sustainable urban development [34]	Improve intellectual capital and human development [3]
Improve collaboration practices and incentivise new ways of working [2]	*At micro-level:*	Support the sharing of social and cultural capital [3, 34]
Support social inclusion [33]	Improve value chain processes increasing supply chain opportunities [3]	Increase cultural sustainability [3]
Improve education practices, skills and knowledge transfer [2, 3, 34, 35]	Improve entrepreneurship initiatives, through changes in organizational structures and reduce operation costs thanks to higher service efficiency [3, 26, 30, 35]	Increase cultural and creative value [3]
Enable the creation of new policies for creativity [3]	Improve service quality and generate innovation [26, 34]	Improve innovative content production [2, 34, 37]
Improve institutional capital [3]	Reduce timing in delivery, using and access services [3]	
	Increase consumers' demand by better targeting their needs and improve business sustainability [16, 26, 34]	
	Increase the implementation of reverse marketing activities [16, 17]	
	Create new business models (i.e. crowdsourcing), and increase access to finance [3, 16, 17, 26, 35]	
	Reduce IPR & copyright issues [2, 33, 35]	

Assessment methodology will be evaluated by a set of experts to identify among the suggested indices the ones can present a well-inclusive picture of the framework. Our synthetic indices are: Social Impact, Economic Impact, at macro and micro-economic level of effectiveness, and Cultural Impact. There is no cross-relation among sub-indices and the table follows a vertical approach. Each index is composed of a number of dimensions or sub-indices and each dimension will be then linked to a specific variable, once the experts will accept these macro-areas (Table 1).

The social impact synthetic index is composed of 8 sub-indices, used to understand the impact on Big Data of several relevant institutional and societal topics intrinsically related to the services and products developed by the Creative Industries at macro-level. Within this context, some scholars have detected the need to focus on the analysis of the impact of Digital Creativity in generating new sharing communities, both of workers and citizens, cooperating for improving living standards, education, skills and knowledge transfer practices [2, 33, 34]. The Creative Industries Council [35] has also assigned a relevant role to them in supporting the openness and diversity of talents within the creative production process. The United Nations [3, 4] have more focused on their impact for improving institutional and knowledge capital, not only for the development of new policies within companies, but also at governance level. The UK Trade and Investments Ministry [29] has detected the importance of improving cross-sectorial communications among engineers, artists and creatives for enhancing social benefits. Galloway et al. [33], and Lee et al. [34] pointed out that Digital Creativity produces a high impact on democracy and freedom, by consequently improving social inclusion at all levels.

These social impact indices are relevant for analysing data flows and their consumption in real-time, thus producing a huge effect on the 'creativity' value of these industries. Indeed, making Big Data more easily accessible to relevant stakeholders and citizens in a timely manner can generate exponential value in supporting openness, and enable talents to develop new business and social activities. This approach also produces a relevant impact enhancing democratic processes, new sharing communities, and improving social inclusion for people that originally were not able to access to these data. Moreover, a significant constraint on creating value from Big Data will be a shortage of talent, particularly of people with deep expertise in statistics and Machine Learning [38]. By using this approach, Big Data improve the diversity of talents and introduce new educational practices, skills and knowledge transfer; they are already generating positive effects on wages, with a huge impact on global better standard of living. As an ever-larger amount of data is digitised and is exchanged across different organisations and companies, there is a set of policy issues that will become increasingly important. This need requires to improve cross-sectorial communication, to create new policies and, by consequence, also institutional capital formation by acquiring new knowledge for different contexts [39].

The economic impact index is composed of 12 sub-indices. In terms of economic impact indices, the United Nations [3] studied it mainly at macro-level, for assessing quantitatively the impact on employment and labour productivity, but also at micro level, analysing the increase of supply chain opportunities and access to finance, the reduction in the time and costs needed to deliver a service. Although a relevant part of the study is on tourism growth, they analysed the economic impact of Digital Creativity on Small and Medium Enterprises' (SMEs)' entrepreneurial activities, the contribution given by digital services to value-added, and subsequent changes in organisational practices. Newbigin [2] and Bahkshi et al. [16, 25], among the others, have identified the economic impact of Digital Creative Industries on gross value-added, too. The latter also depicted the impact of these industries in the marketing approach, by enabling companies to reach more users and better target their needs, by developing new business models. Scholars and institutions agree that impact on Intellectual Property Rights should be carefully considered, since a large amount of revenues is related to the protection of the intellectual and creative work. Other relevant sub-indices are related to the macro-level of effectiveness, to the capacity of Digital Creative Industries to develop international trades [30, 35], and improve business sustainability [16, 17, 34], including urban development. Starting from the analysis of the technological impact on 'creativity'; these sub-indices contribute to investigate mainly how business processes change when Creative Industries use Big Data in their productions.

Indeed, by analysing the effects on creative value chains, companies can also better understand the effects on corporate marketing functions and the change of organisational structures. Relevant information can be extracted by combining some of these sub-indices. For instance, to estimate how the Digital Creativity Industries value and leverage data assets, and if these data can really confer more value than a brand, they should use the specific sub-indices able to assess the impact on improving service quality and generate innovation, and/or enhance business sustainability and entrepreneurial activities by reducing, among others, the time-lag to access or deliver a service and its costs. Within the Big Data market, it is also important to control what existing models are likely to be disrupted, and in this sense the listed sub-indices aim to identify the potential new business models arising by using data-intensive technologies. Some correlated effects are not to be neglected (i.e. increase in consumers' demand by better targeting their needs, in employment rate and supply chain opportunities): IT is useful to analyse also the impact on Intellectual Property Rights and Copyright, especially in terms of reducing or increasing the possibility for Creative SMEs to generate technological transfer activities and new patents.

In terms of cultural impacts, the 8 sub-indices here proposed catch the contribution of Big Data in increasing mainly the creation and sharing of new asset value. In this sense, the United Nations [3] also provides a relevant analysis of the cultural impacts of Digital Creative Industries. Lee et al. [34] focused on the relevance of increasing cultural rights for accessing cultural values per se, thanks to the manipulation of contents by creatives. According to the United Nations [3] and the British Council [2], a pivotal role is also played by Digital Creative Industries in improving cultural diversity practices, and incentivising the creation of new cultural experiences.

Moreover, Digital Creativity produces the highest impact on generating and sharing new intellectual and cultural capital for the development of innovative contents. Consequently, this approach detects a high impact on human development and social capital that effectively enable these type of industries to increase the sustainability of cultural practices. Within this context, Big Data effectively contribute to increase cultural capital, thanks to the creation of new talents, required by the multidisciplinary and interdisciplinary knowledge, a solid base to develop innovative contents. The cultural impact sub-indices clearly show how much all the three main indices are related to each other, as well as the role of Big Data on Creative Industries.

5   The Collaborative Value Network Business Model

A virtual collaborative value network business model is suggested to support Creative Industries in benefitting from Big Data. It aims to support these companies to reverse the traditional value chain by collaboratively creating and selling products through ICT tools. By implementing a new business model, this approach will have impacts on international trade, still one of the major commercial issues for Creative Industries, and public services [21, 22]. The model aims to overcome the current lack of interoperability, Intellectual Property Rights and privacy issues of Creative Industries' products, thus implementing an easier access to creativity services and increasing cross-sectorial innovation of Digital Creativity. The suggested approach will reach a positive impact in terms of social cohesion and innovation, industrial and regional development, social economy, employment and education [5–8, 26, 40].

The business model design-approach is integrating the aforementioned socio-economic impact assessment methodology, to drive startups and SMEs in developing products and services by providing effective economic innovation for society and governments. Taking into account previous analyses of the Creative Industries' user needs and market, it has been possible to identify this model. The Creative Industries' sector is characterized by the absence of medium-sized enterprises, with a relevant gap between micro-SMEs and big companies. Small firms are working mainly locally and only big players benefit from international partnerships. According to Ernst & Young [41], in 2013, Creative Industries employed nearly 30 million people worldwide and generated US$2.25 trillion in revenues. Apart US market, Europe is the second-largest Creative Industries environment, accounting for US$709 billion of revenues (32% of the global total), and 7.7 million jobs (26% of all Creative Industries jobs), mainly relying on a well-structured ecosystem of big players. It has contributed US$200 billion to the digital economy, albeit it is a mature market, showing a weak growth rate in power consumption (0.6%). The users identified are mainly startups and SMEs who need support in accessing the Creative Industries' market and are structurally weak in accessing finance. SMEs have limited business skills and knowledge about market opportunities. Access to innovation is also fundamental and can be achieved only by stimulating the triangulation among governments and Creative Industries' companies, through digital collaboration tools.

Schroeder [39] has investigated through a series of interviews with experts three different business models related to Big Data and has highlighted they are not mutually exclusive, and many companies have adopted more than one simultaneously. The first model, called "data users", refers to companies using data to create value chain or new businesses. The second, "data suppliers", generates marketable data or provides data brokerage services. The third one, named "data facilitators", offers consultancy services, such as analytics, hardware and expertise on making value from Big Data (Fig. 1).

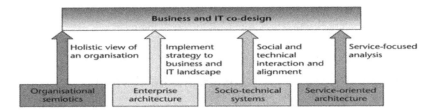

Fig. 1. Theoretical Foundations of Co-Design

Starting from the analysis of these Big Data business models, our hypothesis is that a technological infrastructure for sharing data among different companies can allow to generate a new business model within the Creative Industries framework, taking into account most of the different characteristics of the creative sector. Indeed, our model aims to co-design creative business activities and an Information Technology system to provide an integrated platform enabling the collaboration among Digital Creative Industries working in different contexts and countries on the same service or product. The model is built also upon the approach of the Theoretical Foundations of Co-Design [5–8, 24], which captures business and technical information allowing companies to use information systems' functionalities to meet their business goals. Within this framework, the design is focused on the business system where the technological platform is only a product, which enables enterprises to split complex problems, providing a shared view of the business, and comparative advantages.

The Co-Design Model is based on organisational semiotics, giving a holistic view of the organisation and processes integrated with the Information System. Enterprise Architecture is a strategic planning approach, which improves business effectiveness and performances. Socio-technical systems show the relevance of the interaction between community and technology in the working process. The service-oriented-architecture re-designs business "operations" into "services". In the model, these fields are mutually reliant and operating together to support the co-designing, each providing a complementary perspective. Using these co-relations, our model considers at micro-level the interaction between socio-economic and cultural values to be integrated within the platform for enabling the co-design of creativity services. Indeed, the main goal is to enable Digital Creative Industries to share and extract new value from Big Data.

The model combines the Impact Assessment methodology previously presented with the approach suggested by Liu et al. [24]. It designs a real architecture acting as an intermediary, a bridge between public and private organisations for sharing knowledge, data and infrastructures. Big Data are one of the most powerful drivers to generate new Creative Industries' services. However, they are not relevant if companies do not consider the interrelated values they can bring in terms of social, cultural and economic effects. Before creating new services, they should investigate the potential impact through our methodology by validating its effectiveness and efficacy. After testing it, the digital platform effectively enables Creative Industries to extract 'value' from Big Data, and produce successful results. By implementing and integrating this model, the platform shows different catalogues of available data coming from research infrastructures (i.e. climate data sets, disaster management, smart cities, predictive anal-ysis, environmental information, earth surveillance data), elsewhere not available as open data, and made accessible to companies for developing new products and services (Fig. 2).

Fig. 2. The model

Through the digital platform, companies can also interact with researchers willing to collaborate to develop new products and services, able to offer specific training on data and infrastructures. In this model, organisational semiotics support agents to collab-orate in the co-designing of creative new services and products, thus increasing the value of the network and improving business agility, thanks to the service-oriented architecture [40, 42]. The new business model related to this collaborative approach considers companies as both "users, suppliers and facilitators", supporting themselves for the collaboratively creating new business opportunities on the basis of the sharing economy process [43, 44]. For instance, Creative Industries can use geospatial Big Data for acquiring new information useful to monitor and protect artistic heritage.

This is the case of a pilot project intended to explore the potential of Big Data for producing official statistics and improving the timeliness, the details and, in some cases, the accuracy of commercial statistics on primary cultural destinations [10]. Wikipedia holds 39 million of articles in 246 languages, and it is widely used, recording 21 million page views per hour. The analysis has been based on its page views per month for all articles in its 31 language versions related to the 1.031 World Heritage Sites included on UNESCO's list in 2015 World Heritage Sites. By complementing Big Data with the professional activities of different type of creatives, new technological services for architectural and Cultural Heritage can be developed. Hence, this model enables compa-nies to increase the value coming from Big Data, thanks to the creation of multiple network effects, the reduction of data collection, storing and switching costs, entry barriers and infrastructural costs.

6 Conclusions

Traditional Creative Industries, being heavily labor-intensive, might deserve a lack of productivity [45], but in real world 'creativity' and ICT produce new jobs and economic growth. However, in a globalised and digitised world, they hidden soft and economic power. The link between Creative Economy and its socio-economic and cultural impacts has been transformed by these two forces moving in opposite directions. Globalisation increases market size, where digitisation - sometimes a disruptive innovation - leads powerful economic agents to capture in isolation the new value-added, bringing about a change of the paradigm, where globalisation provides a supra-national framework to these changes.

In this view, the paper explores the impacts of Big Data on service innovation developing a model for Creative Industries assuming that firms who are exposed to more information and data, both new tech-based and creative ones, have more interconnections between industrial structures in their associations. Using a multidisciplinary approach, the main result shows that Big Data can really support Digital Creative Industries in developing new business activities and services, especially in the sub-sectors implementing ICT tools (television & media productions, communications, advertising and marketing, design, finance).

At macroeconomic level, Creative Industries develop disruptive innovation using new technologies, which determine a decrease in value but, adding value to products, they generate new business models. It leads to an enlargement of markets for Creative Industries, and to a significant increase in the degree of choice for consumers, counterbalancing the destructive power of globalisation.

At microeconomic level, they improve the sharing of collaborative value among Digital Creative Industries, since their network forces are wider and denser than in less creative sectors. Digital Creative Industries' producers can reduce the time needed to deliver services and consequently costs, by benefitting from sharing new and valuable contents, not made available before. The next step of this study will be to present the results of the validation of the methodology through its evaluation by a set of experts in the field, thus complementing the model for its application in real contexts.

References

1. Department for Culture Media and Sport: Creative Industries Mapping Documents 2001. DCMS, London (2001)
2. Newbigin, J.: The Creative Economy: An Introductory Guide. Creative Economy Unit. British Council, London (2010)
3. United Nations: Creative Economy Report 2013. Widening Local Development Pathways. UNDP/UNESCO, New York/Paris (2013)
4. UNCTAD Stats: Creative Economy. http://unctadstat.unctad.org/wds/ReportFolders/reportFolders.aspx
5. Caves, R.E.: Contracts between arts and commerce. J. Econ. Persp. **17**, 73–83 (2003)
6. O'Connor, J.: The Cultural and Creative Industries: A Literature Review. Creativity, Culture and Education, Newcastle (2010)
7. Bakhshi, H., Throsby, D.: Culture of innovation. An economic analysis of innovation in arts and cultural organisations. Research report, NESTA, London (2010)
8. Campbell, D.F., Carayannis, E.G., Dubina, I.N.: Creativity, economy and a crisis of the economy? Coevolution of knowledge, innovation, and creativity, and of the knowledge economy and knowledge society. J. Know. Econ. **3**, 1–24 (2012)
9. ESSnet Culture: European Statistical System Network on Culture. Final Report, Eurostat, Luxembourg (2012)

10. Eurostat: Culture Statistics, 2016 edn., http://ec.europa.eu/eurostat/web/products-statistical-books/-/KS-04-15-737
11. Florida, R.: The Rise of the Creative Class. Basic Books, New York (2002)
12. Beer, S.: The viable system model: its provenance, development, methodology and pathology. J. Oper. Res. Soc. **35**, 7–25 (1984)
13. Chen, H., Chiang, R.H., Storey, V.C.: Business intelligence and analytics: from big data to big impact. MIS Q. **36**, 1165–1188 (2012)
14. Tsai, C.W., Lai, C.F., Chao, H.C., Vasilakos, A.V.: Big Data analytics: a survey. J. Big Data **2**, 21–53 (2015)
15. McAfee, A., Brynjolfsson, E.: Big Data: the management revolution. HBR **90**, 60–66 (2012)
16. Bakhshi, H., Hargreaves, S., Mateos-Garcia, J.: A Manifesto for the Creative Economy. NESTA, London (2013)
17. Tomczak, P., Stachowiak, K.: Location patterns and location factors in the cultural and creative industries. Quaestiones Geographicae **34**, 7–27 (2015)
18. Chapain, C., Cooke, P., De Propris, L., MacNeill, S., Mateos-Garcia, J.: Creative Clusters and Innovation. NESTA, London (2010)
19. Christensen, C.M.: The Innovator's Dilemma: When New Technologies Cause Great Firms to Fail. Harvard Business School Press, Brighton (1997)
20. Markides, C.: Disruptive innovation: in need of better theory. J. Prod. Innov. Manag. **23**, 19–25 (2006)
21. Nadkarni, A., Vesset, D.: Worldwide Big Data technology and services forecast, 2016–2020. Doc. no. US40803116. IDC FutureScape, Framingham (2016)
22. Larson, J.: Big (Data) insights. HIS Q. **2**, 20–27 (2014)
23. BDVA: European Big data value strategic research and innovation agenda. The New Economic Asset for Europe, Version 2.0, BDVA, Bruxelles (2016)
24. Liu, K., Li, W.: Organisational Semiotics for Business Informatics. Routledge, Abingdon (2015)
25. Bahkshi, H., Davies, J., Freeman, A., Higgs, P.: The Geography of the UK's Creative and High-Tech Economies. NESTA, London (2015)
26. Marasco, A., Masiello, B., Izzo, F.: Client involvement and innovation in creative-intensive business services: a framework for exploring co-innovation in advertising agency-client relationships. Economies et sociétés **47**, 445–478 (2013)
27. Latinović, T.S., Preradović, D.M., Barz, C.R., Latinović, M.T., Petrica, P.P., Pop-Vadean, A.: Big Data in industry. In: IOP Conference Series: Materials Science and Engineering, vol. 144, p. 012006. IOP Publishing, Bristol (2016)
28. Chen, M., Mao, S., Liu, Y.: Big Data: a survey. Mob. Netw. Appl. **19**, 171–209 (2014)
29. European Commission: Green Paper. Unlocking the Potential of Cultural and Creative Industries. COM(2010) No.183. European Commission, Brussel (2010)
30. UK Trade & Investments: UK Creative Industries – International Strategy. Driving Global Growth for the UK Creative Industries. UK Trade & Investments, London (2015)
31. Boardman, A.E., Greenberg, D.H., Vining, A.R., Weimer, D.L.: Cost-Benefit Analysis: Concept and Practice, 3rd edn. Pearson Prentice Hall, Upper Saddle River (2007)
32. Köksalan, M., Wallenius, J., Zionts, S.: An early history of multiple criteria decision making. J. Multi-Crit. Decis. Anal. **20**, 87–94 (2013)
33. Galloway, S., Dunlop, S.: Deconstructing the concept of 'creative industries'. In: Eisenberg, C., Gerlach, R., Handke, C. (eds.) Cultural Industries: The British Experience in International Perspective, pp. 33–52. Humboldt University, Berlin. http://edoc.hu-berlin.de
34. Lee, N., Rodríguez-Pose, A.: Creativity, Cities and Innovation: Evidence from UK SMEs, WP no. 13/10. NESTA, London (2013)

35. Creative Industries Council: Create UK. Creative Industries Strategy. CIC, London (2013)
36. Potts, J., Cunningham, S.: Four models of the creative industries. Int. J. Cult. Pol. **14**, 233–247 (2008)
37. Bakhshi, H., Windsor, G.: The Creative Economy and the Future of the Employment. NESTA, London (2015)
38. Qiu, J., Wu, Q., Ding, G., Xu, Y., Feng, S.: A survey of machine learning for Big Data processing. EURASIP J. Adv. Sign. Proc. **1**, 1–16 (2016)
39. Schroeder, R.: Big Data business models: challenges and opportunities. Cogn. Soc. Sci. 1166924 (2016). http://dx.doi.org/10.1080/23311886.2016.1166924
40. Clarke, R., Nilsson, A.: Business services as communication patterns: a work practice approach for analysing service encounters. IBM Syst. J. **47**, 129–141 (2008)
41. Ernst &Young: Cultural times. The First Global Map of Cultural and Creative Industries. EY Ltd., London (2015)
42. Larson, R.C.: Service science: at the intersection of management, social and engineering sciences. IBM Syst. J. **47**, 41–51 (2008)
43. Matzler, K., Veider, V., Kathan, W.: Adapting to the sharing economy. MIT Sl Man Rev. **56**, 71–77 (2015)
44. Puschmann, T., Alt, R.: Sharing economy. Bus. Inf. Syst. Eng. **58**, 93–99 (2016)
45. Baumol, W.J., Bowen, W.J.: Performing Arts: The Economic Dilemma. Twentieth Century Fund, New York (1966)

More Observations, More Variables or More Quality? - Data Acquisition Strategies to Enhance Uncertainty Analytics for Industrial Service Contracting

Björn Schmitz[1]([⊠]), Gerhard Satzger[1], and Ralf Gitzel[2]

[1] Karlsruhe Institute of Technology, Karlsruhe, Germany
{bjoern.schmitz,gerhard.satzger}@kit.edu
[2] ABB Corporate Research, Ladenburg, Germany
ralf.gitzel@de.abb.com
http://www.ksri.kit.edu

Abstract. Service business models expose industrial service providers to an increasing amount of uncertainties. In order to design profitable offerings, providers need to understand how uncertainties affect contract profitability. Both, access to data and algorithms are key requirements for accurate analyses.

While current research focuses on *developing algorithms* to derive insights from data that already exist, the need for *strategically acquiring relevant data sets* has been neglected so far. In this article, we develop a method for defining data acquisition strategies to improve uncertainty analyses for industrial service contracting. We explain how lacking observations, variables and quality of data affect uncertainty analyses, propose data acquisition strategies as a systematic plan to acquire relevant data and develop an approach for ranking acquisition strategies by measuring their acquisition effort and business benefit.

The method is applied in an industrial use case to demonstrate its benefit for assessing cost uncertainties in full-service repair contracts.

Keywords: Strategic data acquisition · Uncertainty analysis · Service contracting · Industrial services

1 Introduction

The shift towards service business models has led to massive changes in the design of value propositions and contracts in the industrial sector. Manufacturing companies consider selling functional outcomes of products or services in long-term contracts rather than marketing distinct offerings in spot transactions [1,2]. As a result, new types of contracts such as extended warranty, full-service [3], availability [2] or performance-based contracts [4] are evolving.

From a contracting perspective, this development towards long-term relational engagements poses a considerable challenge with regard to contract design

© Springer International Publishing AG 2017
S. Za et al. (Eds.): IESS 2017, LNBIP 279, pp. 159–172, 2017.
DOI: 10.1007/978-3-319-56925-3_13

and pricing [1,5,6]. It manifests itself in the question how to handle uncertainty and how to assess its impact on contract profitability. Quantifying uncertainty is subject to two requirements: First, decision makers need access to data (e.g. historical data about repair cost of equipment) and they need algorithms and statistics to model uncertainty and to assess its impact on business (e.g. describe repair cost through probability distribution).

The digitization of the industrial sector and the maturity of data analytics both promise to realize considerable improvements in the assessment of uncertainty and thus in the dissemination of new business models and contracts. Currently, the focus of research centers around the question what we can learn from analyzing data that is already available. A rich body of literature deals with the development of diagnostic and prognostic algorithms for industrial applications like maintenance optimization or production scheduling. However, practitioners and researchers have hardly taken up the perspective which data will be needed in the future and how to define and prioritize strategies to acquire it [7,8].

In contracting this has led to the problem that companies refrain from innovating their business models [9] or have failed in servitizing their business [10] - due to lacking skills in managing uncertainties, but also due to scarcity of data [5,11]. As a result, there is no comprehensive body of literature on the strategic acquisition of data, on the added value of acquiring data from different sources within and across organizations and on putting it into use for the assessment of uncertainties in industrial contracting. Therefore, the objective of this article is to support researchers and industrial decision makers in defining and prioritizing data acquisition strategies to enhance the assessment of uncertainties in industrial service contracting.

The research problem is addressed through conceptual research that is evaluated by applying our approach in an industrial use case. Introducing the concept of uncertainty we provide a systematic analysis of related literature on data acquisition in contracting and data analytics in Sect. 2. On this basis, we evaluate how the assessment of uncertainty can contribute to an improved design of industrial service contracts and outline how data-related problems impede such analyses (Sect. 3). In Sect. 4, we introduce the concept of data acquisition strategies and develop a method to rank different acquisition strategies by trading off their acquisition effort against their business benefit. In order to show the benefit of this approach, the method is applied in an industrial use case to improve the assessment of cost uncertainties in the design of full-service repair contracts (Sect. 5). Section 6 concludes with a summary of the contributions, their managerial implications and provides an outlook for future research.

2 Uncertainty Analysis in Service Contracting

Business relationships between industrial service providers and their customers are usually administered through contracts. Contracts are formal written agreements to govern relationships between two contracting parties [12]. As focal elements, contracts formalize what is being delivered (*value proposition* [13]) and how providers are compensated for this offering (*revenue mechanism* [13,14]). Together, delivery costs and revenues determine the profitability of a contract.

When pricing contracts, providers often face the challenge that a contract's delivery costs and/or revenues are uncertain - especially if they engage into long-term relationships. For instance, in industrial maintenance contracts, the cost of repairing production equipment after failures may be influenced by the specific spare parts used, by the required effort of repairing the equipment, etc. Revenues, on the other hand, may be certain or uncertain as well (e.g. fixed price vs. cost-based price [14]), which will ultimately determine how much profit is being earned from a contract. Thus, in order to design profitable contracts, it is essential to understand how uncertainty affects contract profitability.

Formally, uncertainty is the absence of certainty. That is, at the time a decision is being made, the decision maker is not certain about the decision outcome to be realized in the future [15]. Two conditions can be distinguished (ibidem): *Risk* or *measurable uncertainty* [16] describes a condition where the decision maker has knowledge of the probability of decision outcomes to be realized. For instance, the maintenance provider may know how maintenance costs are distributed. *Ignorance*, on the other hand, characterizes a condition of *true uncertainty* [16], where the decision maker is ignorant about what is going to happen in the future. The more informed a decision maker is - i.e. the higher the level of certainty - the easier it is to opt for the optimal decision alternative.

Acquiring data to come to more informed decisions seems an obvious and effective strategy to improve decision making - and we will elaborate on the benefit of strategic data acquisition in the remainder of this article. However, not all uncertainties can be reduced. For instance, considering a dice that generates a sequence of numbers between one and six, it is not possible to predict its exact outcome, regardless of the number of trials observed. Theory therefore distinguishes between *aleatory uncertainty* which is given, and *epistemic uncertainty* which can be reduced through the acquisition of data [17]. This article, investigates the reduction of epistemic uncertainties.

Below, we introduce related work on data acquisition. Articles were identified through a forward/backward keyword search [18] in Google Scholar. Keywords included variations and combinations of the terms *data acquisition, data management, data analytics, uncertainty/risk management* and *industrial service.*

2.1 Related Work on Data Acquisition in Service Contracting

In literature on service contracting, the topic of strategic data acquisition is addressed in articles dealing with the identification, assessment and management of uncertainties in the delivery of long-term service contracts. This research stream is closely related to literature on *product-service systems* [19] and investigates the assessment of uncertainties in full-service, availability-, usage-, performance- or outcome-based contracts.

Herzog et al. [1], Erkoyuncu et al. [2,11] and Schulte and Steven [20] develop frameworks for uncertainty management in service contracting. They propose generic steps for dealing with uncertainties (e.g. identify, assess, analyze, reduce, control), but do not address the question which data is required for reducing uncertainties. Other articles focus on the identification and categorization of uncertainties for particular types of contracts [21,22], but do not discuss how to quantitatively measure them. Schwabe et al. [23] provide an overview of methods and metrics for quantitatively measuring uncertainties, but do not investigate how to collect relevant data - although they emphasize that many methods build on the analysis of quantitative data. Finally, Erkoyuncu et al. [5,17] develop methods for estimating the delivery cost of product-service systems. They discuss the treatment of epistemic and aleatory uncertainties in cost estimation and discuss qualitative and quantitative approaches for their measurement. Although both articles highlight the importance of acquiring data for quantitative uncertainty analyses, they do not discuss data acquisition strategies.

2.2 Related Work on Data Acquisition in Data Analytics

As a second research stream we investigated literature on data analytics, since the collection of data forms an integral part of any data analytics project. Data acquisition is mainly addressed in review articles on data analytics [8,24] and its adoption in the industrial sector [7,25]. Furthermore, we identified articles that develop methods to improve information exchange in and across organizations [26,27] and to assess the inherent uncertainty in data [28].

Kaisler et al. [8] provide a list of data-related challenges in big data analytics. Challenges comprise data storage and transport (data size), its management (data provenance) and processing (computing capacity). Moreover, the authors discuss problems of quality vs. quantity, of data growth vs. data expansion, of speed vs. scale, data ownership, compliance vs. security, value of data, distributed data vs. distributed processing. Their research emphasizes why a strategic and systematic treatment of this topic is so important. The identified problems are confirmed by other authors. Bose et al. [24] regard the availability and sharing of data across organizations as an important challenge and emphasize the need of acquiring relevant target data sets. Klöpper and Schlake [25] discuss big data research challenges in the industrial sector. They state that the concept of storing any amount of data is not economically feasible and point to the necessity of choosing the right data. Baglee et al. [7] confirm the need for differentiating and prioritizing data in order to avoid data overload and to prevent information islands. They also highlight the requirement for assessing data uncertainty and inconsistency in order to enhance data modeling and analysis.

Koronios et al. [26] and Jess et al. [27] propose solutions to improve the data exchange in and across organizations. Koronios et al. introduce internal data markets where data can be exchanged and described through meta data (quality, price, terms of use). Jess et al. discuss a market-based approach to acquire cross-organization data. They develop a framework of characteristics to identify suitable scenarios for using market-based data acquisition strategies.

Finally, Durugbo et al. [28] develop a methodology to assess data uncertainty (in terms of quality and accuracy of data) in product-service systems. The approach helps to identify data quality issues.

3   Towards Strategic Data Acquisition for Analyzing Uncertainties in Industrial Service Contracting

The literature analysis confirms the need for the development of methods for identifying, collecting and acquiring data sets. However, although data acquisition is considered important, it has not been systematically investigated yet. The following section establishes the basis for the development of data acquisition strategies (Sect. 4). We will first explain how the analysis of uncertainties can improve the design of service contracts (Sect. 3.1). Subsequently, we discuss how problems with the input data can impede accurate and meaningful uncertainty analyses which may lead to an adverse contract design (Sect. 3.2).

3.1   Uncertainty Analyses and Their Benefit for Contracting

When designing contracts, providers can benefit from uncertainty analyses in three ways (Fig. 1): First, their analysis can help to derive more *accurate estimates* of costs and revenues. These estimates allow to define prices or contingencies that increase the likelihood of meeting predefined profit targets [17]. Second, if providers understand what type of uncertainties affect their revenues and costs, they can offer *customized contracts* [24] that account for the specific uncertainties induced from offering a service to a particular customer. Third, a better understanding of uncertainties would also allow to better *mitigate the impact of uncertainty* by reacting more effectively to adverse events [29].

Depending on the benefits providers want to realize, they can adjust the focus or objective of uncertainty analyses (Fig. 1): Estimating revenues and costs requires an accurate *measurement of uncertainties*. Ideally, quantitative techniques like statistical or predictive modeling are employed to form accurate estimates (e.g. [5,17,23]). For instance, providers could derive probability distributions from historical data to measure revenue and cost uncertainties. Second, providers may analyze, which factors have an influence on the revenue or cost of a contract (*identify uncertainty drivers*) (e.g. [21,22]). The more profound the understanding of these factors is, the more degrees of freedom do providers have for customizing their offerings [24] and for effectively mitigating uncertainties [29]. For instance, using regression models, one could analyze the impact of factors like operating hours or equipment age on the amount of incurred repair cost. Finally and ideally, providers are able to predict how these factors will develop over time in order to account for possible trends (*forecast development of uncertainty drivers*). For instance, if providers had knowledge of their customers' production plans and hence the operating hours of their equipment, they could more accurately predict the number of equipment failures over the contract duration.

3.2   Data-Related Problems and Their Impact on Contracting

The steps above describe how the joint analysis of uncertainties and contract design is ideally performed. However, relevant data for this analysis is often

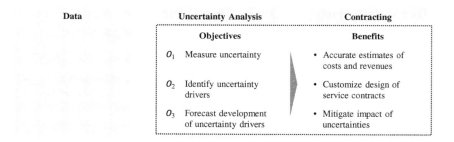

Fig. 1. Objectives of uncertainty analyses and their benefit for service contracting.

not available in sufficient volume, structure or quality. Understanding, how and which data-related problems affect the analysis of uncertainties is essential for addressing data acquisition in a systematic manner.

Three types of problems can impede the reduction of uncertainties (Fig. 2): An *insufficient number of observations* may result in inaccurate uncertainty measurements. A typical example is the determination of failure rates of industrial production equipment, which requires a sufficient amount of failure data to obtain accurate probability distributions. Since failures are rare events, acquiring enough observations is a challenge. The second problem refers to an *insufficient number of measured variables* which may impede providers from identifying factors that affect a target variable of interest. For instance, the failure rate of equipment and hence the repair cost may depend on the particular equipment type or age. Identifying such correlations through data-driven approaches (e.g. [21]) requires a rich set of observations and a rich set of measured variables, since both the direction and the magnitude of an effect are of interest. Some authors refer to this challenge, as collecting the "right" data [7,8,27]. Finally, even if enough data is available, it may have an *insufficient data quality* [8,26,28]. Wrong data can easily lead to prediction errors with a high business impact or may result in drawing wrong conclusions [26]. Problems of data quality may materialize directly - e.g. in form of incomplete, inconsistent or inaccurate data - or indirectly in the sense that the data is not trustworthy or plausible [28]. Methods for assessing data quality have been proposed in [28,30,31].

Thus, in order to decide which data to acquire, providers need to understand what type of data-related problems they face. In the following, we will explain how data acquisition strategies can help to address this challenge.

4   Development of Data Acquisition Strategies

Since flaws in uncertainty analyses are caused by different data-related problems, providers need to decide whether to invest in increasing *observations*, in measuring more *variables* or in improving the *quality* of relevant data sets - or any combination of it. In order to do this in a structured manner, we suggest

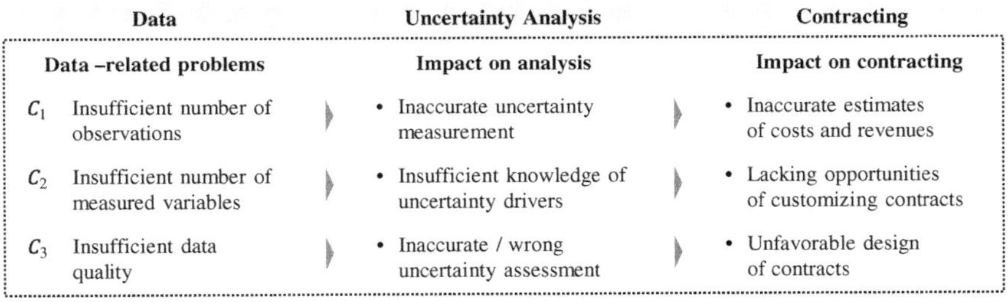

Data	Uncertainty Analysis	Contracting
Data –related problems	**Impact on analysis**	**Impact on contracting**
C_1 Insufficient number of observations	• Inaccurate uncertainty measurement	• Inaccurate estimates of costs and revenues
C_2 Insufficient number of measured variables	• Insufficient knowledge of uncertainty drivers	• Lacking opportunities of customizing contracts
C_3 Insufficient data quality	• Inaccurate / wrong uncertainty assessment	• Unfavorable design of contracts

Fig. 2. Data-related problems and their impact on the analysis of uncertainties.

to define data acquisition strategies and to develop methods that help decision makers to compare and rank different acquisition alternatives. Formally, we regard a *data acquisition strategy* as a *plan to acquire data sets with more observations, more variables or at a higher level of quality for improving the analysis of uncertainties with regard to a specific business objective.* In this section, we develop a structured approach to determine the utility of different acquisition strategies (Sect. 4.1) by trading off their respective acquisition effort (Sect. 4.2) against their potential business benefit (Sect. 4.3).

4.1   Utility Analysis to Evaluate Different Data Acquisition Strategies

In order to compare the benefit of different acquisition strategies, we propose using methods from the domain of *utility analysis* (Fig. 3). The method helps to structure decision making in complex situations by attaching a utility $u(a)$ to each decision alternative $a \in A$. In our case, a decision alternative a may be any combination of strategies S_1 to S_3 for a particular data set of interest. The utility $u(a)$ is computed based on a scoring function $u(a) = \sum_{r=1}^{m} w_r * u_r(a)$ that weights the importance of relevant attributes $r \in R$ with a weighting factor w_r such that the sum of all weighting factors equals 1. We suggest to use attributes that measure the effort and the benefit of acquiring data sets (compare also [7,26,27]). In contrast to similar methods like cost-benefit analysis, utility analyses do not require to express the value of specific decision alternatives and attributes in monetary units. This is beneficial, since it is not always feasible to measure the effort or benefit of acquired data in monetary terms. The attribute weights and the utility attached to specific attributes of decision alternatives may be agreed upon in joint focus groups and expert assessments with sufficient domain knowledge. Overall, the method can be flexibly adapted both in terms of the effort and depth of the analysis (e.g. how many attributes are considered, how the weighting is performed) and with regard to the application case considered. Although the attribute weights may change when analyzing different use cases, we will discuss how to conduct a utility analysis and we will propose attributes that may influence the effort and benefit of acquiring data.

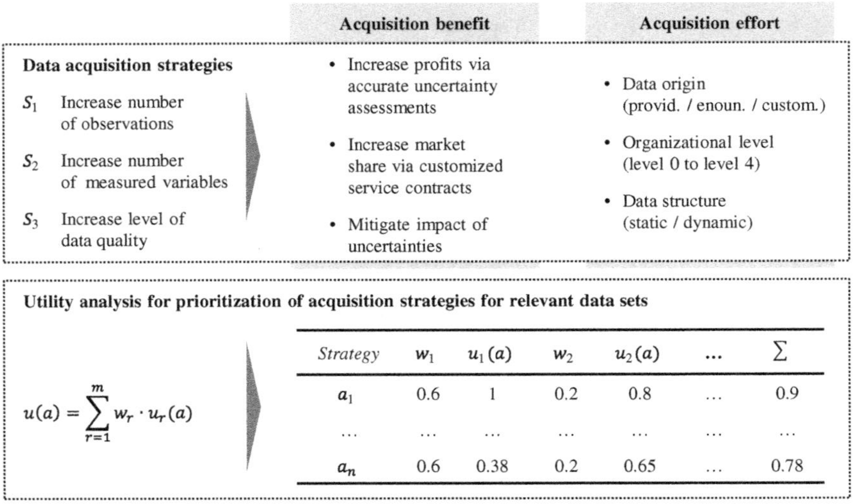

	Acquisition benefit	Acquisition effort
Data acquisition strategies S_1 Increase number of observations S_2 Increase number of measured variables S_3 Increase level of data quality	• Increase profits via accurate uncertainty assessments • Increase market share via customized service contracts • Mitigate impact of uncertainties	• Data origin (provid. / enoun. / custom.) • Organizational level (level 0 to level 4) • Data structure (static / dynamic)

Utility analysis for prioritization of acquisition strategies for relevant data sets

$$u(a) = \sum_{r=1}^{m} w_r \cdot u_r(a)$$

Strategy	w_1	$u_1(a)$	w_2	$u_2(a)$	...	Σ
a_1	0.6	1	0.2	0.8	...	0.9
...	...	...	...	...	...	...
a_n	0.6	0.38	0.2	0.65	...	0.78

Fig. 3. Prioritization of data acquisition strategies for uncertainty analysis.

4.2   Effort of Data Acquisition

This section proposes attributes to measure the effort of acquiring data in the industrial sector. For each attribute, we will discuss under which conditions the acquisition effort increases and we will point to literature that supports this claim. We do not consider the propositions to be generally true, however, we are convinced that they are relevant for many use cases in the industrial sector and beyond. Also, due to the limited scope of this work, we will not discuss the technical effort - e.g. data integration, processing, and storage [7,8,24] - of acquiring data.

The first factor that significantly affects the effort of acquiring data is the *data origin* - i.e. if it is generated internally or by external parties such as customers or suppliers [27,32]. Fromm et al. [32] classify data sources into *provider internal data*, *third-party data* and what they refer to as *encounter data*. Provider data is generated within the provider's internal organization, third-party data is generated and owned by third parties (e.g. customers or suppliers). Encounter data is generated at the time a service is delivered or co-created between provider and customer. According to Bose et al. [24] and Jess et al. [27], acquiring data from third parties is more difficult than collecting provider internal data. Also, we argue that it is usually easier to acquire encounter data than customer data. First, this data is jointly created in the process of service delivery [32]. Second, acquiring data during the service encounter may not require additional effort and clearings from customers. Although customers may refrain from providing access to their IT-systems, they may accept that service technicians document part of this information themselves. Thus, we make the following proposition:

Proposition 1: *The effort of acquiring provider internal data is lower than acquiring encounter data is lower than acquiring customer data.*

Second, data may be owned and generated at different *organizational levels* of an enterprise - from strategic planning to operations [7]. Due to information silos [25–27] or due to the strategic value of data, customers may not be able or agree to share data that is generated at specific organizational levels. For manufacturing companies, the enterprise systems that support these organizational levels are characterized in a norm on enterprise-control systems [33]. Four levels of enterprise systems are distinguished: *Level 4* (enterprise resource planning) captures data for strategic corporate planning. *Level 3* (manufacturing execution system) deals with operational management concerning production planning and optimization. *Level 2* (supervisory control and data acquisition) to *level 0* (input output signals of devices) comprise systems for supervising production processes and for real time monitoring and controlling of production equipment. We argue that the strategic value of data will usually increase from level 0 (shop floor) to level 4 (enterprise resource planning). Thus, the effort of acquiring level 4 data from customers is probably higher than acquiring level 0 data. For instance, while customers may agree to share data of condition monitoring systems (shopfloor), they may not be willing to share data about their planned production volume for the next year (ERP system).

Proposition 2: *The higher the strategic value of data, the higher is the effort to acquire it.*

Third, it is important to account for the *data structure* and to distinguish between what we refer to as *static data* and *dynamic data*. Static data does not change over time and can hence be collected at a one-time expense. An example of static data is master data of industrial production equipment which may comprise information about serial numbers, equipment types, production dates, etc. Dynamic data, on the other hand, evolves over time. An example of dynamic data is failure data, which allows improving failure statistics each time an equipment fails over its lifecycle. Naturally, dynamic data cannot easily be acquired in an ad-hoc fashion which motivates the following proposition:

Proposition 3: *The effort of acquiring static data is lower than the effort of acquiring dynamic data.*

Depending on the use-case, the influence and directions of factors may differ and other factors may need to be considered to determine the acquisition effort of data. The utility analysis and the expert assessments have the flexibility to account for such exceptions. However, we believe that the outlined factors will be relevant for many cases encountered in the industrial sector.

4.3   Benefit of Data Acquisition

Benefits of acquiring data for service contracting have already been laid out in Sect. 3. The value of increasing observations, variables or data quality needs to be assessed for each use-case individually and cannot be easily generalized. However, we will point to opportunities that may bring particular value for reducing uncertainties in service contracting.

Increasing the number of observation helps to derive more accurate estimates of a target variable which may yield significant benefits for contracting: More accurate cost estimates could help providers to charge lower risk premiums for accepting operational risks from their customers. In price competitive markets, providers could hence offer lower prices to increase market shares. In high margin markets, calculating with lower contingencies may directly translate into higher margins and profits. A lot of potential to increase observations of data resides in exploiting installed base information - especially for dynamically evolving data. For instance, providers may not only increase observations of failures by observing a single unit of equipment for a longer period of time, but also by observing the entire installed base of equipment at the same time.

Increasing the number of measured variables helps to identify uncertainty drivers. It supports the design of more customized offerings for specific customer segments, which allows to market more competitive offers and to increase contract profitability. The challenge of acquiring more variables is foremost about identifying relevant candidate variables, and subsequently about determining the direction and magnitude of their impact. In the industrial sector, opportunities come from two directions: First, despite advancements in data analytics, domain knowledge still plays a critical role for identifying the most relevant data sets and variables to acquire (compare [21]). Second, we believe that a lot of potential lies in the analysis of static data such as information about the type and application purpose of equipment. Often, static data is easier to acquire - e.g. it can often be collected during a service encounter. Moreover, it can be more easily used for customizing contracts and for segmenting markets compared to dynamic data, as it does not change over the contract duration.

Finally, providers can increase data quality. Ideally, data collection is already designed such that data quality is accounted for, as it requires enormous effort to improve the quality of data once it has been collected. While measures like data completeness or consistency may be assessed automatically (e.g. [30,31]), detecting problems of data plausibility or trustworthiness is much harder and will usually require a lot of domain knowledge. While cognitive and semantic technologies are still under development, we believe that quality assessments of domain experts may give the most reliable results as to which degree we can trust in analyses made.

In the following, we will exemplary show, how the proposed framework can be put into use to define data acquisition strategies for analyzing cost uncertainties in service contracting.

5   Application of Method for Analyzing Cost Uncertainties in Full-Service Repair Contracts

The following results originate from research that was conducted in cooperation with a discrete automation equipment provider. In order to improve the design of

Table 1. Data acquisition strategies for analyzing repair cost uncertainties

a	S^d	Data set	Effort								Benefit		Σ
			Origin^a		Organ. level^b		Structure^c						
			w_1	$u_1(a)$	w_2	$u_2(a)$	w_3	$u_3(a)$			w_4	$u_4(a)$	
a_1	S_2	Mechanical load	0.3	-1.0	0.3	-0.2	0.3	-1.0			1	1.0	0.34
a_2	S_2	Operating conditions	0.3	-0.5	0.3	-0.2	0.3	-1.0			1	0.8	0.59
a_3	S_2	Commissioning date	0.3	-0.5	0.3	-0.2	0.3	-0.5			1	0.1	-0.26
a_4	S_2	Decommissioning date	0.3	-1.0	0.3	-0.2	0.3	-0.5			1	0.5	-0.01
...	...	...	...	...	...	...	...	...			...	...	...
a_{10}	S_2	Application purpose	0.3	-0.5	0.3	-0.2	0.3	-0.5			1	0.5	0.14

[a] Data origin: provider (-0.2), encounter (-0.6), customer (-1.0)
[b] Organ. level: level 0 (-0.2), level 1 (-0.4), level 2 (-0.6), level 3 (-0.8), level 4 (-1.0)
[c] Data structure: static (-0.5), dynamic (-1.0)
[d] Data problem: observations (S_1), variables (S_2), quality (S_3)

their full-service repair contracts, the company aimed to better understand which factors affect the repair cost of equipment. For this purpose, we performed a data-based analysis of historical repair data with regression analyses, support vector machines and neural networks to identify and measure the impact of different cost drivers on the total repair cost of equipment. In the course of this analysis and in discussions with domain experts we identified data-related problems that impeded further analyses along all three dimensions (*observations, variables, quality*). Exemplary challenges (S_2) along with a simple utility assessment of corresponding data acquisition strategies are listed in Table 1.

The table lists data acquisition strategies a_1 to a_{10}, each referring to a specific data set. The acquisition effort is measured by the attributes *data origin* (*provider, encounter, customer*), *organizational level* (*level 0 to level 4*) and *data structure* (*static, dynamic*) with equal weights $w_1 = w_2 = w_3 = 0.3$. The effort $u_1(a), u_2(a), u_3(a) \in [-1, 0]$ of acquiring each data set was determined by two researchers with input of industrial experts. For each attribute r, a scale of attribute values was determined in order to account for their different acquisition effort (see footnotes in Table 1). The benefit of each acquisition strategy is expressed by a single value $u_4(a) \in [0, 1]$ weighted with factor $w_4 = 1$. The benefit reflects the potential of each data set for improving contracts to increase either market share or contract profitability. The overall utility of each strategy is the weighted sum of effort and benefit. For example, strategy a_1 aims to acquire data about the mechanical load under which equipment is operated. Since abnormal loads are root causes of failures, the data could help to derive more accurate estimates of the number of failures to occur over the contract lifetime. This would enable providers to customize contracts and to market specific offerings to distinct customer segments (*benefit*). For instance, providers could offer a discount for customers with normal load profiles while increasing prices for high-load customers. The acquisition *effort* is determined as follows: Load data is *dynamic* data since it evolves and changes over time. It is generated at the *customer's* production site by sensors and controllers at the field level (*level 0*). Acquiring this data is rather effortful, since it needs to be constantly transmitted and since providers require the consent of their customers for collecting the data. The overall utility of this particular strategy is simply the weighted sum of effort and benefit.

The utility of each remaining acquisition strategy was determined according to the same rationale. A comparison of the derived utility values allows to establish a ranking among all alternatives. Based on this ranking, providers can prioritize investment decisions to acquire the necessary data for achieving their strategic business objectives.

6 Conclusion

In this work, we developed a method for defining and prioritizing data acquisition strategies to improve uncertainty analyses in industrial service contracting.

First, we explained how the analysis of uncertainties enables providers to design more profitable and customized service contracts. We described three objectives of uncertainty analyses (*measure uncertainty, identify uncertainty drivers, forecast uncertainty drivers*) and explained their benefit for contracting (*accurate cost/revenues estimates, contract customization, uncertainty mitigation*). Second, we described how data-related challenges affect uncertainty analyses and introduced the concept of data acquisition strategies (*observations, variables, quality*) to address these problems. Third, we explained how to compare and prioritize data acquisition strategies through utility analyses and proposed concepts to evaluate their effort (*origin, organizational level, structure*) and benefit. Finally, we applied the concepts and method in an industrial use-case to show their benefit for the analysis of cost-uncertainties in full-service repair contracts.

For companies engaging into data analytics projects, our research provides directions for identifying data-related challenges and for acquiring data in a systematic manner. Besides its application in dedicated analytics projects, the method may guide discussions to identify and acquire those data sets that enable companies to gain a strategic competitive advantage on a corporate level [24]. Despite these contributions, there is room for improvements. In the future, we plan to further investigate factors that affect the effort and benefit of acquiring specific data sets. For instance, we are convinced that data sets which are relevant for a broad range of use-cases are probably more valuable than data sets that are only relevant for a single use case [8]. In addition, data acquisition is closely related to technical challenges such as data integration, processing, and storage that may affect the acquisition effort ([7,8,24]). Further research is required to account for such factors.

We are convinced that corporate investments in data analytics technologies and connectivity of devices will pay off even more if companies significantly increase their efforts to acquire those data sets that will allow them to realize competitive advantages in the future. Although proposed concepts and methods are research in progress, they may serve as a first basis to further advance research in this important field.

References

1. Herzog, M., Meuris, D., Bender, B., Sadek, T.: The nature of risk management in the early phase of IPS^2 design. Procedia CIRP **16**, 223–228 (2014)
2. Erkoyuncu, J.A., Roy, R., Shehab, E., Kutsch, E.: An innovative uncertainty management framework to support contracting for product-service availability. J. Serv. Manag. **25**(5), 603–638 (2014)
3. Stremersch, S., Wuyts, S., Frambach, R.T.: The purchasing of full-service contracts: an exploratory study within the industrial maintenance market. Ind. Market. Manag. **30**(1), 1–12 (2001)
4. Hypko, P., Tilebein, M., Gleich, R.: Benefits and uncertainties of performance-based contracting in manufacturing industries: an agency theory perspective. J. Serv. Manag. **21**(4), 460–489 (2010)
5. Erkoyuncu, J.A., Durugbo, C., Shehab, E., Roy, R., Parker, R., Gath, A., Howell, D.: Uncertainty driven service cost estimation for decision support at the bidding stage. Int. J. Prod. Res. **51**(19), 5771–5788 (2013)
6. Huber, S., Spinler, S.: Pricing of full-service repair contracts. Eur. J. Operat. Res. **222**(1), 113–121 (2012)
7. Baglee, D., Marttonen, S., Galar, D.: The need for Big Data collection and analyses to support the development of an advanced maintenance strategy. In: 11th International Conference on Data Mining, pp. 3–9. IEEE, Las Vegas (2015)
8. Kaisler, S., Armour, F., Espinosa, J.A., Money, W.: Big data: issues and challenges moving forward. In: 46th Hawaii International Conference on System Sciences. pp. 995–1004. IEEE, Wailea (2013)
9. Benedettini, O., Neely, A., Swink, M.: Why do servitized firms fail? A risk-based explanation. Int. J. Operat. Prod. Manag. **35**(6), 946–979 (2015)
10. Gebauer, H., Fleisch, E., Friedli, T.: Overcoming the service paradox in manufacturing companies. Eur. Manag. J. **23**(1), 14–26 (2005)
11. Erkoyuncu, J.A., Roy, R., Shehab, E., Wardle, P.: Uncertainty challenges in service cost estimation for product-service systems in the aerospace and defence industries. In: 1st CIRP Industrial Product-Service System (IPS^2) Conference, pp. 200–206. Cranfield University Press, Cranfield (2009)
12. Bolton, P., Dewatripont, M.: Contract Theory. MIT Press, Cambridge (2005)
13. Chesbrough, H., Rosenbloom, R.S.: The role of the business model in capturing value from innovation: evidence from Xerox Corporation's technology spin-off companies. Ind. Corp. Change **11**(3), 529–555 (2002)
14. Van Ostaeyen, J., Van Horenbeek, A., Pintelon, L., Duflou, J.R.: A refined typology of product-service systems based on functional hierarchy modeling. J. Clean. Prod. **51**, 261–276 (2013)
15. Bamberg, G., Coenenberg, A.G., Krapp, M.: Betriebswirtschaftliche Entscheidungslehre, 15th edn. Vahlens Kurzlehrbücher, Vahlen, München, Germany (2012)
16. Knight, F.H.: Risk, Uncertainty, and Profit. Houghton Mifflin, Boston (1921)
17. Erkoyuncu, J.A., Roy, R., Shehab, E., Cheruvu, K.: Understanding service uncertainties in industrial product-service system cost estimation. Int. J. Adv. Manuf. Technol. **52**(9–12), 1223–1238 (2011)
18. Webster, J., Watson, R.T.: Analyzing the past to prepare for the future: writing a literature review. MIS Q. **26**(2), xiii–xxiii (2002)
19. Baines, T.S., Lightfoot, H.W., Evans, S., Neely, A., et al.: State-of-the-art in product-service systems. Proc. Inst. Mech. Eng. Part B J. Eng. Manuf. **221**(10), 1543–1552 (2007)

20. Schulte, J.K., Steven, M.: Risk management of industrial product-service systems (IPS2) – how to consider risk and uncertainty over the IPS2 lifecycle? In: Dornfeld, D.A., Linke, B.S. (eds.) 19th CIRP Conference on Life Cycle Engineering, pp. 37–42. Springer, Berkeley (2012)

21. Erkoyuncu, J.A., Durugbo, C., Roy, R.: Identifying uncertainties for industrial service delivery: a systems approach. Int. J. Prod. Res. **51**(21), 6295–6315 (2013)

22. Sakao, T., Öhrwall Rönnbäck, A., Ölundh Sandström, G.: Uncovering benefits and risks of integrated product service offerings - using a case of technology encapsulation. J. Syst. Sci. Syst. Eng. **22**(4), 421–439 (2013)

23. Schwabe, O., Shehab, E., Erkoyuncu, J.: Uncertainty quantification metrics for whole product life cycle cost estimates in aerospace innovation. Prog. Aerosp. Sci. **77**, 1–24 (2015)

24. Bose, R.: Advanced analytics: opportunities and challenges. Ind. Manag. Data Syst **109**(2), 155–172 (2009)

25. Klöpper, B., Schlake, J.C.: Aufbrechen der Datensilos-Big Data Forschungsfragen aus dem Bereich Industrial Analytics. Lecture Notes in Informatics - INFORMATIK 2014, pp. 79–81. German Informatics Society, Stuttgart, Germany (2014)

26. Koronios, A., Redman, T., Gao, J.: Internal data markets: the opportunity and first steps. In: 4th International Conference on Cooperation and Promotion of Information Resources in Science and Technology, pp. 127–130. IEEE, Beijing (2009)

27. Jess, T., Woodal, P., McFarlane, D.: A framework for identifying suitable cases for using market-based approaches in industrial data acquisition. In: 1st International Data and Information Management Conference, pp. 113–124. Library and Information Statistics Unit, Loughborough University, Loughborough (2014)

28. Durugbo, C., Erkoyuncu, J.A., Tiwari, A., Alcock, J.R., Roy, R., Shehab, E.: Data uncertainty assessment and information flow analysis for product-service systems in a library case study. Int. J. Serv. Operat. Informat. **5**(4), 330–350 (2010)

29. Schmitz, B., Düffort, F., Satzger, G.: Managing uncertainty in industrial full service contracts: digital support for design and delivery. In: 18th IEEE Conference on Business Informatics, pp. 123–132. IEEE, Paris (2016)

30. Gitzel, R., Turrin, S., Maczey, S., Shaomin, W., Schmitz, B.: A data quality metrics hierarchy for reliability data. In: 9th IMA International Conference on Modeling in Industrial Maintenance and Reliability, pp. 1–6. Kent Academic Repository, London (2016)

31. Gitzel, R.: Data quality in time series data - an experience report. In: Proceedings of the 18th IEEE Conference on Business Informatics - Industrial Track, pp. 41–49. CEUR, Paris (2016)

32. Fromm, H., Habryn, F., Satzger, G.: Service analytics leveraging data across enterprise boundaries for competitive advantage. In: Bäumer, U., Kreutter, P., Messner, W. (eds.) Globalization of Professional Services, Chap. 13, pp. 139–149. Springer, Heidelberg (2012)

33. Deutsche Kommission Elektrotechnik Elektronik Informationstechnik im DIN und VDE: International Standard IEC 62264-1:2013 Enterprise-control system integration - Part 1: Models and Terminology. Beuth, Berlin (2013)

From Data Science to Value Creation

Jürg Meierhofer$^{(\boxtimes)}$ and Kevin Meier

Institute of Data Analysis and Process Design,
Zurich University of Applied Sciences, Winterthur, Switzerland
`juerg.meierhofer@zhaw.ch`

Abstract. Value creation with data science methodologies generates important insights. However, these insights do not systematically provide service value to customers. Therefore, we show a systematic approach to use data science for the process of service design. We develop a structure of data science methodologies in the dimensions of their potential to create service benefit. This enables the mapping of the value contribution of the data science tools on the different perspectives and phases of the service design process. Based on this mapping, a direct link can be established between the outcomes of the data science methodologies and the value drivers for the customer. The resulting new methodology allows the systematic value creation from insights generated by data science.

Keywords: Service science · Data science · Service design · Data product design · Data-driven value creation

1 Introduction

This paper describes a new methodology for creating value from data science. Specifically, the goal is to develop data-based services and products starting with insights generated by data science tools. According to [1] data science tools provide insights which support decision making. However, these insights do not yet make up data products for which user or customers are ready to pay. For new data science-based results the appropriate business applications often are not obvious. In other words, data science tools may provide solutions for which we need to find a suitable application. Therefore, this article describes a way how to systematically design services given data science-based outcomes.

The generic value chain for data-intensive user services is shown in Fig. 1. We assume that we have data available as well as the data science tools to create insight from this data. However, we lack a methodology how to create service value for users out of the insights. The literature provides numerous sources describing how to extract value from data. [1] describes how data science can be used as a driver for analytical engineering and how organisations can incorporate data science in their business strategy. [2] provides an long list of data science applications across various industries and use cases. These sources focus on the generation of insights from data, which represents an essential prerequisite for the creation of value from data. However, it is not clear whether these insights

© Springer International Publishing AG 2017
S. Za et al. (Eds.): IESS 2017, LNBIP 279, pp. 173–181, 2017.
DOI: 10.1007/978-3-319-56925-3_14

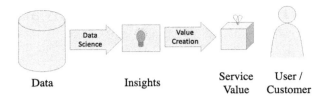

Fig. 1. Creating user value out of data science-based insight

can be considered providing service value from the user's perspective or if there are customers willing to pay.

According to [6] data products are not about data, but about enabling users to do what they want to do. Data products are to deliver results rather than data and data is invisible in the product. [7] points out that data products must always be designed based on the user's objectives in order to create actionable value and not just more data. The so-called drivetrain approach is proposed to find the optimum actionable outcome. Both [6,7] make clear that data products need to be designed to meet the user's objectives and needs.

In order to make sure that the resulting services and products generate value for users, the concepts of service design and service science can be applied, which provide a systematic procedure to create service value for users and customers. The service concepts are based on the assumption that the beneficiary of the service there is always a human being [3]. This holds also for industrial services (e.g., in the Industry 4.0 context), in which human actors are creating value typically supported by technical machines. Services are usually provided in service systems, which are configurations of people, technologies, organisations, and information that create and deliver value to all stakeholders in the system [3,4].

Therefore, we now put the focus on the problem of understanding the user's needs and designing value propositions to meet those needs. When talking about value creation for humans, there are the established tools for service design [11–13]. These tools point out that the value needs to cover both functional as well as emotional user needs (in [11] so-called jobs, pains and gains).

The challenge "How to use data science for service innovation?" is addressed in the literature. In [8] an approach is shown to combining service design with business analytics to improve the service innovation process. However, the approach is focussed on the very specific aspect of identifying patterns of customer behaviour and product usage by customers. The literature [1] discusses the application of data science for business and states that business problems need to be decomposed into sub-problems that can be solved by analytics. The approach is primarily focused on the solution design (the phase "Design of Value Proposition and Processes" in Fig. 3). Also [9,10] apply analytics or artificial intelligence for the segmentation of customers or understanding customer behaviour. Although these approaches highlight important potentials to improve service innovation with the help of data science, there are many additional aspects of the service design process which can leverage the benefits of data science. The literature about service design [11–13] describes the phases of the process for engineering customer-centric services. It becomes clear that in various steps of this process, analytics can help finding better solutions. However, we lack a systematic approach for this support.

According to [4] the application of data's value for smart service systems represents a current research challenge. [5] suggests that the application of advanced technology like smart systems or cognitive computing to advance service is put on the research agenda.

The approach discussed in this paper differs from the approaches in the literature in two ways: First, it considers the application of analytics for the end-to-end service design process, i.e. comprising all phases of Fig. 3. In particular, the phases preceding and following the design of the value proposition are considered. Second, the approach discussed here explicitly takes into account a literature-based structure of data analytics w.r.t. the potential benefits (Fig. 2).

In Sect. 2 we elaborate a structure of data science tools which can be mapped on the structure of the service design process. The latter is discussed in Sect. 3. The mapping scheme of the data science tools to the service design process is presented in Sect. 4. Section 5 provides application examples for the new approach.

2 Structure of Data Science Tools and Their Value Contribution

In this section, we structure the data science-based tools according to their potential contribution to create service value (Fig. 2). The nine boxes in Fig. 2 represent fundamental data mining concepts according to [1]. We use them as basic elements of data science-based insights. They can be combined to larger data science models targeting at value creation for specific service design constellations. Thus, we have a toolbox available for the creation of service value.

The data science concepts can be differentiated into unsupervised and supervised data mining methods (left-hand side of Fig. 2). In case of open, explorative problem statements without a clear goal the unsupervised methods can be applied (e.g., not yet clear how to segment the users or which are the stages of the customer journey). On the other hand, if insights for specific problem statements with specific variables are required, supervised methods are selected (e.g., whether a specific service will be used and paid by customers in a specific situation).

The examples on the right hand side of Fig. 2 are adapted and generalised from [2] and represent an abstract of all potential applications. They suggest elements of potential data products. Figure 2 helps to find elements of services that can be created when we have a specific set of data science-based methodologies available. Therefore, this approach provides a means to find potential data products.

The challenge that we now have to solve is the following: the data science-based outcomes shown in Fig. 2 represent solutions. However, we do not yet

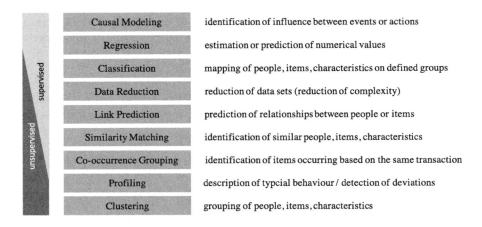

Fig. 2. Structure of data science-based insights with their benefits for services

know for which user this solution may be useful and in which context. I.e., generating data science-based insights provide solutions for yet unknown user needs. Therefore, a process for modelling the user is presented and discussed in Sect. 3.

3   Design of Service Value for Users

Service science and service design provide us with a rich toolbox of methodologies for creating value for users [3,11–13]. There are many variants of the service design process, but they have in common that they generate user insights and an understanding of user needs at a very early stage of the process and then iterate in designing, testing, and improving the value proposition [11–13]. We define here the service design process as shown by the steps in Fig. 3. We differentiate the overarching phases "explore" (for understanding the user and his needs), "create" (for designing the solution), "reflect" (for testing and improving the solution in co-creation with users), and "implement" (for the operational and market deployment of the service).

The process starts with setting the application field describing the situation and context in which the user's challenge is located (e.g., the application field may be "user wants to travel from A to Z"). Next, the phases "Customer Insight Research" and "Customer Profile Modelling" help to understand the user's needs (jobs, pains, gains) in the different contexts and along his user journey. Upon this, in the phase "Design of Value Proposition and Processes" the service and the service processes are designed to fit with the user needs in terms of the jobs, pains and gains. In the phase "Test and Improvement", the service is tested with users (e.g., "Does the value proposition fit the user's needs?", "Do users understand and like the service?", etc.). Note that from this phase the service design process usually jumps back to the previous phases for re-designing the

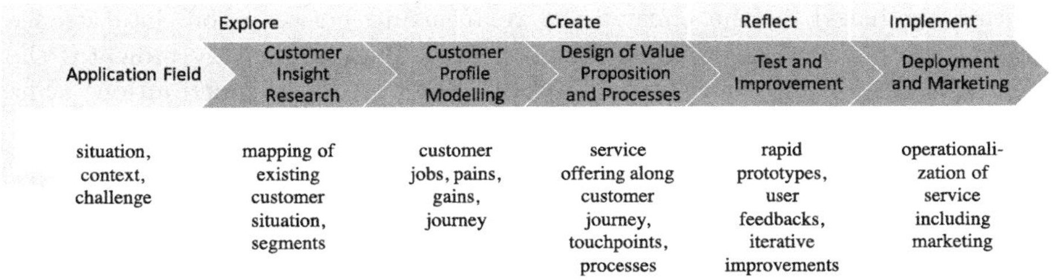

Application Field	Explore		Create	Reflect	Implement
	Customer Insight Research	Customer Profile Modelling	Design of Value Proposition and Processes	Test and Improvement	Deployment and Marketing
situation, context, challenge	mapping of existing customer situation, segments	customer jobs, pains, gains, journey	service offering along customer journey, touchpoints, processes	rapid prototypes, user feedbacks, iterative improvements	operationali- zation of service including marketing

Fig. 3. Phases of the service design process

value proposition, or even for adapting the customer profile. I.e., there are several iterations of "design – test – improve" until the service is found to be ready for deployment. In the last phase of the service design process the service is operationalised and marketed.

In each phase of this process there are design and engineering challenges to be solved. In the phase "Customer Insight Research", for example, the challenge is to capture and quantify the functional and emotional needs and to classify them into segments for the next phase "Customer Profile Modelling", in which the typical customer profiles are described. In the phase "Design of Value Proposition and Processes", the task is to identify service elements which provide relevant value to the user. In the phase "Test and Improvement", the hypothesis for the value proposition needs to be assessed quantitatively. If it is falsified (e.g., the target customer does not sufficiently appreciate the service), a new hypothesis for the value proposition needs to be designed (iteration of "design – test – improve").

4   Identifying the Value Drivers for the Data Tools

In Sect. 2 the fundamental concepts of data science are structured and mapped to potential data products. In Sect. 3 the steps of the service design process and the corresponding design challenges are discussed.

A coupling matrix is now developed bringing the service design phases of Fig. 3 in relation with the data science-based insights discussed in Fig. 2. The approach is now to map the available data science tools to the design phases such that they provide value. The mapping has been developed as follows: for each phase of the service design process of Fig. 3, the typical problem statements as described in Sect. 3 haven been put in relation with the outcomes of the data science tools (Fig. 2). For instance, in the phase "Customer Insight Research", the identification of customer segments with similar needs is one of the most relevant challenges, to which clustering can contribute substantially. In this way, the mapping of the data science tools on the different phases of the service design process has been constructed. The result is shown in Fig. 4. The dots in the matrix show to which phase of the service design process the different data science tools can contribute value. The size of the dots has been assigned qualitatively based on the constructed relationship between the data science tool and the phase of the service design process. It qualitatively indicates the strength of this value contribution (small, medium, large contribution). The empirical validation of this constructed mapping is subject to further studies.

Generic application pattern (dots with numbers 1 to 5 in Fig. 4): In a given business situation in a consumer service context, we may have data available about our customer's past behaviour and how they used various products. This data allows us to cluster customers into groups depending on different attributes (dot 1 in Fig. 4), i.e. to define our customer segments. We can describe a typical customer behaviour using profiling algorithms (dot 2 in Fig. 4). In the value proposition design phase, we may apply almost the full spectrum of data science tools to design our product or service (cluster of dots 3 in Fig. 4): for instance, we may have external data sources about environment conditions (e.g., weather and traffic) allowing us to apply co-occurrence algorithms to analyse in which context the customer is most likely to have which specific problem or pain and design the service features accordingly. Or we may use classification or link prediction to assign service features to specific users or contexts. Additionally, for an early estimation of a business case for our new service, we may have data for regression models indicating the intensity of the service usage and the willingness to pay. In the test phase (dot 4 in Fig. 4), we can apply causal modelling to avoid test insights based on correlations, for example. And, for the go-to-market-phase, we may have data to apply classification for target marketing (dot 5 in Fig. 4).

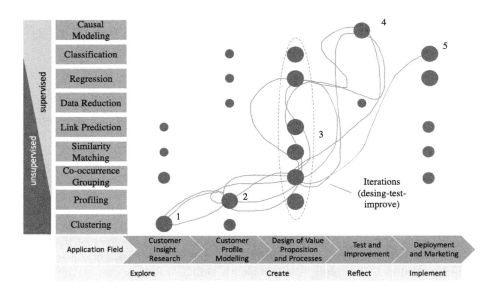

Fig. 4. Mapping the data science-based insights on the service design process

Like the service design process, the procedure to use the data science tools for finding data-driven value propositions is not a linear one. The design process is highly iterative and goes forth and back (again, iterations of "design – test – improve"). These iterations are indicated by the squiggle in Fig. 4.

5   Application Examples

The mapping scheme of Fig. 4 and the generic application pattern of the previous section were applied in first case studies. We discuss two different kinds of cases in this section.

A first case study was conducted with a service provider striving for a new offering in the area of "connected home". Thus, the application field was given, but it was not yet clear which value proposition this service should provide to which customer segments. Various demographic and behaviour data were available. A clustering of customers with different attributes can show whether customers fall into natural groups (e.g., based on the place of residence, patterns of presence and energy consumption etc.). Co-occurrence grouping can be applied to look for customers who already use similar services. The typical behaviour of customers on the company's website can be investigated by profiling. Deviations from these patterns of behaviour may indicate an opportunity for the new service. The value proposition then refers to the customer segments and the customer profile. For example, by a combination of classification and regression, the energy consumption can be predicted for the identification of potential savings. Finally, classification may support personalised marketing campaigns.

In contrast to the previous example for a new service development, many application examples do not follow the service design process in a linear way. For the second example, a case study was investigated in which technical tools for speech-to-text and natural language processing were available and the application field was in the context of customer service representatives (CSRs). The task was then to elaborate how the benefit of these technical tools can be leveraged for the daily work of the CSRs. Thus, considering the CSRs as the customers of the service, the technical tools were considered contributing to the service design phase "Design of Value Proposition and Processes" in Fig. 4. To start, we went back to the phase "Customer Insight Research" to understand the functional and emotional needs of the CSRs and to cluster them into segments. In the case study, this step was partly done using qualitative methods since limited quantitative data was available. Next, the typical journeys and profiles of the CSRs could be described. This allowed us to return to the phase "Design of Value Proposition and Processes" and shape a precise value proposition based on the technical tools (i.e., based on speech-to-text and natural language processing). The value proposition was then focused on the support of the CSR during the dialog with the customer and comprised the following elements: the automatic retrieval of customer data records, the estimation of the customers context situation, the presentation of offers to be made, as well as the recording of the conversation with the customer in a structured way.

6   Conclusions and Further Development

In this paper the question how to get value out of data science-based insights was discussed. First, a structure for the data science principles was shown with their potential contribution to service value creation. Then, the service design process was outlined. The design challenges in the different phases of the service design process were discussed, thus revealing where the data science-based insights can plug in for deploying their value. Finally, the data science and the service science approaches were combined in a new structure which outlines the application of data science along the process of service design.

This new structure has been developed and validated with several practical use cases so far. Complementary research will validate the applicability of the scheme in different industries and different service constellations and adapt it based on the outcomes. Additionally, the constructed mapping scheme of Fig. 4 needs to be verified with more and different empirical application examples. Moreover, the approach discussed in this paper started on the technical side, i.e., the data science-based insight, and then searched for user value contributions fitting to this. Future research will additionally take into account starting on the user side, i.e., the application field and the user's jobs, pains, and gains, and then look for appropriate data science contributions to support the targeted value proposition design.

References

1. Provost, F.P., Fawcett, T.: Data Science for Business. O'Reilly, Sebastopol (2013)
2. Siegel, E.: Predictive Analytics. Wiley, Hoboken (2016)
3. Lusch, F.L., Vargo, S.L.: Service-Dominant Logic. Cambridge University Press, Cambridge (2014)
4. Peters, C., Maglio, P., Badinelli, R., Harmon, R.R., Maull, R., Spohrer, J.C., et al.: Emerging digital frontiers for service innovation. Commun. Assoc. Inf. Syst. **39**(1), 136–139 (2016)
5. Spohrer, J., Demirkan, H., Lyons, K.: Social value: a service science perspective. In: Kijima, K. (ed.) Service Systems Science. TSS, vol. 2, pp. 3–35. Springer, Tokyo (2015). doi:10.1007/978-4-431-54267-4_1
6. Loukides, M.: The Evolution of Data Products. O'Reilly, Sebastopol (2011)
7. Howard, J., Zwemer, M., Loukides, M.: Designing Great Data Products. O'Reilly, Sebastopol (2012)
8. Scherer, J.O., Kloeckner, A.P., Duarte Ribeiro, J.L., Pezzotta, G., Pirola, F.: Product-service system (PSS) design: using design thinking and business analytics to improve PSS design. In: Procedia CIRP, vol. 47, pp. 341–346. Elsevier, Amsterdam (2016)
9. Wang, B., Miao, Y., Zhao, H., Jin, J., Chen, Y.: A biclustering-based method for market segmentation using customer pain points. Eng. Appl. Artif. Intell. **47**, 101–109 (2016)
10. Kwong, C.K., Huimin, J., Luo, X.G.: AI-based methodology of integrating affective design, engineering, and marketing for defining design specifications of new products. Eng. Appl. Artif. Intell. **47**, 49–60 (2016)

11. Osterwalder, A., Pigneur, Y., Bernarda, G., Smith, A.: Value Proposition Design: How to Create Products and Services Customers Want. Wiley, Hoboken (2014)
12. Polaine, A., Løvlie, L., Reason, B.: Service Design: From Insight to Implementation. Rosenfeld Media, Brooklyn (2013)
13. Brenner, W., Uebernickel, F. (eds.): Design Thinking for Innovation Research and Practice. Springer, Cham (2016)

Towards a Unified Approach to Identify Business Model Patterns: A Case of E-Mobility Services

Fabian Hunke[(✉)], Ronny Schüritz, and Niklas Kuehl

Karlsruhe Service Research Institute (KSRI), Karlsruhe Institute of Technology (KIT),
Kaiserstr. 89, 76133 Karlsruhe, Germany
{fabian.hunke,ronny.schueritz,niklas.kuehl}@kit.edu

Abstract. The introduction of new technologies creates the pursuit for innovative business models. In order to compare and evaluate such models in new markets, business model patterns can support the provision of new insights. However, so far there is no agreed upon transparent approach that helps to identify these patterns. Based on a combination of established statistical methods we propose a systematical approach that allows to identify business model patterns of any given domain. In order to validate the approach, we apply it on a data set of 58 e-mobility projects and as a result identify five distinct and semantically meaningful business models types. This paper contributes on the one hand by suggesting a new approach to identify different patterns of business models and on the other hand provides a valuable insight of the current state of e-mobility service business models that can further drive the adoption.

Keywords: Business models · Patterns · Clustering · E-Mobility services

1 Introduction

Electric mobility (e-mobility) gains increasing interest in academic research. Great potential is recognized within the automotive industry and the energy sector. From the sustainability perspective it is furthermore believed to contribute to a positive climate change and to overcome global resource shortages [1]. Regardless the obvious advantages of e-mobility, the adoption of electric vehicles in the German market falls short on the expectations [2].

Different authors state a rise of new business opportunities in the field of e-mobility in the near future [1, 3–6]. These opportunities derive – amongst others – from service business models that are widely believed to have the power to overcome the general disadvantages of e-mobility, such as the total cost of ownership, a shorter driving range, a significant thinner charging network and a lack of standardization and, thus, drive a broader adoption [4, 5]. This is not surprising as Chesbrough [7] has already stated the important role of the business model (BM) in capturing the value from a new technology. However, two questions remain: What are the BMs that capture the value of e-mobility services and by what means can we distinguish them?

In order to reduce complexity, enable a simple understanding of a matter and create a level of comparison, patterns can be used [8]. This notion of using patterns to describe

© Springer International Publishing AG 2017
S. Za et al. (Eds.): IESS 2017, LNBIP 279, pp. 182–196, 2017.
DOI: 10.1007/978-3-319-56925-3_15

and compare BMs is not entirely new. Several authors have already contributed work on *business model archetypes* [9] or *business model patterns* [10–12]. However, so far the identified patterns in research are developed based on personal experience and therefore are lacking a transparent method [9–11, 13]. Apart from that, other researchers may apply concrete methods, but do make them transparent during their research process by deriving patterns manually through intuition [14] or choosing manually the relevant dimensions that are used for discriminating the pattern [15].

Following the call for greater methodological sophistication in BM research [16–18] this research aims to tackle the lack of a standardized, observable way of pattern identification in any given domain by suggesting an approach that is replicable and takes advantages of accepted statistical methods. In order to show its general feasibility, we apply the approach to the domain of e-mobility, evaluate its performance and thereby further contribute by identifying patterns in a set of e-mobility service BMs.

The remaining paper is structured as follows: Since clustering BMs in the field of e-mobility services is related to the topics of e-mobility services in general, BM research and analytical pattern identification, Sect. 2 provides context from a literature review of each topic to support the understanding. Section 3 introduces step-by-step the general approach to identify BM patterns. Section 4 applies the approach to the domain of e-mobility services—and reveals five distinct BM patterns. Section 5 evaluates the introduced method, summarizes the contributions, provides implications, illustrates the limitations and addresses future research.

2 Related Work

2.1 E-Mobility Services

As a field of research, the topic of e-mobility has consequently gained increasing interest by a wide range of authors within the last decade [19]. Especially in Germany, the field gained popularity since 2011, when the German government announced the goal to achieve one million electric vehicles on the road in Germany by 2020 [20]. Services play a key role concerning the success of e-mobility, since they serve as potential contact points with users. They help to involve consumers easily which eventually lowers the barrier of entry [21]. In this sense, services act as an e-mobility multiplier. This ultimately accelerates the market penetration of e-mobility related products or services [5]. In order to allow a common understanding, we define e-mobility as "(…) a highly connective industry which focuses on serving mobility needs under the aspect of sustainability with a vehicle using a portable energy source and an electric drive that can vary in the degree of electrification" [19].

2.2 Business Models

Research in BMs has gained momentum within the last decade [16, 18, 22]. Despite the apparent interest in the BM as a concept, the research field presents itself as an intangible topic. Researchers offer a wide range of explanations on what a BM represents, since different research motivations lead to divergent roles and functions assigned to the individual BM concept [18]. Therefore, existing definitions only partly overlap in the literature [16]. That leads to a variety of representations that aim to describe a BM. The representation supports explaining the business, running the business, and developing the business [17]. The commonly used component-based approach is a textual representation of a variety of dimensions and characteristics of the BM [23]. These dimensions describe certain elements of the BM such as key resources, key activities, and revenue model. While there is wide range of possible dimensions, at its core the BM describes how value is created (value creation), what value is offered to the customer (value proposition) and how this value is then captured (value capturing) [24].

BM research in the field of e-mobility has a tendency to only conduct analysis of BMs in a particular and isolated context. For example, there are approaches that develop BMs around the electric vehicle using it as the central and solely element of the business [1]. Others suggest BMs which are focused on a certain industry like the energy or automotive sector [25]. Some approaches even aim to develop BMs for an even more narrowed area of e-mobility, for example for the fast commercialization of plug-in cars [26]. Kuehl et al. introduce a unified BM framework for e-mobility services to support comprehension and make services comparable [19]. It describes the essential dimensions of a BM value creation (key resources, key activities, cost structure), value capturing (revenue streams, customer segments) and value proposition.

2.3 Pattern Identification

The term pattern was decisively coined by the architectural theorist Christoph Alexander [27]. It is universally applicable to different topics [8]. An analysis of existing BMs with regard to underlying patterns has been carried out by several researchers. Abdelkafi et al. [4] conduct a research on publications dealing with BM patterns and identified 200 patterns in total, which were partly overlapping. Weill et al. [9] aim to distinguish different BM configurations in order to analyze the business performance of the largest 1000 companies in the US economy on a two dimensional taxonomy. Andrew and Sirkin [13] investigate different forms of businesses which successfully manage to achieve a payback from their investment on innovation and differentiate between three types of BM patterns. Johnson [10, 28] provides a categorization of BMs based on his own definition of the BM and gives an overview of 19 patterns. A similar work is conducted by Osterwalder and Pigneur [11], who introduce five patterns based on the BM canvas proposed by Osterwalder [29]. Gassmann et al. [12] carries out the most extensive research on BM patterns in the reviewed literature as they observe 55 different patterns which are meant to serve as innovative concepts for developing new business ideas upon. These authors do not apply a formal method to develop the patterns but suggest that patterns can be formed by a personal observation of firm's BM configurations [9–11, 13]. However, finding patterns within BM configurations by close observation highly depends on the individual analyzing the data set. This eventually leads to a non-replicable and possibly biased result. Other researchers apply systematic approaches to unveil BM patterns in a specific industry. Echterfeld et al. [14] introduce a pattern-based BM design methodology to exploit disruptive technologies. Hartmann et al. [15] analyze a data set of 100 BMs from start-up firms that rely on data as a resource of major importance for their business. They configure the data according to a pre-defined BM framework and apply the k-Medoids clustering algorithm. Kuehl et al. [19] undertake an investigation of BM patterns in the field of e-mobility services, where data is analyzed by two separate methods (k-Means and a greedy-like algorithm).

Although these authors contribute to seizing the business model pattern identification discipline, certain shortcomings need to be mentioned. Kuehl et al. [19] applies two very transparent algorithms, but does not include any form of validation. Hartmann [15] and Echterfeld [14] introduce a method that could be generally applied in any domain. However, both require unguided decisions during the research process in order to identify patterns. Hartmann [15] decides at one point for the dimensions that are used for the algorithm. Echterfeld [14] requires a personal decision by revealing patterns manually in a graphic.

3   Methodology

In order to develop an approach for extracting BM patterns the Knowledge Discovery in Databases (KDD) process is used as a framework that composes of the elementary steps for identifying patterns in data [30]. Based on previous work by Halkidi (2001) [32] the introduced approach uses the four steps data preparation, data analysis, validation as well as interpretation (shown in Fig. 1).

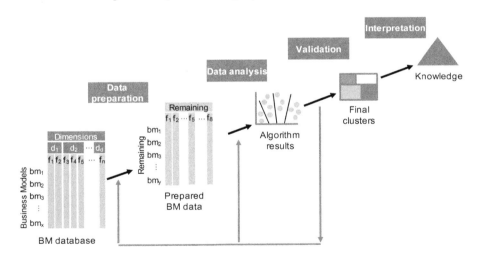

Fig. 1. The Knowledge Discovery Process based on [32].

The identification of BM patterns starts with a database of BMs which are coded according to a BM framework. A BM framework consists of multiple dimensions (e.g. key resources, key activities, etc.) with multiple characteristics (e.g. key activity: providing, aggregating, etc.). For a single BM i therefore follows: $bm_i = \{f_1, f_2, f_3, \dots, f_n\}; f_m = \{0;1\}; m = \{1, 2, 3, \dots, n\}.$, where f represents the BM characteristics described in the BM framework. 1 indicates the presence, 0 the absence of a characteristic within a BM. The initial database consequently consists of a binary matrix in which columns represent the potential BM characteristics and rows represent individual BMs.

3.1 Data Preparation

This phase aims to create a target data set for a pattern identification process. Any information within the data which is not of relevance for a data mining task need to be filtered since the may interfere the results [32]. Therefore, the data preparation phase is twofold.

First a selection of BM framework characteristics is include in the pattern identification process. Only those can be further considered which describe a distinct incident. Characteristics being absent in every BM are neglected since they do not add value to the process of finding patterns of similar BMs. Accordingly, characteristics which are only considered by a single BM are neglected since they comprise a BM's unique characteristic.

Second is projecting the remaining data set. This step aims to reduce the effective number of characteristics under consideration by finding useful components to represent the data [30]. The projection is applied using the Principal Components Analysis (PCA). The PCA is favored over similar statistical methods due to the precise order of priority concerning the principal components. This allows to define distinct decision rules for the selection of relevant principal components during the research process which allows to reduce dimensionality by neglecting the irrelevant ones.

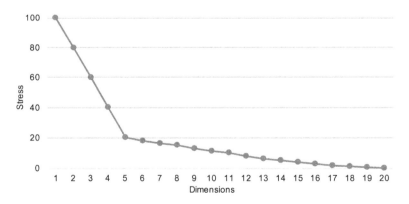

Fig. 2.   Ideal scree plot for applying the elbow criterion; adapted from [33].

The method computes a set of k orthogonal vectors (principal components) which are linear combinations of all original variables (BM characteristics) and function as a new set of axes for the data explaining the same information as the original ones. The relevant principal components are selected using a scree plot and the 'elbow' criterion [33]. The scree plot graphs the variance explained by each principal component in a decreasing order over the number of the corresponding principal components (exemplarily illustrated in Fig. 2). The elbow marks the graph's dramatic change in the stress of variance and leads to the selection of the principal components to the left of this elbow which can equally represented the data objects with little distortion [33]. In this way, components which only provide a low contribution to the explained variance of the original data set can be eliminated. In the case of Fig. 2 the criterion would select the first five components and therefore reduce the dimensionality from 20 to 5 dimensions for further analysis.

The data preparation phase ends with the selection of the relevant principal components and a non-binary target data set is passed to the next process phase.

3.2 Data Analysis

This phase aims to apply a data mining task on the non-binary target data set. Since labels for classes of patterns with similar BM configurations are not previously known in the data set, the research objective turns into an unsupervised learning data mining task [34]. It requires to group the data into classes of similar objects which is called clustering. It hereby singles out characteristics, which coin different groups distinctively.

Different types of clustering algorithms exist in the literature [34]. In the context of this clustering task a partitioning method is appropriate for the analysis since the research intends to unveil mutual exclusive clusters. More precisely, the k-medoids algorithm is chosen to cluster the data set at hand. This algorithm groups n objects into c clusters by minimizing the sum of dissimilarity between each object p within cluster i and the cluster's representative object o_i [36]:

$$\min \sum_{i=1}^{c} \sum_{j \in C_i} \text{dist}(p, o_j) \tag{1}$$

The k-medoids algorithm is chosen over the contemplable k-means algorithm due to the cluster's corresponding point o_i (medoid), an actual data object which is most centrally located in the cluster. It brings a twofold advantage over the k-means which uses the mean value of the cluster's objects to represent the center of the cluster. On the one hand, medoids form a more meaningful cluster's center in this context. They can serve as archetypes for clusters later in the investigation process. On the other hand, k-medoids is less sensitive to outliers in the data set in general [34].

After the data projection is performed only assumptions can be made on how the data looks likes, since it retrieves a non-binary data set in k-dimensional Euclidean space. Hence, the Euclidean distance as the conventional distance measure in this space is used to measure dissimilarity between data objects [36]:

$$\text{dist}(x, y) = \sqrt{\sum_{i=1}^{k} (x_i - y_i)^2} \tag{2}$$

Afterwards, the number of clusters k is determined. The parameter influences the granularity of the results since it manages a trade-off between the accuracy of each object to fit its assigned cluster and the compression of the data set to a small number of groups. A rule of thumb is provided in the literature by setting the number of clusters to $\sqrt{n/2}$ for a data set of n objects [37]. In case the resulting number does not provide a clear suggestion on which number to choose both the lower and upper discrete bound are further examined. Additionally, the silhouette coefficient is plotted over the number of clusters. The number of clusters with the highest value then represents the best solution for k. There is, however, no right procedure to determine the correct number of clusters. Instead, the clustering result needs to be validated and eventually conducted again if the cluster analysis reveals poor results. This iterative procedure unveils the optimal number of clusters as the one with the best overall clustering result.

3.3   Validation

The clustering results are presented as a list of clusters with their assigned objects. The silhouette coefficient is then used to validate the clustering results. It offers the advantage of only depending on the partition of the data set, but not on the clustering algorithm itself. This makes the validation process independent from the previous research steps and allows a comparison to other clustering results from different studies on the data set [38]. The silhouette coefficient $s(i)$ measures how well an object i has been clustered compared to closest neighbor cluster on a ratio $-1 \leq s(i) \leq 1$ [38]. Accordingly, the average silhouette width denotes the average s(i) for all objects i within a cluster and overall silhouette width the average $s(i)$ for all clusters.

The validation phase is two-stepped. First, the overall silhouette width will be examined. Clustering results featuring an overall width of less than 0.25 will be neglected. According to Kaufmann and Rousseeuw [35], a silhouette coefficient under 0.25 suggests that no substantial structure in the data exists. In this case, the research process returns to the data analysis phase. Second, if the overall width indicates an underlying clustering structure ($s \geq 0.25$), the average silhouette width for each cluster will be consulted. Only clusters with at least an underlying structure will be further interpreted. In order to validate the retrieved results, the clustering process is carried out several times. In case the clustering solution show consistency, the clustering result can be considered reliable [39].

4   Application to E-Mobility Services

In the following section, we present the results from applying the approach to e-mobility services. As already pointed out, services play a key role concerning the success of e-mobility and, thus, provides an ideal opportunity to apply the introduced method in order to search for patterns of BMs exploiting e-mobility as a bundle of new technologies. According to the proposed methodology, we give insights about the data source, the data preparation, data analysis as well as the validation. Finally, we interpret the clustering results.

4.1   Data Source

The initial data set is provided from a government-funded web platform. It collects BM information of 58 e-mobility services, which are directly entered by the e-mobility projects themselves [19]. The unified e-mobility services BM framework serves as its basis (see Fig. 3). Applying this framework to real projects results in the required format to process it for the proposed BM clustering (c.f. Sect. 3).

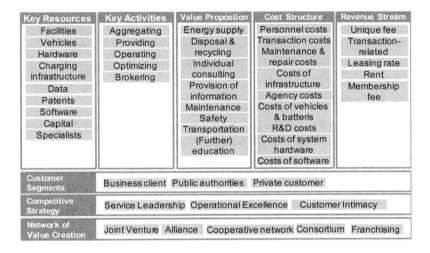

Key Resources	Key Activities	Value Propostion	Cost Structure	Revenue Stream
Facilities	Aggregating	Energy supply	Personnel costs	Unique fee
Vehicles	Providing	Disposal &	Transaction costs	Transaction-
Hardware	Operating	recycling	Maintenance &	related
Charging	Optimizing	Individual	repair costs	Leasing rate
infrastructure	Brokering	consulting	Costs of	Rent
Data		Provision of	infrastructure	Membership
Patents		information	Agency costs	fee
Software		Maintenance	Costs of vehicles	
Capital		Safety	& batteris	
Specialists		Transportation	R&D costs	
		(Further)	Costs of system	
		education	hardware	
			Costs of software	

Customer Segments	Business client Public authorities Private customer
Competitive Strategy	Service Leadership Operational Excellence Customer Intimacy
Network of Value Creation	Joint Venture Alliance Cooperative network Consortium Franchising

Fig. 3. BM framework for e-mobility services based on [19]

4.2 Data Preparation

As part of the data preparation, we remove 39 characteristics in total for the analysis. Note that this does not mean these are irrelevant in terms of business model research. Instead, they could not add value to the data mining process. The web platform offered the option for the projects to select 'unclear' for the dimensions *cost structure*, *revenue stream*, *customer segments*, *competitive strategy* as well as *network of value creation*. In order to only regard completely filled BM, we neglect these dimensions. Because the

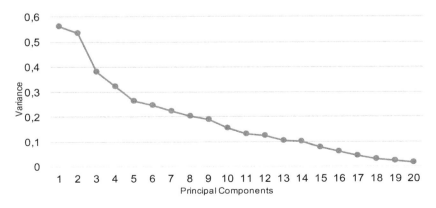

Fig. 4. Scree plot resulting from PCA.

characteristics *patents* and *hardware* (both key resources) are continuously absent, they are also neglected. The data reduction leaves 20 characteristics to describe the collected BMs. They consist of the dimensions *value proposition*, *key resources*, and *key activities* with the characteristics depicted in Fig. 3 (without *patents* and *hardware*).

The resulting scree plot from the PCA unveils that the additional variance provided by each component beginning from component five lies on a similar, low level (see Fig. 4). Because there is no clear elbow component, both potential candidates (three and five) are included for the following steps until the validation.

4.3   Data Analysis

The k-Medoids algorithm is initialized with a random seed for both PCA component cases. The clustering is run several times with different random seeds. The rule of thumb for determining the number of clusters suggests to search for k $= \sqrt{58/2} \approx 5, 39$ clusters. Therefore, five and six clusters are examined closer in the overall silhouette width plot for validation.

4.4   Validation

Table 1 shows the average silhouette coefficients for the relevant cases. It indicates that six clusters with three PCA components are the superior choice with s($c = 3$, $k = 5$) = 0.32.

Table 1.   Average silhouette coefficient for combinations of PCA components and clusters.

Average silhouette coefficient s(c, k)		
# Clusters ▶	k = 5	k = 6
# PCA components ▼		
c = 3	0.31	**0.32**
c = 5	0.30	0.27

This indicates a weak, but existing clustering structure. Based on the average silhouette width of 0.16 one cluster with 10 BMs is not further considered. The rest of the clusters show an average silhouette width of 0.25 or higher and are therefore further considered in the research process. The clustering result therefore consists of five vectors entailing the assigned objects' names. Subsequently, the entries of each cluster's vector are matched with the original row names of the binary matrix. By doing this, a binary data subset for each cluster is obtained. Each subset now reflects the binary BM configuration again which describes the presence or absence of the respective characteristic in the BM. This fragmentation of the initial data set consequently represents the final clusters of the data analysis.

4.5   Interpretation

In the following the clusters are described with regard to the BMs characteristics' frequency. This is carried using a bar plot for each BM cluster. It is exemplarily shown in Fig. 5. Afterwards, the cluster's prevalent characteristics are identified and used to interpreted the clusters as BM types. Again, the approach is not supposed to unveil components being irrelevant to a business model. Instead, insights of different business model types and their core components are supposed to be derived.

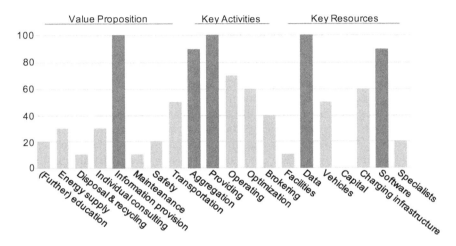

Fig. 5. Characteristics' frequency within BM Type 1: analytics as a service.

BM Type 1: Analytics As A Service. This cluster comprises of 10 BMs which all offer a provision of information to the customer as the value proposition. Additional value proposition is only used in rare frequency. The key resource is represented by data. Additionally, 90% of the BMs also use software as a key resource. The key activity is described by providing information. Moreover, 90% within this cluster perform aggregation as a key activity. Thirdly, 70% of the cluster's BMs consider operating a key activity. Value proposition providing information seems consistent in the context of key resource data from which the relevant information offered seems to be derived. The collection of data is performed through an aggregation by the majority of BMs. Resource software indicates that the businesses derive the necessary information from the data by themselves. Furthermore, operating indicates that the aggregation is conducted by an own infrastructure or platform.

Concluding, data and the distribution of gathered knowledge seems to be the core business of this cluster. In order to collect necessary data, an own infrastructure or platform is mostly operated. The gathered data is then aggregated and analyzed by an own software to gain the information. Therefore, this group is referred to as the analytics as a service cluster.

BM Type 2: Charging As A Service. The 7 BMs in this cluster all use a charging infrastructure as a key resource (100%). Additionally, vehicles and software are also used as key resources by respectively 57% of the BMs. The key activity is expressed by operating (100%). It is enriched through three additional activities: providing (71%), brokering (86%), optimization (71%). The value proposition is represented by the characteristics energy supply (71%) and individual consulting (57%). Summarizing, this BM type addresses the importance of charging infrastructure. They build the core business around the operation of a charging infrastructure. Energy supply as the dominant characteristic for the value proposition seems plausible in combination with the operation of the charging infrastructure. The seven entailed BMs are referred to as the charging as a service type.

BM Type 3: Data As A Service. 15 BMs this cluster mainly offer a provision of information to the customer (80%). The major key resources are presented by data (87%) and enriched through specialists (53%). The aggregation expresses the main key activity (87%). Providing is also noteworthy, but it is less considered since only 67% use this sort of activity. Taking into consideration the main value proposition and key activity of this cluster, it seems very similar to the Analytics as a service BM type. Both seem to use occurring data as the core element of their business and the central linkage between the BM elements. However, the strong differences are revealed when looking at the business characteristics which are consciously not considered in this cluster. The resource software only plays a minor role in this cluster, whereas it is regarded a key resource in the analytics as a service business type. Instead, specialists play a considerable role in the cluster at hand, whereas specialists are not considered crucial at all in the first type. Also, the activity of operating any sort of infrastructure or platform does not reflect an important issue of this BM type. Taking into consideration the previously made observations, this cluster needs to be considered as an own BM type. Since any sort of software is not considered crucial in this cluster, it can be presumed that a transformation process of data does not take place. It is rather the pure aggregation of data which constitutes the core element of the business. Therefore, the 15 BMs in this cluster are referred to as data aggregation services. The characteristics (further) education and safety slightly stand out as characteristics of the value proposition. This gives an indication that the data aggregation and provision service takes place in a wide range of different themes throughout the field of e-mobility.

BM Type 4: Transportation Service. The 6 BMs in this cluster mainly offer transportation as the value proposition (83%). In fact, only one BM does not provide this kind of offering. The key resource is mostly represented by vehicles (67%). Additionally, software is used by 50% of the BMs. The key activity is mainly described by operating (67%). Additionally, providing and brokering, are carried out by respectively 50% of the BMs. Summarizing, the value proposition and the key resources suggest a traditional transportation service. Accordingly, key activity operating seems reasonable in the context of a transportation service offering an own electrical powered vehicle fleet. However, the key activities providing and brokering transportation also suggest a car sharing business type as a transport services which uses a software-based business realizing a brokering of vehicles. Therefore, this BM type is referred to as the transportation service.

BM Type 5: Knowledge As A Service. The 10 BMs mainly builds their core business upon specialists as a key resource (70%). Other key resources are represented by a significantly lower percentage. The value proposition is provided by an individual consulting (50%) and (further) education (40%). Both seem weak regarding the percentage of BMs entailing this characteristic. However, they seem to be the superior choice as opposed to the remaining characteristics. Key activities are represented by providing (60%), brokering (60%) and optimizing (50%). This BM type creates value by addressing the need for e-mobility related knowledge. On the one hand, the business configurations suggest the existence of a consulting service provision, carried out by specialists. On the other hand, value seems to be alternatively offered through an e-mobility related education, carried out by specialists. Therefore, this BM type is referred to as knowledge as a service (Fig. 6).

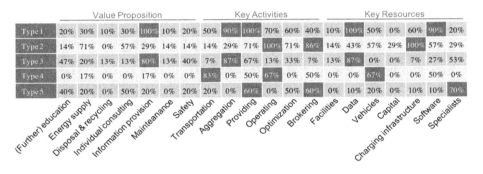

Fig. 6. Summary of the identified business model types' core components.

5 Conclusion

The paper at hand proposes a general approach for finding BM patterns by applying established, mathematical methods. The approach combines the Principal Components Analysis, a k-Medoids clustering and the silhouette coefficient as a quantitative evaluation measure. This ensures that the results are replicable and the approach is deterministic—and therefore easily automatable and scalable.

The approach is applied to a dataset from the domain of e-mobility services. We are able to show its general feasibility as we are able to identify five semantically meaningful clusters, namely *analytics as a service, charging as a service, data as a service, transportation service* as well as *knowledge as a service*. Interestingly, an emphasized role of ICT is discovered within e-mobility services as two BM clusters explicitly offer a provision of information to their customers in different ways. One noteworthy aspect is that *analytics as a service, data as a service* and *transportation service* are partially confirmed by the work of Kuehl et al. [19] who already derive a *data-driven* and a *fleet-related* service. It can be seen that the proposed approach creates a more detailed clustering result since the BM clusters are more specified.

The approach has several limitations. First, the analysis was only applied to one domain with a small data set of 58 business models—and while the silhouette coefficient showed structure of the clusters, it could be higher for a strong underlying clustering structure. Second, a bias in the research remains regarding the initial data set since the entered framework information could not be reflected critically. Hence, subsequent single case studies should be conducted for every identified cluster in order to evaluate and illustrate the different business model types.

Nonetheless, the application of the approach was able to deliver interesting results for the field of e-mobility services and future work will apply it to different domains with a higher volume of available data to enable new, quantitative-based insights for the field of business model research.

References

1. Kley, F., Lerch, C., Dallinger, D.: New business models for electric cars: a holistic approach. Energ. Policy **39**, 3392–3403 (2011)
2. Statista. http://de.statista.com/statistik/daten/studie/265995/umfrage/anzahl-der-elektroautos-in-deutschland/
3. Spath, D., Rößler, A., Rose, H., Rothfuss, F., Satikidis, D., Scheffler, G.: SYSTEMANALYSE BW e MOBIL 2013: IKT- und Energieinfrastruktur für Innovative Mobilitätslösungen in Baden-Württemberg, 2nd edn. Fraunhofer, Stuttgart (2013)
4. Abdelkafi, N., Makhotin, S., Posselt, T.: Business model innovations for electric mobility: what can be learned from existing business modell patterns. Int. J. Innov. Manage. **17**, 1–42 (2013)
5. Cocca, S., Fabry, C., Stryja, C. (eds.): Dienstleistungen für Elektromobilität: Ergebnisse einer Expertenstudie. Fraunhofer, Stuttgart (2015)
6. Budde Christensen, T., Wells, P., Cipcigan, L.: Can innovative business models overcome resistance to electric vehicles? Better place and battery electric cars in Denmark. Energ. Policy **48**, 498–505 (2012)
7. Chesbrough, H., Rosenbloom, R.S.: The role of the business model in capturing value from innovation: evidence from Xerox Corporation's technology spin-off companies. Ind. Corp. Ch. **11**(3), 529–555 (2002)
8. Bushmann, F., Meunier, R., Rohnert, H.: Pattern-Oriented Software Architecture: A System of Patterns. Wiley, Chichester (1996)
9. Weill, P., Malone, T., D'Urso, V., Herman, G., Woerner, S.: Do some business models perform better than others? A study of the 1000 largest US firms. MIT Center of Coordination Science, Working Paper No. 226, pp. 1–39 (2005)
10. Johnson, M.W.: Seizing the White Space: Business Model Innovation for Growth and Renewal. Harvard Business Press, Boston (2010)
11. Osterwalder, A., Pigneur, Y.: Business Model Generation: A Handbook for Visionaries, Game Changers, and Challengers. Wiley, New York (2010)
12. Gassmann, O., Frankenberger, K., Csik, M.: Geschäftsmodelle Entwickeln: 55 Innovative Konzepte mit dem St. Galler Business Model Navigator. Hanser, München (2013)
13. Andrew, J., Sirkin, H.: Cashquelle Innovation: Wie aus Ideen Gewinne Sprudeln. Hanser, München (2007)
14. Echterfeld, J., Amshoff, B., Gausemeier, J.: How to use business model patterns for exploiting disruptive technologies. In: 24th International Conference on Management of Technology, International Association for Management of Technology, pp. 2294–2313 (2015)

15. Hartmann, P.M., Zaki, M., Feldmann, N., Neely, A.: Capturing value from big data: a taxonomy of data-driven business models used by start-up firms. Int. J. Oper. Prod. Man. **36**(10), 1382–1406 (2016)
16. Zott, C., Amit, R., Massa, L.: The business model: recent developments and future research. J. Manage. **37**, 1019–1042 (2011)
17. Schneider, S., Spieth, P.: Business model innovation: towards an integrated future research agenda. Int. J. Innov. Man. **17**(1), 1–34 (2013)
18. Spieth, P., Schneckenberg, D., Ricart, J.E.: Business model innovation: state of the art and future challenges for the field. R&D Manage. **44**, 237–247 (2014)
19. Kuehl, N., Walk, J., Stryja, C., Satzger, G.: Towards a service–oriented business model framework for e-mobility. In: European Battery, Hybrid and Fuel Cell Electric Vehicle Congress, pp. 1–9 (2015)
20. Bundesministerium für Wirtschaft und Technologie Bundesministerium (eds.) Regierungsprogramm Elektromobilität. Bundesregierung, Berlin (2011)
21. Dudenhöffer, K.: Why electric vehicles failed. J. Man. Control **24**(2), 95–124 (2013)
22. Ghaziani, A., Ventresca, M.J.: Keywords and cultural change: frame analysis of business model public talk. Sociol. Forum **20**(4), 523–559 (2005)
23. Burkhart, T., Werth, D., Krumeich, J., Loos, P.: Analyzing the business model concept: a comprehensive classification. In: 2011 International Conference on Information Systems, pp. 1–19. Association for Information Systems, Shanghai (2011)
24. Schüritz, R., Satzger, G.: Patterns of data-infused business model innovation. In: 18th Conference on Business Informatics Proceedings, pp. 133–142. IEEE Press, Paris (2016)
25. Handberg, K., Owen, G.: Electric vehicles as grid support. In: Beeton, D., Meyer, G. (eds.) Electric Vehicle Business Models: Global Perspectives. Lecture Notes in Mobility, pp. 129–146. Springer, Cham (2015)
26. Williander, M., Stålstad, C.: Four business models for a fast commercialization of plug-in cars. In: Beeton, D., Meyer, G. (eds.) Electric Vehicle Business Models: Global Perspectives. Lecture Notes in Mobility, pp. 129–146. Springer, Cham (2015)
27. Alexander, C.: The Timeless Way of Building. Oxford University Press, New York (1979)
28. Johnson, M., Christensen, C., Kagermann, H.: Reinventing Your Business Model. Harv. Bus. Rev. **86**, 57–68 (2008)
29. Osterwalder, A.: The Business Model Ontology: A Proposition in a Design Science Approach. Dissertation, University of Lausanne (2004)
30. Fayyad, U., Piatetsky-Shapiro, G., Smyth, P.: From data mining to knowledge discovery: an overview. In: Fayyad, U., Piatetsky-Shapiro, G., Smyth, P., Uthurusamy, R. (eds.) Advantages in Knowledge Discovery and Data Mining, pp. 37–54. American Association for Artificial Intelligence, Menlo Park (1996)
31. Halkidi, M.: On clustering validation techniques. J. Intel. Inf. Syst. **17**, 107–145 (2001)
32. Miligan, G.: Clustering Validation: Results and Implications for Applied Analyses. World Scientific, New York (1996)
33. Pyle, D.: Data Preparation for Data Mining. Morgan Kaufmann, San Francisco (1999)
34. Han, J., Kamber, M.: Data Mining: Concept and Techniques, 2nd edn. Morgan Kaufmann, San Francisco (2006)
35. Kaufman, L., Rousseeuw, P.J.: Finding Groups in Data: An Introduction to Cluster Analysis. Wiley, New York (1990)
36. Rajaraman, A., Ullman, J.D.: Mining of Massive Datasets. Cambridge University Press, Cambridge (2012)

37. Han, J., Kamber, M., Pei, J.: Data Mining: Concept and Techniques, 3rd edn. Morgan Kaufmann, San Francisco (2011)
38. Rousseeuw, P.J.: Silhouettes: a graphical aid to the interpretation and validation of cluster analysis. J. Comput. Appl. Math. **20**, 53–65 (1987)
39. Ketchen, D., Shook, C.: The application of cluster analysis in strategic management research: an analysis and critique. Strateg. Manage. J. **17**, 441–458 (1996)

Towards Requirements Analytics: A Research Agenda to Model and Evaluate the Quality of Unstructured Requirements Specifications

Patrick Kummler[(✉)]

Karlsruhe Service Research Institute (KSRI), Karlsruhe Institute of Technology (KIT),
Kaiserstraße 89, 76133 Karlsruhe, Germany
patrick.kummler@kit.edu

Abstract. Communication between actors in a service system can be based on unstructured text. The quality of this text is crucial for the effort and output of a service system. The paper presents an approach to evaluate and model the quality by using requirements from automotive development projects as practical example. The aim is to define quality by using relevant attributes and quantifiable measures. First results include the development of an assessment tool and an initial analysis of the available dataset.

Keywords: Service systems · Software requirements · Analytics · Machine learning · Automotive software · Requirements quality · Quality metrics · Quality measurement

1 Introduction

The main economic activities in developed countries are service activities [1, 2]. The increase of the service sector is based on the tremendous outsourcing of activities, such as maintenance or tasks in research and development. According to Vargo and Lusch [3], a service is defined as *"the application of competences for the benefit of another, meaning that service is a kind of action, performance, or promise that's exchanged for value between provider and client."* In contrast to the goods-dominant logic, a service activity is based on communication and exchange with the client: *"[T]he more knowledge-intensive and customized the service, the more the service process depends critically on client participation and input"* [4, p. 72].

The main goal of a service system is to create reciprocal value – "service is exchanged for service" [1]. Therefore, a service system concludes the co-creation of value through the configuration of actors and resources. An actor is defined as a person including their knowledge and skills while resources are described as technology, information and physical artifacts. Thus, a service system is a *"socio-technical system that enables value co-creation guided by a value proposition"* [5, p. 399].

Service systems can increasingly be observed in different industries. For example, in the development of automotive software products the actors of a service system can be recognized as automotive manufacturers and suppliers. In this case the service

© Springer International Publishing AG 2017
S. Za et al. (Eds.): IESS 2017, LNBIP 279, pp. 197–209, 2017.
DOI: 10.1007/978-3-319-56925-3_16

interactions are based on communication and exchange of information and requirements between these actors.

Generally, services increasingly influence the automotive industry [6–8]. The importance of configurable services as part of current and future mobility services increases the relevance of effective co-creation of value between car manufacturers and suppliers. The creation process of service solutions, that are provided by developing software-based functionalities in products, is crucial for future competitiveness. Besides the importance of these services, the complexity of software products is growing [9, 10]. Several established quality approaches exist to describe and specify software requirements [11]. However, industrial practice is still suffering from a lack of structure regarding the interaction and cooperation between actors in a development project.

In the development process of a software product the specification of requirements is the first concrete and tangible output [12]. However, it is also the most critical part as the specification is the major source of failures for upcoming activities [13]. At the same time the specification and adjustment of requirements is an ongoing and highly agile process [14]. Thus, it must be ensured, that the quality of requirements is on a high level from the beginning of a software project.

In this work, we develop an approach to analyze communication and exchange of information based on software requirements that are specified as unstructured text by using machine learning algorithms. The target is to implement an automated solution to measure and monitor requirements quality. We contribute at two stages during the specification process: First, the quality of the specification can be checked and analyzed at the point of creation and improvements are shown directly; second, the output can be evaluated from different roles.

We portray the research in cooperation with an international automotive engineering company and relevant organizational units of car manufacturers. The research should not only improve effectiveness and efficiency of software-based services in the automotive industry. It should also offer a generalizable solution where the analysis and improvement of unstructured text is necessary for the successful communication between actors of a service system.

The quality assurance is based on the use of metrics that are analyzing the requirements. Before metrics can be applied to the text, the term *quality* must be defined and concretized. Many definitions of quality with regards to software requirements are already established [15–17]. The work in progress analyzes and aggregates these definitions and enhance the approaches with the input from experts, that are creating and receiving software requirements either for testing or implementing. The input is used in different phases of our research. An assessment phase delivers necessary data to get insights into the qualitative assessment of requirements. Based on these results we develop quantitative measures as input for machine learning techniques.

2 Related Work

The analysis of software requirements was initially conducted in the aerospace sector [12]. Due to high safety and legal provisions, software developers were forced early to guarantee reliable results. The experiences from the aerospace sector for software development should be considered in the automotive industry as the automotive manufacturers are increasingly faced with similar challenges [18, 19].

General problems, solutions and challenges of the requirements engineering are summarized and categorized in [20]. The authors analyze and evaluate different software projects in the automotive industry and indicate various results for the following categories: general issue, requirement engineering process and technical issues. These results can be used as input for the development of relevant quality attributes.

In [21], a framework to measure and improve the quality of textual requirements is presented. Especially the challenge regarding the derivation of quantitative measures from quality attributes is considered. Similar approaches of the requirements quality prediction are based on this research. However, despite sufficient approaches defining relevant quality attributes of software requirements, the industrial practice is rarely asked about the relevance and importance of the quality attributes and the assignment of quantitative measures.

Parra [22] build a machine learning technique to assess the quality of requirements automatically. The authors use the tool RQA (Requirement Quality Analyzer) [21] to provide quantitative criteria. They show that their methodology manages to evaluate the quality of requirements written in natural language in the same way an expert would do with 86.1% average accuracy. Their work focuses on correctness metrics. Other kind of metrics are left aside which offers the possibility for further and comprehensive research.

Mund [17] follows a similar research approach and investigates the influence of requirements quality in the lifecycle of software-intensive systems. This research and the subsequent contribution offer useful information especially regarding the derivation of quantitative measures from quality attributes.

Davis [15] provides a list with different quality attributes for requirements. We plan to enhance this list with current research. We analyze different approaches of quality attributes referring to requirements quality and compare and aggregate the information. In this context, we focus especially on weak phrases used in the specification of requirements [23].

3   Research Methodology

The aim of our research is to analyze and improve service interactions based on unstructured text for the co-creation of value. By using service interactions between automotive manufacturers and suppliers we want to develop an automated solution that analyzes, measures and monitors software requirements to support the specification process in a software project.

A literature review according to vom Brocke [24] helps us to systematically find relevant literature. Although a literature review is described as *"a summary of a subject field that supports the identification of specific research questions"* [25, p. 31], we already defined potential questions. Our research topic is based on personal experiences made during the interaction and collaboration with organizations and enhanced by analyzing relevant practice literature [26, 27]. In Webster [28] it is mentioned, that such experience helps to justify and support a proposition. In the following we propose our research questions:

RQ1 How can the quality of unstructured text improve the communication between actors in a service system?

RQ2 How is it possible to measure the quality of unstructured text by using qualitative attributes and quantitative measures?

RQ3 How is the quality of unstructured software requirements defined?

RQ4 Which quality attributes are relevant for the interaction between actors and which of them are consciously adapted in the development of software functions for the automotive industry?

RQ5 Which quality attributes influence the quality perception?

RQ6 Is it possible to automatically recognize the quality of unstructured requirement text?

The proposed research questions are focused on the quality of unstructured text. Thus, it is necessary to propose a resilient definition of the term quality and to evaluate it throughout the research progress. RQ1 analyzes the relation between different actors of a service system where the quality of unstructured text communication is a crucial part. We analyze how the communication quality can be improved by considering the relevant unstructured text.

To increase the communication quality in a service system different quality metrics are necessary. In the second research question, we evaluate and implement quality attributes that analyze unstructured text. Established approaches are presented and applied to the available data. The implementation of enhanced attributes is also a relevant part.

In a specific way, RQ3 gives an overview of existing quality notions regarding software requirements. For the term quality, especially in conjunction with software requirements, many different definitions exist. In our research, we want to aggregate these definitions and develop a comprehensive approach by considering input from expert interviews as well.

RQ4 is concerned about which quality attributes are relevant for the communication and interaction between actors in a service system. Referring to our practical example, we consider different actors of software development projects in the automotive industry. This research question studies which quality attributes from literature and standards are consciously adapted in the industrial practice.

Following the previous research question, RQ5 aims to identify the relevant quality attributes that influence the quality perception. The goal is to analyze quality attributes that are relevant for the actors in a service system.

Based on the previous results RQ6 questions whether an automatically recognition of requirement quality is possible. By gathering sufficient data, we want to answer this question using machine learning techniques to predict the communication quality.

In the field of quality measurement for requirements, a lot of earlier work exists. Due to the number of similar approaches regarding the quality characteristics of software requirements, such as IEEE Std. 830 [29] or ESA PSS-05 [30], it is planned to create an overview and compare the different concepts later in the research. Our research builds upon previous findings and past research and enhances these results with input from experts. This information is used to adapt an approach that focus on industrial practice.

We plan a systematic and guided literature review defined as a *"structured approach to identifying, evaluating and synthesizing research [and to] address a specific research question that guides data collection, extraction and aggregation process"*. [31, p. 9] The purposes of our review are to comprehensively understand the domains of requirements engineering, quality measurement and machine learning techniques. Additionally, we want to uncover research and feature gaps for the measurement of requirements quality. A relevant part is the identification of research methods and strategies that are commonly used for the quality measurement of requirements.

Following vom Brocke [24], we focus on different phases during our literature review. As the goal of the review should be clear in the beginning we adopt the taxonomy according to Cooper [32] in the initial phase. We start with analyzing handbooks and seminal literature about requirements engineering [26, 27]. The next phase focuses the relevant database, keywords and back- and forward search as well as an ongoing evaluation of sources. It is planned to analyze relevant journals and doing an author- and topic-based search simultaneously. We evaluate important IS conferences [33, p. 121] for similar topics and research approaches. After sufficient literature [31] is collected we analyze and synthesize relevant content into a concept matrix to evaluate relevant concepts.

In our research, we consider the Cross-Industry Standard Process for Data Mining (CRISP-DM) as standard reference model for our data mining approach [34]. Initially we start with the business understanding phase and define objectives for our data analysis. As we already have relevant data for our research topic we proceed with activities that *"enable [us] to become familiar with the data, identify data quality problems, discover first insights into the data, and/or detect interesting subsets to form hypotheses regarding hidden information"* [34, p. 10]. The initial analysis of our dataset is presented in Sect. 6.2 and shows basic syntactical measures. This analysis helps us to get familiar with the available data and to identify potential tasks regarding the pre-processing of our dataset. From the final dataset, we randomly choose requirements for the assessment tool. For the modelling phase, we choose the application of a support vector machine (SVM). According to Kivinen [35], SVM algorithms are well suited for problems with dense concepts and sparse instances. Moreover, SVM have the ability to learn independently from the dimensionality of the feature space. *"[SVMs] allow fully automatic parameter tuning without using expensive cross-validation"* [36, p. 3]. Additionally, we follow the approach of Hsu et al. [37] that offers a practical guide for support vector classification. Nonetheless, other techniques are considered to analyze the optimal approach according to the dataset.

Simultaneously to the literature review we conduct exploratory interviews with different roles: requirements engineer, developer and tester. A requirements engineer is the person that specify the requirements. This person creates the unstructured text we want to analyze and improve. A developer or a tester receive requirements and capture the unstructured text. We conduct the interviews by meeting our participants "face-to-face" [38]. A group basis is not desired, as the single participants could be biased by the answers of other participants especially regarding the questions about quality. As there are various persons with different years of profession we prefer a semi-structured interview as this type refers to qualitative research and is non-standardized [39].

With our research, we contribute in different parts of relevant scientific areas. We analyze existing approaches that assess the communication quality based on unstructured text. We use input from experts' perspective to enhance these approaches and aim at creating a tool that automatically analyzes text according to defined quantitative measures. Our solution helps to reduce misunderstanding between actors and reduces unnecessary exchange of information due to inadequate quality. We conduct the quality analysis of the unstructured text by using quality attributes acknowledge from the experts. Additionally, we conduct a structural analysis to find quantitative measures representing a well-defined requirement. By using machine learning algorithms, we want to define appropriate rules that classify requirements automatically.

4   Research Process

The research is planned in three phases: Quality Definition, Assessment & Analysis and Tool Development. Figure 1 shows the research process.

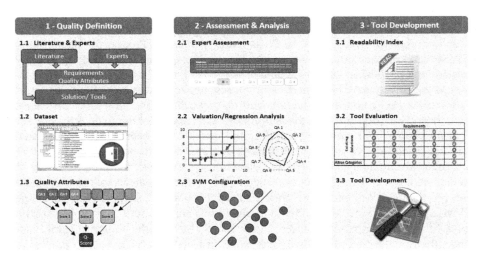

Fig. 1. Research process for requirements analytics

The first phase *Quality Definition* includes tasks to gather data and create attributes to measure the quality of unstructured text in requirement documents. We gain information about the quality measurement of requirements by considering established attributes from literature and standards. Additionally, we plan to conduct around 20 expert interviews to gain enhanced information and best practices about necessary quality attributes and quantitative measurements. The experts for the interview sessions have different roles in software projects (requirements engineer, software developer and tester). Simultaneously we capture data consisting of unstructured requirements that are used for the interaction and communication between actors of the service system. The next task covers the definition of relevant quality attributes. The outcome of the first phase contains an overview about established quality attributes and insights about necessary characteristics of qualitative requirements in the industrial practice. Besides gathering relevant data, the first phase provides a set of defined and crucial quality attributes. This step is the most important and challenging part of the research as the following phases are based on its output.

During the *Assessment & Analysis* phase, we gather additional data through the assessment of requirements to analyze the quality in two ways: first, we want to determine which quality attributes influence the overall quality perception of requirements; second, we want to implement a solution to predict the quality of requirements by using machine learning techniques. The first task in this phase contains the tool development to assess the unstructured requirement text with the defined quality attributes from the previous phase. After reaching a level of saturation - when more assessments would not introduce significantly new findings and insights [31] - the results are used to determine dependent quality attributes by analyzing the correlation for each quality attribute. We want to identify relevant quality attributes and the influence of each attribute regarding the overall quality assessment of a requirement.

The following task includes the derivation of quantitative measures from the quality attributes defined in the phase *Quality Definition* by using the input from the expert interviews as well. With these measures, it is possible to implement a support vector machine (SVM) that classifies the quality of unstructured document text.

The results of the second phase contain the assessed requirements and the application of a regression analysis which is used to gain detailed insights about the data. To predict the quality of requirements, quantitative measures are derived and assigned to each relevant quality attribute. In this way, a support vector machine analyzes and uses the data as input to predict requirements quality.

In the last phase (*Tool Development*), the tool evaluation and implementation is focused. In the first task, we consider and analyze different readability indices. It is planned to adapt syntactical features to implement and evaluate an approach for a software requirement readability index. Based on already established readability indices, such as Flesch, Dale-Chall and SMOG, a similar approach would help to identify the complexity and understandability of a software requirement. An evaluation of available tools for the qualitative measurement of software requirements based on features and metrics is necessary. The last task in this phase contains the development of a solution based on evaluated features to improve the quality of software requirements.

5   Research Issues

The analysis of textual requirements can be an *"aid for writing the requirements right, but not for writing the right requirements"* [21, p. 26]. A measurement can be an approach to improve the correct understanding of requirements. However, an overall challenge in the analysis of requirement text is to define attributes that offer insights about the content quality. In [21], the quality measurement is mainly based on the analysis of countable characteristics. Generally, a quantitative measurement approach does not necessarily give insights about the content quality as the text is only the carrier of the content. However, we are convinced that understandable, clear and unambiguous content lead to a better communication between actors and, referring to our automotive example, lead to a higher quality of software products.

In our research, requirements are assessed without detailed context. This could lead to a different analysis from experts and thus the creation of distorted rules from our machine learning algorithm. However, we believe that intelligibility is one important criterion on its own as well. Our methodology is dependent on the quantitative attributes that we define to analyze the requirements. The definition of the right and accurate metrics is a crucial point for the success of our predictions. On the other hand, our research method depends a lot on experts' inputs and evaluation. Given the numerous features in a text analysis, it might take a while to reach the saturation level of our algorithm. A challenge is also to find enough experts that are willing to share their time and knowledge for these analyses.

The development of this tool is based on several hypotheses: quality can be made objective and measurable, metrics can portray experts' quality analysis and metrics are sufficient to interpret natural language. We refer to empirical results to confirm these hypotheses. Moreover, we plan to use one set of chosen requirements to develop this tool, which might not be sufficient to apply the deduced classifying rules to every other requirement.

As mentioned in [17] another possible issue is the unawareness of related work. Metrics, its application and natural language analysis is a broad subject. Thus, we might be tempted to use others disciplines hints and at the same time we might miss some important inputs from distant fields we are unaware of.

6   First Results

In this section, we present first results regarding the development of the assessment tool and the analysis of the dataset.

6.1   Development of Assessment Tool

The assessment tool is constructed as a web survey using JavaScript and PHP scripts connected to a SQL database. Initially general information is asked: gender, role in the service system and years of experience in this role.

After a necessary data preprocessing, a random set of requirements is selected for the assessment. We want to make sure that the requirement set is evaluated several times by different roles. Thus, we take the given role as input for our SQL queries to display the requirements to be analyzed.

For each text, the expert first decides whether it is a requirement, an information or another type (e.g. headings to structure the requirements document) that is used in the specification process of requirements. In case *requirement* is selected, the assessor can detail the concrete requirement type according to the V-model [40]. This standard development process for software development offers different requirements types in each development phase. However, despite of established procedures and standards to create different documents for each requirement type, industrial practice is still suffering from a lack of clear separation between these types. Therefore, and due to reasons of simplicity requirements engineers often combine different requirement types in one document. This situation urges to ask about the concrete type in the assessment phase, such as customer requirements, architectural requirements, (non-) functional requirements or software requirements. After selecting the type, the expert assesses the overall quality of the requirement and the quality attributes that are defined in the first phase of our research process. The assessments are saved in our database through SQL queries.

6.2 Data Analysis

The assessment tool provides information about how requirements are perceived by the experts and leads to the assessment of requirements based on defined quality attributes. Besides this, we analyze objective information about the available dataset. We plan to define metrics based on lexical indicators (connective terms, imprecise terms) and analytical indicators (verbal tense) [20].

The available dataset is based on an export from several development projects and contains 32.956 object types in 54 requirements documents. Objects types are defined as headings, information and requirements. As headings are not relevant for our analysis we exclude this type from the dataset. The remaining object types are enhanced with additional information, such as language of the object text and a differentiation between software and hardware related content. Further we reduce the dataset by excluding empty content and double spaces in the relevant object text to 26.492 objects. We analyze the dataset by defining the following basic syntactical measures:

- Number of characters
- Number of words
- Average length of words

In Table 1, the analysis of the number of characters reveals that there is a remarkable difference between software and hardware related content. The mean value of hardware referring to the number of characters is 121.99 whereas for software it is 138.56. Such characteristics can also be observed in the standard deviation where the value for software content (120.68) is one third higher than for hardware content (88.32). The values for minimum and maximum number of characters lead to the necessity to invest sufficient time for a successful pre-processing of the data. As example, an object (requirement or information) consisting of only ten characters can hardly transport sufficient information. In contrast, for an object with 3900 characters it should be possible to split the content in several objects. Throughout the measures software objects contains more characters in comparison to hardware objects.

Table 1. Analysis of number of characters

Objects	Mean value	Standard deviation	Minimum value	Maximum value
Hardware	121.99	88.32	10	1876
Software	138.56	120.68	18	3900

In Table 2 the results from the analysis of number of words are shown. The mean value is similar for hardware (18.02) and software (19.50). However, the outliers for software objects (650) have to be analyzed separately. The analysis of the average length of words in Table 3 confirms the necessity of pre-processing before using the dataset for further research. The mean value for the average length of words is similar for hardware (6.00) and software (6.29). The standard deviation for both object categories corresponds as well (1.29 for hardware and 1.56 for software). Again, the software related content reveals outliers referring to the maximum value of the average length of words (hardware with 17.58 and software with 24.81).

Table 2. Analysis of number of words

Objects	Mean value	Standard deviation	Minimum value	Maximum value
Hardware	18.02	12.95	5	166
Software	19.50	16.00	5	650

Table 3. Analysis of average length of words

Objects	Mean value	Standard deviation	Minimum value	Maximum value
Hardware	6.00	1.29	1.20	17.58
Software	6.29	1.56	2.00	24.81

The first analysis of the dataset leads to the necessity to invest sufficient time in pre-processing the data in order to create a qualitative dataset. This analysis shows the range of outliers and differences between hardware and software related content and contains first results to get familiar with the available dataset. Additional tasks are necessary and presented in the work in progress.

7   Conclusion and Outlook

The paper sets up to improve communication quality between actors in a service system. We refer to the relationship between actors in development projects of automotive manufacturers and present related work that investigates metrics to assess the quality of requirements. We propose a research agenda to analyze software requirements that are specified as unstructured text by using machine learning algorithms. Quality attributes based on experts' input and relevant literature help to create an assessment tool to analyze the perceived quality of requirements. With these results, we plan to develop quantitative measures that serve as input for a support vector machine. We aim to offer a tool that automatically predicts the quality of unstructured text in the communication between actors of a service system.

The next step of our research contains a comprehensive literature review to gain information about fundamental and current research referring to communication in service systems and quality measurement of requirements. We plan to finalize the interview questionnaire near-term and start interview sessions with relevant actors soon. Simultaneously we analyze the dataset and identify potential and relevant tasks for pre-processing the data.

Acknowledgments. The author would like to thank Léa Vernisse for her general support regarding the implementation of the assessment tool and her input for the research issues as well as Gerhard Satzger for his continuous support.

References

1. Maglio, P.P., Vargo, S.L., Caswell, N., Spohrer, J.: The service system is the basic abstraction of service science. Inf. Syst. E-Bus. Manage. **7**(4), 395–406 (2009)
2. Tien, J.M., Berg, D.: On services research and education. J. Syst. Sci. Syst. Eng. **15**(3), 257–283 (2006)
3. Lusch, R.F., Vargo, S.L.: The Service-Dominant Logic of Marketing: Dialog, Debate, and Directions. M.E. Sharpe, New York (2006)
4. Spohrer, J., Maglio, P.P., Bailey, J., Gruhl, D.: Steps toward a science of service systems. Comput. (Long. Beach. Calif.) **40**, 71–77 (2007)
5. Böhmann, T., Leimeister, J.M., Möslein, K., Leimeister, J.: Service systems engineering. Bus. Syst. Eng. **6**(2), 73–79 (2014)
6. Gaiardelli, P., Songini, L., Saccani, N.: The automotive industry: heading towards servitization in turbulent times. In: Servitization in Industry (2014). http://doi.org/10.1007/978-3-319-06935-7-4
7. Verstrepen, S., Deschoolmeester, D., Berg, R.J.: Servitization in the automotive sector: creating value and competitive advantage through service after sales. In: Mertins, K., Krause, O., Schallock, B. (eds.) Global Production Management. ITIFIP, vol. 24, pp. 538–545. Springer, Boston (1999). doi:10.1007/978-0-387-35569-6_66
8. Westphal, I., Nehls, J., Wiesner, S., Thoben, K.-D.: Steigerung der Attraktivität von Elektroautomobilen durch neue Produkt-Service-Kombinationen. Industrie Manage. **5**, 19–24 (2013)
9. Broy, M.: Automotive software and systems engineering. In: Proceedings of the Second ACM and IEEE International Conference on Formal Methods and Models for Co-design, pp. 143–149 (2005)
10. Grimm, K.: Software technology in an automotive company: major challenges. In: Proceedings of the 25th International Conference on Software Engineering, pp. 498–503. IEEE Computer Society (2003)
11. Kneuper, R., Sollmann, F.: Normen zum Qualitätsmanagement bei der Softwareentwicklung. Informatik Spektrum. **18**(6), 314–323 (1995)
12. Wilson, W.: Writing effective natural language requirements specifications. In: Software Technology Conference, pp. 1–14 (1997)
13. Alshazly, A., Elfatatry, A., Abougabal, M.: Detecting defects in software requirements specification. Alexandria Eng. J. **53**(3), 513–527 (2014)
14. Juristo, N., Moreno, A.M., Silva, A.: Is the European industry moving toward solving requirements engineering problems? IEEE Softw. **19**(6), 70–77 (2002)
15. Davis, A., Overmyer, S., Jordan, K., Caruso, J., Dandashi, F., Dinh, A.: Identifying and measuring quality in a software requirements specification. In: IEEE Software Metrics Symposium, pp. 141–152 (1993)

16. Lochmann, K., Fernández, D.M., Wagner, S.: A case study on specifying quality requirements using a quality model. In: APSEC, pp. 577–582 (2012)
17. Mund, J.: Quality Assessment of Requirement Specifications Using Metrics–A Research Proposal (2013)
18. Bellotti, M., Mariani, R.: How future automotive functional safety requirements will impact microprocessors design. Microelectron. Reliabil. **50**(9), 1320–1326 (2010)
19. Papadopoulos, Y., Parker, D., Grante, C.: Automating the failure modes and effects analysis of safety critical systems. In: Proceedings of the Eighth IEEE International Symposium on High Assurance Systems Engineering, pp. 310–311 (2004)
20. Weber, M., Weisbrod, J.: Requirements engineering in automotive development-experiences and challenges. In: Proceedings of the IEEE International Conference on Requirements Engineering, pp. 331–340 (2002)
21. Génova, G., Fuentes, J., Llorens, J., Hurtado, O., Moreno, V.: A framework to measure and improve the quality of textual requirements. Requir. Eng. **18**(1), 25–41 (2013)
22. Parra, E., Dimou, C., Llorens, J., Moreno, V., Fraga, A.: A methodology for the classification of quality of requirements using machine learning techniques. Inf. Softw. Technol. **67**, 180–195 (2015)
23. Femmer, H., Méndez Fernández, D., Wagner, S., Eder, S.: Rapid quality assurance with requirements smells. J. Syst. Softw. **123**, 190–213 (2015)
24. Brocke, V., Simons, A., Niehaves, B., Riemer, K., Plattfaut, R., Cleven, A.: Reconstructing the giant: on the importance of rigour in documenting the literature search process. In: Proceedings of the 17th European Conference on Information Systems, Verona, pp. 2206–2217 (2009)
25. Rowley, J., Slack, F.: Conducting a literature review. Manage. Res. News **27**(6), 31–39 (2004)
26. Ebert, C.: Systematisches Requirements Engineering: Anforderungen ermitteln, dokumentieren, analysieren und verwalten. dpunkt.verlag, Heidelberg (2014)
27. Rupp, C., Klaus, P.: Basiswissen Requirements Engineering: Aus-und Weiterbildung nach IREB-Standard zum Certified Professional for Requirements Engineering Foundation Level. dpunkt.verlag, Heidelberg (2015)
28. Webster, J., Watson, R.T.: Analyzing the past to prepare for the future: writing a literature review. MIS Q. **26**(2), xiii–xxiii (2002)
29. IEEE Std. 830-1998: Recommended Practice for Software Requirements Specification. IEEE (1998)
30. Kaindl, H., Lutz, B., Tippold, P.: Methodik der Softwareentwicklung: Vorgehensmodell und State of the art der professionellen Praxis. Springer Fachmedien Wiesbaden GmbH, Wiesbaden (1998)
31. vom Brocke, J., Simons, A., Riemer, K., Niehaves, B., Plattfaut, R., Cleven, A.: Standing on the shoulders of giants: challenges and recommendations of literature search in information systems research. Commun. Assoc. Inf. Syst. **37**(1), 205–224 (2015)
32. Cooper, H.: Organizing knowledge syntheses: a taxonomy of literature reviews. Knowl. Technol. Policy **1**, 104–126 (1988)
33. Walstrom, K.A., Hardgrave, B.C.: Forums for information systems scholars: III. Inf. Manage. **39**(2), 117–124 (2001)
34. Chapman, P., Clinton, J., Kerber, R., Khabaza, T., Reinartz, T., Shearer, C., Wirth, R.: CRISP-DM 1.0 step-by-step data mining guide (2000)
35. Kivinen, J., Warmuth, M.K., Auer, P.: The perceptron algorithm versus winnow: linear versus logarithmic mistake bounds when few input variables are relevant. Artif. Intell. **97**(1–2), 325–343 (1997)

36. Joachims, T.: Text categorization with support vector machines: learning with many relevant features. In: Nédellec, C., Rouveirol, C. (eds.) ECML 1998. LNCS, vol. 1398, pp. 137–142. Springer, Heidelberg (1998). doi:10.1007/BFb0026683
37. Hsu, C., Chang, C., Lin, C.: A practical guide to support vector classification. Department of Computer Science and Information Engineering, National Taiwan University (2003)
38. Saunders, M., Lewis, P., Thornhill, A.: Research Methods for Business Students, 6th edn. Pearson Education, Harlow (2012)
39. King, N.: Using interviews in qualitative research. In: Cassell, C., Symon, G. (eds.) Essential Guide to Qualitative Methods in Organizational Research, pp. 11–22. Sage, London (2004)
40. Dröschel, W., Wiemers, M.: Das V-Modell 97: der Standard für die Entwicklung von IT-Systemen mit Anleitung für den Praxiseinsatz. Walter de Gruyter GmbH & Co. KG (1999)

A GIS-Based Decision Support System for Locating Primary Care Facilities

Melanie Reuter-Oppermann[1(✉)], Daniel Rockemann[1], and Jost Steinhäuser[2]

[1] Karlsruhe Service Research Institute (KSRI),
Karlsruhe Institute of Technology (KIT), Karlsruhe, Germany
melanie.reuter@kit.edu
[2] Institute of Family Medicine, University Medical Center Schleswig-Holstein,
Campus Lübeck, Lübeck, Germany

Abstract. Keeping up a high level of primary care services for a whole country or even a federal state is a very challenging task due to the demographic change and many other reasons, also for Germany. In the future it is expected that even more general practitioners (GP) are necessary to cover close to come healthcare. Therefore, an efficient use of resources and an optimized planning is crucial. Mathematical models and approaches can help facing the challenge by determining optimal locations for practices, shift schedules or appointment strategies, for example. To use these in practice, decision support systems (DSS) are necessary that link the input data to the approaches and display the results. The outline of such a decision support system for optimally locating GP practices is presented in this paper.

Keywords: Primary care services · General practitioners · Decision support tool · Location planning

1 Introduction

Due to the demographic change people in many countries are getting older and are facing more medical conditions. Therefore, the demand for primary care is increasing while the number of general practitioners (GPs) is decreasing. This is currently also the case in Germany. The situation is especially critical in rural areas and social hotspots in cities as these are often less attractive for GPs to live in [1]. From a system point of view, a functioning primary care service network is crucial for medical and financial reasons [2]. Without its continuity of care e.g. more and more patient are then using the emergency services or out-of-hour doctors. By doing so they reduce the capacity for the actual emergency patients and in addition, emergency services result in higher costs for the insurance companies as well as the health care providers, e.g., hospitals [3].

In a survey from 2014 only 8.9% of the medical students were planning on becoming a GP [4]. While different programs try to increase the attractiveness of the GP profession [5], it is also important to use the resources efficiently.

© Springer International Publishing AG 2017
S. Za et al. (Eds.): IESS 2017, LNBIP 279, pp. 210–222, 2017.
DOI: 10.1007/978-3-319-56925-3_17

In Germany, every inhabitant can freely choose their general practitioner (GP) and about 90% did so [6]. The median number of contacts with a GP in Germany is 10 [7]. This makes the planning even more difficult. Therefore, Operations Research methods and approaches can be applied to this situation. Those methods and approaches can be used to optimize the primary care services, for example by determining the best locations for new practices in order to minimize driving times for the patients.

Determining locations for facilities is a well-studied problem in the area of Operations Research. In healthcare, models and approaches have been successfully applied to preventive or acute care location problems, ambulance planning and hospital layout planning, for example. In [8,9] overviews over healthcare facility location problems are given. In another review Afshari and Peng also discusses the challenges for locating healthcare facilities, as well as measures and criteria. [10] These reviews do not address primary care facilities directly, though. Overall, the literature on that topic is still quite scarce.

GIS-based systems have been successfully applied in healthcare, see for example [11,12]. Goli et al. presented a GIS for locating road emergency stations [13], Schuurmann et al. defined hospital catchments [14].

We have developed a GIS-based decision support system for determining (optimal) locations for GP practices in order to increase the service level for the patients. It allows for interaction with decision makers as it displays the chosen locations on a map which can then be easily discussed. In this paper we present the design of the system and explain its use. While our current focus is on Germany, it can be easily applied to other countries.

2 Decision Support Tool

In this section we present the design of a decision support tool capable of gathering data, using this data to calculate locations of primary care medical practices and visualizing the locations to make them easily understandable. By doing so we give an insight into the opportunities arising from an integration of operations research (OR) methods into a geographic information system (GIS).

In case of locating facilities without using a GIS, the mathematical model is the core application. Especially in the health care sector, most models are location-allocation models, trying to locate a subset of a set of given facilities in a way that demand points are supplied most efficiently. Thus, location-allocation models simultaneously locate facilities and allocate demand points to these facilities. Input parameters for these models are usually the population figures and a distance matrix including all the distances from each facility to each demand point. The distances can be captured one by one using Google Maps and saved in an Excel file, for example. With this Excel file and the population figures as inputs, an optimal solution of the location-allocation model aiming for a comprehensive medical supply and a minimum ride time for each patient to their doctor is found by a solver. The results in the form of concrete locations are usually delivered as numbers. Let us consider an example in which the authorities want to locate two new hospitals in New York City. The goal is to supply medical care to as many people as possible. The hospitals may be located in two of the city's five boroughs (*1*st Manhattan, *2*nd Brooklyn, *3*rd Queens, *4*th Bronx, *5*th Staten Island). Taking into account the population figures of each borough, the already existing medical care supply and the mean distances among characteristic points representing the boroughs, the solver may determine *2* and *5* as the optimal solution. This indicates that Brooklyn and Staten Island might be the best places to locate the hospitals according to the model. However, a *2* and a *5* are very unspectacular and inexpressive. Only those experts who developed the model can deal with the solution. Yet, experts often are not the ones who make the final decision. Then, solutions are often illustrated manually in order to make them easily understandable for non-experts and final decision makers. Unfortunately, finding locations can often be an iterative process and therefore a GIS-based decision support tool that automatically displays the solutions can reduce the effort significantly.

The GIS-supported decision support tool includes the following tasks:

– display a mathematical location-allocation model (1),
– gather data for the model (2),
– solve the model (3), and
– visualize the results of the solved model (4).

Two tools have been combined and merged to the decision support tool which is presented in this paper. The first tool is Quantum GIS (QGIS), a GIS used to gather data (2) and visualize the results of the solved model (4). The second tool is IBM's ILOG CPLEX Optimizer, a software that is able to model mathematical issues (1) and solve them with powerful algorithms (3). Both QGIS and the IBM ILOG CPLEX Optimizer are standalone applications. However, for our purposes, the two tools must exchange data, as QGIS supplies data whereas the IBM ILOG CPLEX Optimizer consumes data. The connection of both can be achieved by using a Python program: QGIS as well as IBM ILOG CPLEX offer Python application programming interfaces (APIs) that allow accessing the logic and algorithms of the two tools. Thus, the added Python program (which was named Python Main) first addresses the QGIS Python API to load the data found with QGIS into the program. Second, it passes the data to another added Python program named Python CPLEX, containing the encoded mathematical model, and using the CPLEX Python API to trigger the IBM ILOG CPLEX Optimizer to solve the model. After that, Python Main receives the solution from Python CPLEX and saves it in txt-files. The last step consists in loading the results data of the txt-files into QGIS to visualize the locations and allocations. For this task, a Python QGIS plugin was developed. It addresses the QGIS Python API to manipulate the QGIS map canvas, which is responsible for how data is shown on the map. Figures 1 and 2 show the structure of the decision support tool.

Having presented the interaction of the components of the decision support tool to give the readers an idea about its setup, a more detailed insight into the tool's components can be given. Step by step, every component (QGIS, Python

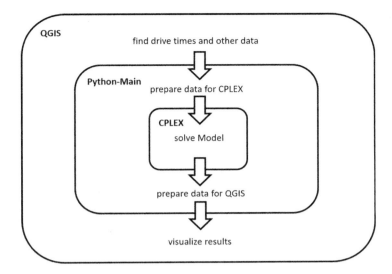

Fig. 1. The three modules QGIS, Python-Main and CPLEX and their tasks. Data is passed from the outside to the inside and back to the outside.

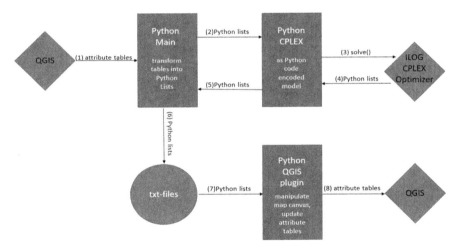

Fig. 2. The Python programs and the data formats they work with.

Main, Python CPLEX with the ILOG CPLEX Optimizer, Python QGIS plugin) is described comprehensively.

It is the model (and therefore the tool's component Python CPLEX, as it contains the encoded model) that is the core of the tool. With its objective function and its constraints, the model defines which data is to be used. A comprehensive overview about common models for locating health care facilities is given by Guneş and Nickel [8] as well as by Daskin and Dean [9].

The tool currently operates a Maximum Covering Location model (MCLP), which was originally designed by Church and ReVelle [15] in 1974 and has since been adapted to different problems. However, due to the fact that the mathematical model was separated from QGIS using IBM ILOG CPLEX as the solver, every mathematical model (or heuristic when omitting the solver) could be used to find locations without harming the operability of the tool. In order to run another model type (than an MCLP) or a heuristic, all that is to be done consists in adjusting Python CPLEX by typing in the new encoded model. The objective of the implemented version of the MCLP is to supply medical care to as many people of the surveyed place as possible by minimizing the ride times of each patient to their primary care practice. So, the model needs data about both the population of the considered region and drive times of the patients to the practices. In addition to that, the model gives a recommendation about how many doctors should work in each practice, insomuch needing data about the current number of doctors working in each practice and the number of patients each doctor can treat.

Finding appropriate data is often more complex than finding a solution to the model. The population data must consider the age composition of the considered region. This is because of the fact that older people have a significantly higher rate of family doctor visits per year than younger people. Thus, districts where the population is of a higher average age need more doctors than districts with the same population figure but a lower average age. Therefore, the population data is a key factor for the allocation of doctors to districts, determining the extent to which the population of a city is supplied with primary medical care.

Data about how many doctors are currently working in the practices can be collected from the practices' websites, for example.

The last important aspect concerning data is determining the ride times between each facility/demand node pair, as the secondary objective of the mathematical model is to minimize the journey of each patient to their family doctor. Ride times can only be determined when the shortest routes from each district to each primary care practice have been calculated. The basis of route calculations within GIS (QGIS, Google Maps, GPRS) is a routable road network. A routable road network is the abstraction of a physical road network into a digitalised network of nodes and edges with information about speed limits, distances, road conditions, the presence of one-way streets and many more restrictions. Calculations upon this routable road network yield shortest roads and drive times. QGIS itself does not provide a routable road network, yet it offers the possibility to access Google Maps. Note that in the current version of the decision support tool we focus on ride times by car and also include walking times, if applicable.

Geo-related observations are always conducted for exactly one place. A *place* is a very relative concept, as it highly varies in its shape: it can be a district, a city, a region, a country or even the whole earth. In our case, the place is the city or a region for which the primary care service network is to be improved. However, a place itself is characterized by its features. *Features* are objects of the surveyed place, for example houses, rivers, streets, mountains or cities. They are represented on maps as geometric symbols like polygons, points or lines. In our case, the features of the place are first the population districts that need primary care supply (expressed in the model as demand nodes), second the existing primary care practices or locations for new ones (facility nodes), and third the routes between the primary care practices and the districts. Features are stored in a *layer*, a collection of like-minded features. The developed tool uses three types of layers (demand nodes, facility nodes and routes). Features are shaped by their *attributes*. A route, for example, has three attributes: the distance of the route, the drive time and the directions. Attributes are saved in *attribute tables*. An attribute table is a common table consisting of rows and columns. Each column represents one attribute. The column values of one row add up to one feature. There exists exactly one attribute table for each layer. Figure 3 shows the attribute table of the routes layer.

Attributtabelle - Teutendorf_D.shp :: Objekte gesamt: 10, gefiltert: 10, g...   —   ⊏

	distance_t	distance_v	duration_t	duration_v	end_addr	start_addr	ID
0	2.7 km	2745.00000	6 mins	366.00000	Auf dem B...	Teutendorf...	0
1	3.9 km	3912.00000	7 mins	419.00000	Dreilingsbe...	Teutendorf...	6
2	3.0 km	3028.00000	5 mins	298.00000	Rã¶dsaal, ...	Teutendorf...	8
3	3.3 km	3279.00000	29 mins	1727.00000	Mecklenbu...	Teutendorf...	3
4	4.9 km	4889.00000	9 mins	545.00000	Bertlingstr...	Teutendorf...	5
5	3.4 km	3386.00000	6 mins	366.00000	Grã¶nland...	Teutendorf...	4
6	4.4 km	4379.00000	8 mins	473.00000	Moorredde...	Teutendorf...	1
7	4.0 km	4046.00000	31 mins	1842.00000	Pã¶tenitze...	Teutendorf...	9
8	3.1 km	3138.00000	6 mins	369.00000	Rã¶nnauer...	Teutendorf...	7
9	4.4 km	4387.00000	8 mins	473.00000	Moorredde...	Teutendorf...	2

Fig. 3. The attribute table of a routes layer. It contains the distances as string (distance_t) and double (distance_v) values, drive times as string (duration_t) and double (duration_v) values as well as the string values of the start and end addresses (start_addr, end_addr) of the routes. Each row represents one route.

The data of the attribute tables is visualized on the QGIS *basemap*, a digital map of the surveyed place.

QGIS itself provides a set of basic functions for visualizing, administrating and acquiring geo-related data. However, from the very beginning on, the QGIS functionality was planned to be extended by plugins written by users. Plugins are small programs that fulfill special tasks within the QGIS framework. For this reason, QGIS has been equipped with Python and C++ APIs. Every user can therefore write a plugin developed to fulfill one or a few special tasks within QGIS. These plugins can then be published in the QGIS plugins repository and can then be used by every other QGIS user. Due to the growing number of QGIS users, also the number and variability of existing plugins is increasing.

As mentioned above, QGIS is able to access Google Maps to do route calculations and to locate addresses. This access is guaranteed by one of these QGIS plugins, a plugin called *GeoSearch*, which is used for the following tasks:

– Locating demand points on the basemap:
  The user can type an address into the GeoSearch UI, triggering the plugin to calculate the location of the address and return it in the form of a point on the QGIS map as well as a string in the layer's attribute table. Thus, every demand point (district) can be located on the QGIS basemap and captured in the demand point layer's attribute table. Adding the age-adjusted population data (e.g. provided by the department of statistics of the considered city or region) as a second column to the attribute table, a for our purposes fully functioning demand point layer is generated.
– Locating the primary care practices:
  GeoSearch is used to locate the addresses of the primary care practices on the basemap and in the attribute table of the facilities layer. The second column of the attribute table captures the separately investigated number of doctors working in each primary care practice. As a result, a fully functioning facilities layer is obtained.
– Calculating routes:
  GeoSearch can not only be used to locate addresses, but also to find routes from a starting point to a target point. Thus, filling in the addresses of a demand point (starting point) and a primary care practice (target point) into the plugin's UI, a route is returned in the form of a line on the basemap (see Fig. 4). On top of that, the attribute table of the layer contains the route as a feature with the attributes distance, ride time and start and end address. This function of GeoSearch is used to calculate the route from each demand point to each practice to complete the routes layers. Figure 3 shows a part of the attribute table of this layer.

All the data that is relevant for the model has so far been stored in QGIS attribute tables. Now the model can be solved. First, as shown in Fig. 2, the attribute tables must be loaded into the decision support tools second component, Python Main. Python Main's task consists in reformatting the data, hence transforming the attribute tables into Python lists, as only this data structure is understood by the solver IBM ILOG CPLEX. For this task, QGIS provides a Python library called qgis.core which is able to transform the C++ data formats of QGIS (like attribute tables) into Python data formats (like Python lists). Using qgis.core, every layer's attribute table can be represented as a Python list. Finally, Python Main invokes the program Python CPLEX, committing the Python lists to the encoded model.

The third component of the tool, Python CPLEX, is a Python program representing the encoded mathematical model. It addresses the Python API of the IBM ILOG CPLEX Optimizer to express all the variables, constraints and the objective function as Python code. The data of the attribute tables, which have been transformed to Python lists by Python Main and committed to Python

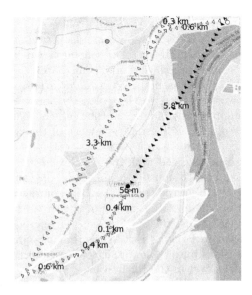

Fig. 4. A route visualized on the QGIS basemap. The yellow and red lines are the output of a request to the GeoSearch plugin. The raw data are saved in attribute tables as like as in one row of the table in Fig. 3 (Color figure online).

CPLEX, serve as the coefficients of the objective function and the constraints. The herby encoded and completed variables, constraints and objective function are then committed to a so called CPLEX entity, the Python object that stores all the information of the model. It is this CPLEX entity that calls the Optimizer's solve-function, leading the Optimizer to solve the model and return the model's results in the form of Python lists. These lists are then passed back to Python Main.

Python Main saves the Python lists containing the results serialized in txt-files.

The last step finally fulfills the visualization goal of the decision support tool. The results of the solved model (containing the locations of the practices to be opened as well as the allocation of districts to the practices) are added to the QGIS attribute tables of the layers. For this purpose, a function which is able to read the content of the txt-files and add it as new columns to the attribute tables is necessary. On top of that, all the practices to be opened (a subset of the features of the facilities layer) as well as all the allocations of the districts to the practices must be highlighted on the basemap. To highlight several features of layers, the QGIS map canvas must be manipulated. The best way to do this is to write a QGIS plugin (see for example [16]). As explained above, plugins are a set of Python programs running in the QGIS framework to fulfill special tasks like manipulating the QGIS map canvas, which determines how the data of the attribute tables is shown on the basemap.

A Python QGIS plugin consists of two Python classes. One class is responsible for the setup of the plugin's UI, a window where several parameters, specifying the data the plugin needs to fulfil its task, can be typed in. The second class represents the plugin's function, thus what it calculates, adds to the attribute tables and shows on the QGIS basemap. With *Plugin Builder*, QGIS offers a tool that builds the framework of the two classes as Python code on its own. This facilitates writing a plugin, as the plugin's author can concentrate on implementing the function of the plugin, while the framework of the two classes is already given.

The plugin's UI can be changed with the *Qt Designer*, a drag-and-drop tool to change Python UIs. The UI must contain input options for specifying the paths to the txt-files containing the results, as their content must be loaded into the plugin. The function-representing class can then read these paths from its UI and save it in strings. Thus, the txt-files can be referenced and its content can be saved deserialized in Python lists. After that, the different QGIS layers are referenced and their attribute tables are extended by new columns filled with the results of the model saved in the Python lists.

The facilities layer's attribute table is extended by a column which can only contain the numbers 0 or 1. Each feature, i.e. primary care practice, which is chosen receives the column value 1, whereas each location that is not chosen receives the column value 0. On the basis of this procedure, the plugin can highlight exclusively the features (primary care practices) that have a column value of 1 on the basemap, as displayed for example in Fig. 5. In addition to that, a second column is amended and filled with the number of doctors that should work in each practice according to the solution of the model.

The routes layer is responsible for representing the allocation of the demand points to the primary care practices. Remember that the routes of every district to every practice have been stored in an attribute table. The routes are visualized

Fig. 5. Visualization of the practices to be opened. White points represent practices to be opened (opening status of 1), whereas grey practices (opening status of 0) are not opened in the solution.

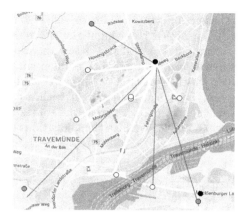

Fig. 6. Visualization of the allocations of one demand point. A line means that the black marked demand point is assigned to the white marked practice located at the end of the line.

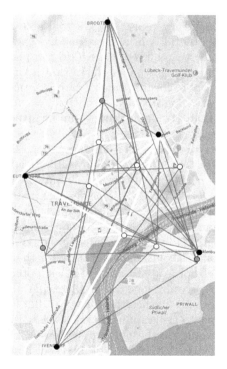

Fig. 7. Visualization of the allocations of all the demand points. A line means that the black marked demand point at the beginning of the line is assigned to the white marked practice located at the end of the line.

as straight lines on the basemap. Now the idea to visualize the allocation of one district to a primary care practice is to highlight the route between the two points (which is only a straight line) on the basemap. Thus, the route layer's attribute table is extended by one column that is filled with the model's values about the proportion to which a district is supplied with medical care by a primary care practice. Only the features (routes) with a column value greater than 0 are highlighted on the basemap, whereas the routes with a column value of 0 are not shown. Figures 6 and 7 visualize the allocations of one and of several demand points.

3   Conclusions and Outlook

The potentials of handling more data than ever in minimum time, brought by the rapid advance of information technology, have led to a rethinking of academic approaches in every data-driven science. Locational analysis, a scientific problem characterized by the participation of many different stakeholder (e.g. decision makers, researchers, interest groups), involves a great risk of conflicts arising by the urge to balance contrary goals. Thus, a tool able to communicate valid data as well as possible solutions in a user-friendly way in order to visualize dependencies and outcomes is required. Here, GIS can play an important role.

The approach presented in this paper shows that a GIS can be transformed into a decision support tool to locate primary care practices. It helps to acquire and visualize spatial data, input as well as output, when locating facilities for example. As a result of that, even non-experts can be given an understanding of how a solution has originated.

The decision support tool presented in this paper does not use locational algorithms included in the GIS (although many GIS already provide approaches to solve distinct location problems on their own), but solves an external location model by using IBM ILOG CPLEX. This brings maximum flexibility regarding the problem formulations (models), as basically all linear objective functions, constraints and parameters, independent of the locational abilities of the underlying GIS, can be used. In addition, optimal solutions obtained by a solver are usually preferable (as long as solution times allow for that).

The flexibility of the tool leads to a lot of other potential applications: the tool could be adapted to solve also location problems in totally different domains. Thus, it is suitable for not only the health care sector, but for every sector in need of mathematical solutions to geographically related issues. The sole component in need of adjustment would be the model (Python CPLEX). Except for this, the procedure of finding data (QGIS can still be used, it simply must find other data than drive times), processing data (Python Main) and visualizing data (QGIS plugin) remains the same - providing a viable foundation for solving many location problems.

As a next step, the decision support tool shall be tested in case studies with different German regions. For that, additional location models will also be implemented. While in the beginning it will only be used internally by researchers to determine and show possible locations, it is of interest to further develop the tool to make it accessible and practicable for many different users. In order to reach this goal, a user-friendly interface and an assistance component must be developed. In future work, additional components will be developed to build a generalized decision support system for the organization of primary care services, for example determining the staff scheduling based on the opening hours and the offered services.

References

1. Stock, C., Szecsenyi, J., Riedinger-Riebl, U., Steinhuser, J.: Projection of general practitioner care demand at the community level. Gesundheitswesen (Bundesverband der Arzte des Offentlichen Gesundheitsdienstes (Germany)), **77**(12), 939–946 (2015)
2. Starfield, B., Shi, L., Macinko, J.: Contribution of primary care to health systems and health. Milbank Q. **83**(3), 457–502 (2005)
3. Breslin, J.M., MacRae, S.K., Bell, J., Singer, P.A.: Top 10 health care ethics challenges facing the public: views of Toronto bioethicists. BMC Med. Ethics **6**(1), E5 (2005)
4. Kassenärztliche Bundesvereinigung: Berufsmonitoring Medizinstudenten 2014. http://www.kbv.de/media/sp/2015_04_08_Berufsmonitoring_2014_web.pdf
5. Flum, E., Magez, J., Aluttis, F., Hoffmann, M., Joos, S., Ledig, T., Oeljeklaus, L., Simon, M., Szecsenyi, J., Steinhuser, J.: Verbundweiterbildung (plus) Baden-Wrttemberg: Development of educational meetings and implications for the implementation of family medicine training programmes in Germany. Zeitschrift fur Evidenz, Fortbildung und Qualitat im Gesundheitswesen 112 (2016)
6. Scherer, M., Abholz, H.H., Chenot, J.F., Gerlach, F.M., Kochen, M.M.: Allgemeinmedizin und Familienmedizin. In: Lehrbuch der Versorgungsforschung, pp. 337–341. Schattauer Verlag, Stuttgart (2011)
7. Riens, B., Erhard, M., Mangiapane, S.: Arztkontakte im Jahr 2007. Zentralinstitut fr die kassenrztliche Versorgung in Deutschland (Zi). Versorgungsatlas-Bericht Nr. 12/02, Berlin (2012)
8. Günes, E.D., Nickel, S.: Location problems in health care. In: Laporte, G., Nickel, S., Saldanha da Gama, F. (eds.) Location Science, pp. 555–579. Springer, Cham (2015)
9. Daskin, M.S., Dean, L.K.: Location of health care facilites. In: Operations Research and Health Care: A Handbook of Methods and Applications, pp. 43–76. Kluwer, New York (2004)
10. Afshari, H., Peng, Q.: Challenges and solutions for location of healthcare facilities. Ind. Eng. Manag. **3**, 127 (2014)
11. Forgionne, G.A., Gangopadhyay, A., Adya, M., Tan, J.K.: BCAS: A Web-enabled and GIS-based Decision Support System for the Diagnosis and Treatment of Breast Cancer. e-Publications@Marquette (2001)
12. Patel, A., Waters, N.: Using geographic information systems for health research. In: Alam, B.M. (ed.) Application of Geographic Information Systems. InTech (2012)
13. Goli, A., Ansarizade, N., Barati, O., Kavosi, Z.: Location of road emergency stations in fars province, using spatial multi-criteria decision making. Bull. Emerg. Trauma **3**(1), 8–15 (2015)

14. Schuurman, N., Fiedler, R.S., Grzybowski, S.C.W., Grund, D.: Defining rational hospital catchments for non-urban areas based on travel-time. Int. J. Health Geograph. **5**(1), 43 (2006)
15. Church, R., ReVelle, C.: The maximum covering location problem. Papers Reg. Sci. Assoc. **32**(1), 101–118 (1974)
16. Sherman, G.: The PyQGIS Programmer's Guide - Extending QGIS 2.x with Python. Locate Press LLC, Chugiak (2014)

Combining Data Analytics with Layout Improvement Heuristics to Improve Libraries' Service Quality

Diogo V. Silva and Vera L. Miguéis[✉]

Faculdade de Engenharia da Universidade do Porto, Porto, Portugal
diogosilvatai@gmail.com, vera.migueis@fe.up.pt

Abstract. Currently, many libraries, either academic or public, possess information systems to support their operations. Although libraries are becoming more aware of the potential of data analytics in supporting library management decisions, there is still a long way to go to take plenty advantage of the information collected. This paper proposes a prescriptive analytics solution to enhance the service provided by libraries, by optimizing libraries layout. The quantitative method introduced aims to identify layout configurations that minimize the time spent by clients in picking books from the library. A new multi-floor layout optimization algorithm is developed, based on the pairwise exchange method heuristic. A real data sample of approximately 66.000 loans, taken from the information system of a European Engineering School's library, was analyzed and processed. The method proposed was used to improve the library's current departments configuration, achieving an improvement of 13.2% in terms of walking distance to collect the books. The results corroborate the effectiveness of the method proposed and its potential in supporting library management decisions.

Keywords: Service enhancement · Multi floor layout · Combinatorial optimization · Heuristics · Data analytics · Library management

1 Introduction

Nowadays, in order to catalogue items and keep track of material circulation and acquisitions, more and more academic and public libraries possess their own library management system (LMS) [1]. However, many libraries are still not making adequate use of the information gathered by their LMS about users' habits and preferences [2]. At the moment, the use of library management systems is mainly focused on achieving better control of the library's processes and not in exploring the collected data for operational efficiency improvement and service improvement based on prescriptive analytics approaches [1]. Although information systems significantly increase the value of information as suggested by Hayes [3], when used poorly, the impact on the quality of the service will not be significant. The declining usage of physical collections paired with the

© Springer International Publishing AG 2017
S. Za et al. (Eds.): IESS 2017, LNBIP 279, pp. 223–234, 2017.
DOI: 10.1007/978-3-319-56925-3_18

increasing demand for electronic networked resources [4] suggests that physical libraries should look for ways to improve their services so they can remain active and competitive. In order to become more efficient in an increasingly digital era, it is important that libraries consider the potential of data analytics and big data in supporting library management decisions.

This paper seeks to make use of the information stored in a LMS by applying an improvement-type layout algorithm to the data collected by it, in order to reduce the time spent by the library's clients in picking their books, thus reducing the traffic flow and improving client satisfaction. The method proposed is based on the traditional pairwise exchange method, usually applied in manufacturing facility layout analysis, and is evaluated using a distance-based objective. In this case, the library's departments correspond to the departments or areas considered in the original application of the pairwise exchange method and the capacities for departments distribution in each floor correspond to the shelves capacity. The method proposed enhances the traditional pairwise exchange method by including capacity restrictions for each floor in order to prevent unfeasible exchanges. Also, "dummy" departments are included in the configurations tested in order to reach solutions that allow for different number of departments in each floor.

The structure of the remainder of the paper is as follows. Section 2 includes a general revision on the use of data analytics in library management and explores some of the different heuristic-based layout optimization techniques applied to facility layout planning problems. Section 3 introduces the approach followed in this project, explaining how the library management system's information was analyzed and processed and how the heuristic was structured in order to adequately carry out a multi-floor optimization with departments or subjects and floors of unequal areas. Section 4 describes the application of the proposed method on a specific case study related to an academic library and presents the results obtained. The paper ends with the conclusions reached and some suggestions for future research.

2 Literature Review

2.1 Data Analytics in Library Management

Nowadays, the evolution of technology and the increasing importance of data analytics is starting to impact more and more the way we manage our information and how we make use of it [5]. This growing tendency has also began to shape the way libraries are managed and what the future holds regarding library management practices. For example, in Singapore, the National Library Board (NLB) was awarded the Best Practice Award in Resource Management at the 2014 Public Services Awards for applying data analytics in order to improve their understanding on borrowing behaviour and improve resource management [6]. Also, other works, such as Lakos [7] and Showers [8], provide a new perspective to the way that data analytics can be used to support decision-making in libraries and how data analytics and metrics can be exploited to help make better decisions, improve services and increase user satisfaction. However, even though libraries throughout the world are starting to invest more in data analytics as a way to identify patterns and trends in customer behaviour [9], the path towards integration between library management and data analytics is not yet fully understood, which represents a big opportunity for research [10]. In this paper, we bring a new prescriptive analytics approach to library management by applying a layout optimization technique that facilitates the loaning process to both the client and the library.

2.2 Heuristic-Based Layout Optimization Techniques

According to Xie and Sahinidis [11], a good facility layout can reduce up to 50% the total operating costs, explaining in a way why layout optimization techniques have been a subject of extensive study in academic research. Determining the optimal placement of facilities within a plant is defined as a facility layout problem (FLP). The facility layout problem is commonly formulated as a generalization of the quadratic assignment problem (QAP), which falls under the NP-hard (non-deterministic polynomial-time hard) category [12]. The QAP is a combinatorial optimization problem that for a set of N facilities and N locations, uses the distance between each pair of locations, the flow between each pair of facilities and the cost of flow per unit distance, to provide an assignment of all facilities to different locations with the goal of minimizing total cost [13]. In this case, total cost is a function of the distance and the flow between facilities. This approach to the FLP's formulation is the most recurrent in literature, in which the objective function to be minimized follows quantitative criteria, such as total distance, cost or time spent. However, layout optimization approaches that allow for both quantitative and qualitative criteria are also popular, resulting in solutions that can better meet the plant's requirements [14]. An approach that mixes multiple optimization objectives, either quantitative or qualitative factors (the second being weighted, for example, by a subjective closeness rating between facilities) falls into the category of the Multi Objective Facility Layout (MOFL) problems [15]. Due to the fact that there is no known efficient way to reach an optimal solution for FLP problems, several heuristics and meta-heuristics have been developed to seek near optimal solutions at reasonable computational time [16]. The pairwise exchange method, the CRAFT algorithm and the CORELAP are examples of heuristical approaches to the facility layout problem [17].

   The pairwise exchange method is an improvement-type layout algorithm that can have both an adjacency-based or distance-based objective function. This heuristic requires the initial layout, the distance between locations and the flow between departments as its main inputs. In each iteration, the method tests all feasible department pair exchanges and calculates, for each new configuration, the value of the associated objective function. If the objective is to minimize the total distance or the material handling costs then, from the resulting list of objective function values, the heuristic selects the configuration that mostly reduces the value of the objective function ("steepest descent" method) as the

new layout configuration [18]. This process is repeated until no tested configuration can decrease the value of the objective function in comparison to the
current one. A significant limitation of the pairwise exchange method is that the
final solution is conditioned by the given initial layout configuration, meaning
that the final configuration may be a local optimum of the problem, with no
insight on its closeness to the global optimum [18].

In this paper, we adapt the traditional pairwise exchange method in order
to account for departments and floors of unequal areas (inputs similar to
CRAFT's), as well as a multi-floor layout (by using "dummy" departments that
are used to "reserve" space in each floor). This heuristic was chosen due to its
formulation's simplicity, allowing us to adapt it adequately to the case study
presented.

3   Method

The typical department layout of a library is not usually designed taking into
account quantitative data regarding loan records throughout time. However,
client's demand for each subject within a library is, arguably, one of the best
criterion in deciding how close to place each subject to the entrance, in order
to reduce the distance walked by clients. The method developed to address this
optimization problem is an improvement type algorithm that rearranges the
subjects or departments within a library, considering the number of occurrences
of each path containing either solo or paired departments. For example, the
number of occurrences of the path "Entrance → Subject X → Exit" or path
"Entrance → Subject X → Subject Y → Exit". In order to calculate the
frequency of each path in book picking, the information of the loan records
stored in a library's LMS is processed and analyzed, thus providing the flow
matrix required for the algorithm.

The remainder of this section explains the steps needed to replicate the algorithm, process the loan record information and analyze the heuristic's results (see
Fig. 1). These steps are described in the following sections. Section 3.1 describes
how a multi-floor library with floors of unequal capacities and subjects with different capacity requirements can be represented and improved using a solution
based on the traditional pairwise exchange method. Section 3.2 explains how the
collected data was transformed into a flow matrix compatible with the heuristic's
formulation.

3.1   Layout Optimization Heuristic

Base Formulation. In libraries context, in order to provide better service
quality, the total distance travelled by clients should be minimized, meaning that
the total number of paths travelled when picking books from one, two or more
departments multiplied by the distance that is associated to each one of those
paths should be diminished. More specifically, library managers are interested
in guaranteeing that the total distance associated with going from the library's

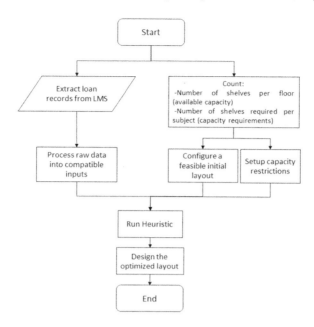

Fig. 1. Holistic representation of the main steps of the proposed method.

entrance to the consecutive departments' locations and back to the entrance is minimized. The information related to the total number of paths travelled for every combination of subjects required by a client is available in the loan records stored in a library's management system. The data associated with the distance between department locations can be obtained by on-site measurements.

For example, considering a simple four-floor library where the three top floors contain a single department of the same size (Fig. 2), and with the following sample of three loan records (Fig. 3), we can derive the distance (d_{ij}) and flow (f_{ij}) matrices between departments i and j (Figs. 2 and 4, respectively).

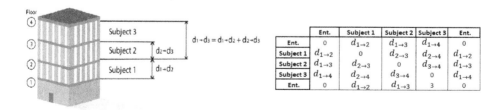

	Ent.	Subject 1	Subject 2	Subject 3	Ent.
Ent.	0	$d_{1\to2}$	$d_{1\to3}$	$d_{1\to4}$	0
Subject 1	$d_{1\to2}$	0	$d_{2\to3}$	$d_{2\to4}$	$d_{1\to2}$
Subject 2	$d_{1\to3}$	$d_{2\to3}$	0	$d_{3\to4}$	$d_{1\to3}$
Subject 3	$d_{1\to4}$	$d_{2\to4}$	$d_{3\to4}$	0	$d_{1\to4}$
Ent.	0	$d_{1\to2}$	$d_{1\to3}$	3	0

Fig. 2. Representation of the example's library and the correspondent distance matrix.

Fig. 3. Transformation of a sample of loan records into book picking paths.

	Ent.	Subject 1	Subject 2	Subject 3	Ent.
Ent.	-	1	2	0	-
Subject 1	-	-	0	1	1
Subject 2	-	1	-	0	1
Subject 3	-	0	0	-	1
Ent.	-	-	-	-	-

Fig. 4. Flow matrix containing the book picking flows for the given sample of loans.

Let the objective function (z) be defined as:

$$z = \sum_i \sum_j f_{ij} d_{ij} \tag{1}$$

Function z represents the total "cost" of the current library layout. According to Eq. 1, the value of this function is equal to the total distance that needs to be travelled in order to replicate all loans accounted in the flow matrix f_{ij}. In this illustrative case, the total "cost" of the library's layout would be equal to 14, considering that the distance between consecutive floors is the same and unitary. This formulation laid the basis for the translation of book picking activities into matrices and for the approach to the multi-floor layout optimization problem. As mentioned before, it was then required to improve it by considering departments and floors of unequal sizes and capacities, respectively. Also, a way to have a different number of departments in each floor had to be developed in order to avoid unnecessarily constrained solutions.

Main Assumptions. The main assumption of this formulation is that the distance d_{ij} between the library departments i and j corresponds to the distance travelled in the vertical direction. This implies that the horizontal distance travelled within each floor when picking books is not considered, resulting in a focus on the number of floors climbed to reach the department locations.

Furthermore, we assume that most of the loans only encompass one or two different subjects. This assumption is particularly relevant because for loans of one or two subjects, the order that the subject locations are visited does not affect the total travelled distance. For example, if a client is going to pick up books of Subject X and Subject Y, this client can either follow one of the two paths: Entrance $\rightarrow$ Subject X $\rightarrow$ Subject Y $\rightarrow$ Entrance or Entrance $\rightarrow$ Subject Y $\rightarrow$ Subject X $\rightarrow$ Entrance. In either case, even if the subjects are on different floors, the total distance travelled is the same. However, if we have a loan record containing 3 subjects, the total distance travelled can differ depending on the path selected and the layout currently employed. In terms of the problem's formulation, this would imply that the flow matrix would have to be recomputed in every iteration of the algorithm, because the layout changes in every iteration. The exclusion of all loans containing more than two subjects is, however, not very compromising due to the fact that usually in a library those loans account for a significantly small percentage of all the total loans. This assumption will be verified in Sect. 4 to validate the applicability of the model proposed.

Another assumption is related to the future behaviour of the library's clients. By adjusting the distribution of the library subjects according to the past loans registered, we are assuming that future loans pattern will be similar. This assumption is not particularly compromising in the case of university libraries, as most clients represent students who borrow books to assist their study in courses that are taught every year.

Final Formulation. Although the initial formulation provides a simple approach to the library's layout optimization problem, its application is limited to basic cases like the example provided. In order to account for subjects and floors of unequal sizes and capacities respectively, while at the same time enable to reach solutions that allocate a different number of subjects to each floor, some additional formulation changes had to be made.

Firstly, for any given iteration, not all pairwise subject exchanges are feasible because not all floors have the same available capacity (number of shelves available) and not all subjects have the same capacity requirements (number of shelves needed). Figures 5 and 6 illustrate these two inputs for the algorithm.

Floor	Capacity
1	C_1
2	C_2
3	C_3
⋮	⋮
N	C_N

Subject	Capacity Req.
A	R_1
B	R_2
C	R_3
⋮	⋮
M	R_M

Fig. 5. Floor-Capacity.   **Fig. 6.** Subject-Capacity.

In each iteration, the algorithm checks if the current pairwise subject exchange under consideration is feasible or not using the inputs shown above. If it is, the corresponding distance matrix is computed, as well as the layout cost. If it is not, then the algorithm will ignore this exchange and will evaluate the next one.

Secondly, the base formulation cannot account for floors with a different number of subjects. A direct consequence of this constraint is that the number of configurations tested will be very small in comparison to the real set of feasible solutions. To overcome this limitation, "dummy" subjects are assigned to every floor in order to reserve position slots for incoming subjects during the pairwise subject exchanges. The "dummy" subjects have null capacity requirements and null flows coming in and leaving out of them because they do not really exist. The most common application of "dummy" departments in Facility Layout Planning problems aims to eliminate excess facility space [19] or reach desired layout shapes [20].

By having their inbound and outbound flows set to zero, "dummy" subjects are not considered preferable to occupy a location when compared against real subjects, as the goal of the algorithm is to minimize the facility layout total cost. This fact promotes the switch of the "dummy" subjects when a better solution is found and enables to configure multi-floor layouts with different number of subjects in each floor. Figure 7 illustrates an example of a flow matrix with "dummy" subjects. The D tags correspond to "dummy" subjects and the S tags correspond to the real subjects.

		Ent.	S_1	S_2	D_1	S_3	S_4	S_5	S_6	D_2	D_3	Ent.
			Floor 1			**Floor 2**			**Floor 3**			
	Ent.	-	$f_{1\to2}$	$f_{1\to3}$	0	$f_{1\to5}$	$f_{1\to6}$	$f_{1\to7}$	$f_{1\to8}$	0	0	-
Floor 1	S_1	-	-	$f_{2\to3}$	0	$f_{2\to5}$	$f_{2\to6}$	$f_{2\to7}$	$f_{2\to8}$	0	0	-
	S_2	-	$f_{3\to2}$	-	0	$f_{3\to5}$	$f_{3\to6}$	$f_{3\to7}$	$f_{3\to8}$	0	0	-
	D_1	-	0	0	0	0	0	0	0	0	0	-
	S_3	-	$f_{5\to2}$	-	0	-	$f_{5\to6}$	$f_{5\to7}$	$f_{5\to8}$	0	0	-
Floor 2	S_4	-	$f_{6\to2}$	$f_{6\to3}$	0	$f_{6\to5}$	-	$f_{6\to7}$	$f_{6\to8}$	0	0	-
	S_5	-	$f_{7\to2}$	$f_{7\to3}$	0	$f_{7\to5}$	$f_{7\to6}$	-	$f_{7\to8}$	0	0	-
Floor 3	S_6	-	$f_{8\to2}$	$f_{8\to3}$	0	$f_{8\to5}$	$f_{8\to6}$	$f_{8\to7}$	-	0	0	-
	D_2	-	0	0	0	0	0	0	0	0	0	-
	D_3	-	0	0	0	0	0	0	0	0	0	-
	Ent.	-	-	-	-	-	-	-	-	-	-	-

Fig. 7. Flow matrix for a multi-floor layout, using "dummy" subjects.

3.2 Data Treatment

In order to construct the flow matrix based on the loan records of a library management system, it is required to process the raw data, taking into account the formulation set described in the previous sub-section. A typical loan record comes with the specification of multiple attributes, such as: client's id, loan date, return date, book's id and number of overdue notices. The construction of the flow matrix only requires the loan information regarding the client's and book's id as well as the date of the loan.

In libraries, a common way to identify a book is by using the universal decimal classification (UDC). This classification system supports complex content indexing and facilitates information search by providing an organized arrangement of all departments or areas of knowledge [21]. The specific areas of knowledge are described by a number code (see Fig. 8), where the first digit corresponds to the main knowledge class code associated to it.

Book's identification is usually specified by combining the area of knowledge UDC code, with an additional sub-code.

5	**Mathematics. Natural Sciences**
...	...
502	The environment and its protection
504	Threats to the environment
51	Mathematics
511	Number Theory
512	Algebra
...	...

Fig. 8. Example of a UDC table.

The first step in data treatment is to extract the area of knowledge associated to the loan, in order to have a simplified list of all subjects that are required during the period of analysis. Then, it is required to determine the number of loans containing one or two subjects, as explained in Sect. 3.1.

Figure 9 illustrates the expected results of this stage.

After selecting the loans referring to one or two subjects, it is necessary to group that data using the flow matrix formulation described in Sect. 3.1. This implies the definition of the paths taken by the clients, starting in the entrance, passing through the respective subjects' positions and going back to the library's entrance/exit. This results in a matrix similar to the one represented

Loan_ID	Client	Date	Subject
1	Client X	14/10/2012	Subject 1
2	Client X	14/10/2012	Subject 3
3	Client Y	14/10/2012	Subject 2
4	Client Z	15/10/2012	Subject 2
...	...	...	...

Subject 1	1
Subject 2	2
Subject 3	0
Subject 4	0
Subject 5	0
Subject 6	0
Subject 7	0
...	...
Subject N	0
	3

	Subject 1	Subject 2	Subject 3	...	Subject N
Subject 1	0	0	1	...	0
Subject 2	0	0	0	...	0
Subject 3	0	0	0	...	0
...	...	...	...	...	...
Subject N	0	0	0	...	0

Fig. 9. Representation of the results expected from the initial "data crunching" process.

in Fig. 7. The "dummy" subjects are added afterwards, according to the need for additional positions slots.

4   Case Study

In this section we present the library used as a case study and the heuristic's results obtained for this case study.

4.1   Library

The library used as case study is a European engineering faculty's library that uses a library management system to track its book circulation and users' activity. The library studied has six floors, having its subjects distributed from the first to the fourth floor. The loan records for the last five years of the library's activity were made available by the library's management team.

In order to apply the method proposed, some inputs had to be specified. The value defined for the number of available position slots in each floor was 11. This value was defined taking into consideration the average occupation of a subject and the average floor capacity. More position slots could be added to each floor, though that would only unnecessarily increase the heuristic's run time and the input matrices' size. The number of floors was 4, as explained above. The flow and distance matrices' size was 46 (result of the product between the number of floors and the number of position slots per floor plus the library's entrance and exit slots).

Additionally, the initial assumption of excluding all loan records containing more than two subjects was tested. In the case of the library used as case study, these loans accounted for only 0.83% of all the loans made, thus validating the assumption.

4.2   Results

Starting from the current layout configuration of the library, the results obtained using the method proposed to enhance this configuration showed that it would have been possible to reduce 12.1% of the total distance travelled by a client, when considering all the past loan records made available. This represents a significant impact on the service quality perceived by clients and on the image of the library.

In order to look for alternative solutions, an initially optimized department layout configuration was created, based on the demand for each library department. This time, instead of using the current library department configuration as the heuristic's starting point, this new configuration was used. In comparison to the current layout's total distance travelled value, the final results showed a reduction of 13.2% in the total distance travelled by a client when simulating the past loans analyzed, meaning that a client would save, on average, 13.2% of time while picking the books.

The total number of exchanges in each run of the algorithm, corresponding to different initial configurations, was less than 10 and the utilization of the lowest floors increased in both solutions, as expected. Table 1 summarizes the results obtained, with the distance measured in terms of number of floors climbed.

Table 1. Summary of the results obtained.

	Total distance (start)	Total distance (end)	Reduction (%)
Current layout	233930	205624	12.1
Initially optimized layout	206819	203119	13.2

These results reveal the potential of the method proposed and reinforce that the combination of data analytics with optimization procedures may constitute a powerful approach to support decision makers in the process of services improvement.

5 Conclusions and Issues for Future Research

The purpose of this study is to develop a tool that may support libraries in the enhancement of the service provided, by adjusting the library's layout and consequently facilitating the process of picking the books required by the clients. This paper contributes to the literature by proposing a quantitative approach to library management that explores the potential of combining data analytics with layout improvement heuristics. The method described in this paper, although more directed to multi-floor buildings, can easily be transposed to any library setting, thus making it a valuable tool for any library manager wishing to improve library's service. Being a quantitative-based approach, the savings resulting from its implementation are more tangible and objective. The results obtained corroborate the effectiveness of the proposed approach, since significant picking time reductions were achieved.

This study is only limited by the main assumptions made regarding: the insignificance of the loan records containing more than two subjects, and the analysis of vertical distances instead of "intra-floor" distances during the optimization process. For the case study presented, the first assumption was verified, as expected, because in libraries the total number of loans containing books from more than two areas of knowledge represents a small portion of the total number of loans made.

Taking this study as a basis, much more can be explored regarding the mixture between data analytics and optimization techniques. The approach presented in this paper could be further developed by implementing, for example, a sensitivity analysis to the floors' capacity restrictions in order to account for situations where adding a small number of book shelves to a floor would greatly expand the set of feasible solutions, thus enabling the specification of better layout configurations.

References

1. Kochtanek, T., Matthews, J.: Library information systems: from library automation to distributed information access solutions. Libr. Inf. Sci. **3**(4), 690–692 (2003)
2. Sero Consulting Ltd., KCCL Glenaffric Ltd.: Library Management Systems Study: An Evaluation and Horizon Scan of the Current Library Management Systems and Related Systems Landscape for UK Higher Education. Sero Consulting (2008)
3. Hayes, R.: Economics of Information. Routledge, London, New York (1997)
4. Martell, C.: The absent user: physical use of academic library collections and services continues to decline 1995–2006. J. Acad. Librarianship **34**(5), 400–407 (2008)
5. EY Group: Big Data: changing the way businesses compete and operate. Tech. report (2014)
6. Ong, G.: A smarter library: using data analytics to improve resource management and services at NLB. Civil Service College (2014)
7. Lakos, A.: Evidence-based library management: the leadership challenge. Portal Libr. Acad. **7**(4), 431–450 (2007)
8. Showers, B.: Library Analytics and Metrics: Using Data to Drive Decisions and Services. Facet Publishing, London (2015)
9. Vizard, M.: Libraries pool big data analytics investments (2016)
10. Emerald Group Publishing Limited: Call for Papers: Library Management and Innovation in the Big Data Era (2016)
11. Xie, W., Sahinidis, N.V.: A branch-and-bound algorithm for the continuous facility layout problem. Comput. Chem. Eng. **32**(4), 1016–1028 (2008)
12. Sahni, S., Gonzalez, T.: P-complete approximation problems. J. Assoc. Comput. Mach. **23**(3), 555–565 (1976)
13. Koopmans, T.C., Beckmann, M.J.: Assignment problems and the location of economic activities. Econometrica **25**(1), 53–76 (1957)
14. Ramtin, F., Abolhasanpour, M., Hojabri, H., Hemmati, A., Jaafari, A.: Optimal multi floor facility layout. In: Proceedings of the International MultiConference of Engineers and Computer Scientists (2010)
15. Chen, C.W.: A design approach to the multi-objective facility layout problem. Int. J. Prod. Res. **37**(5), 1175–1196 (1999)
16. Drira, A., Pierreval, H., Hajri-Gabouj, S.: Facility layout problems: a survey. Annu. Rev. Control **31**(2), 255–267 (2007)
17. Singh, S.P., Sharma, R.R.K.: A review of different approaches to the facility layout problems. Int. J. Adv. Manuf. Technol. **30**(5), 425–433 (2006)
18. Tompkins, J., White, J., Bozer, Y., Tanchoco, J.: Facilities Planning. Wiley, Hoboken (2010)
19. Sarker, R., Mohammadian, M., Yao, X.: Evolutionary optimization. In: International Series in Operations Research and Management Science, vol. 48 (2006)
20. Simkins, T.: On the application of mixed integer programming to the facility layout problem: a case study. Lehigh University Lehigh Preserve (2011)
21. McIlwaine, I.C.: The universal decimal classification: some factors concerning its origins, development, and influence. J. Am. Soc. Inf. Sci. **48**(4), 331–339 (1997)

Service Organizations Case Studies and Practices

A Return on Our Experience of Modeling a Service-Oriented Organization in a Service Cartography

Gorica Tapandjieva[1]([✉]), Giorgio Anastopoulos[2], Georgios Piskas[1],
and Alain Wegmann[1]

[1] Systemic Modeling Laboratory LAMS, École Polytechnique Fédérale de Lausanne,
Station 14, 1015 Lausanne, Switzerland
{gorica.tapandjieva,georgios.piskas,alain.wegmann}@epfl.ch
[2] Information Systems SI, École Polytechnique Fédérale de Lausanne,
Station 14, 1015 Lausanne, Switzerland
giorgio.anastopoulos@epfl.ch
http://lams.epfl.ch/, http://vpsi.epfl.ch

Abstract. We present a longitudinal project using action design research, which is a four-year collaboration between two EPFL enti-ties: The research Laboratory for Systemic Modeling (LAMS) and EPFL's IT department, called the VPSI. During that time the VPSI was going through a transformation into a service-oriented organization. The research project began as an open-ended modeling of some of the VPSI processes. It slowly matured into the design and development of a visualization tool we call *service cartography*. During this research, we learned that, to successfully apply service-orientation, focusing purely on IT architecture and end-customer value is not enough. Attention must be given to the exchange of internal services between the service organi-zation members and their alignment with the services expected by the external stakeholders. In this paper we present the evolution of (1) our understanding of what services are, and (2) our conceptualization of how the service cartography facilitates the service-oriented thinking.

Keywords: Action design research · Service-orientation · Service car-tography · SEAM · Enterprise architecture

1 Introduction

The IT department of EPFL, called the VPSI for Vice Presidency of Informa-tion Systems (SI in French), provides IT infrastructure and development services to the entire EPFL community. Beginning around 2012, the IT department of EPFL began to transform from a traditional IT organization, developing appli-cations and maintaining infrastructure, into a so-called service organization as envisioned by frameworks such as the Information Technology Infrastructure Library (ITIL) [1,2]. An EPFL research laboratory, called LAMS, collaborated

© Springer International Publishing AG 2017
S. Za et al. (Eds.): IESS 2017, LNBIP 279, pp. 237–250, 2017.
DOI: 10.1007/978-3-319-56925-3_19

with the VPSI during the transformation project. LAMS specializes in Enterprise Architecture research. LAMS provided methodological advice to the VPSI members while improving its methods and publishing the results [3–7]. The first author began her involvement in this project by attempting to model business processes that support specific services offered by the VPSI. Gradually this involvement shifted to the mapping of services and their dependencies in an interactive tool that is called a *service cartography*.

Collaborating with the VPSI members, the LAMS researchers became aware that there is lack of clear guidelines: (1) to manage the internal service exchange and (2) to align the internal services with the VPSI's external stakeholders' expectations. The existing service management frameworks, such as ITIL [1], convey a clear vision of services, but they only give abstract guidelines concerning the implementation of this vision. For example, ITIL defines peoples' roles with different responsibilities in service management, but it does not provide information on how these roles collaborate, i.e., exchange services. Accordingly, the researchers learned that to apply service-orientation, focusing purely on IT architecture and end-customers is not enough.

The concept of a service is not new [8]. In the past two decades, services have been researched in the domains of service-oriented thinking [9], service-dominant logic [10], service systems [11], servitization [12], service-oriented architecture [13], and others. Most research on services "focuses on the interaction between the firm and the customer" [14], but an open question still remains: *How do members of a service organization collaborate in the implementation of a certain service?*

We believe that successful service-orientation is characterized by empowering employees to collaborate by exchanging services and aligning the results of this collaboration with the expectations of external stakeholders. Employees' work includes responsibilities in a given context, as well as services they use and they provide to other employees, systems, applications, organizations and end-users. Our ongoing efforts are towards designing and building a service cartography tool. It is envisioned that in this tool, the VPSI members will store and visualize the services exchanged and resources used, and by doing so, they will build a shared understanding of the internal collaboration.

This project can be categorized as action research because of its duration (four years) and the active operational role assumed by the researchers [15,16]. In addition, the research output is a designed artifact, namely the service cartography tool, which makes the project compatible with design-science frameworks [17]. The research method we use is action design research (ADR) [18].

In Sect. 2 we give details of our research method. We describe the context of our ADR project, as well as our project iterations in Sect. 3. In Sect. 4, we explain the limitations and challenges researchers face when conducting an action-research project of this magnitude. In Sect. 5 we list the related work, and in Sect. 6 we present our conclusions.

2   Research Method

Action Research

We believe that "knowledge is created through transformation of experience" [19]. As experience is gained by doing, we focus more "on what practitioners do, rather on what they say they do" [16]. The research undertaken in real organizational context, aiming to solve immediate problem situation in collaboration with practitioners is called action research. As described by Avison et al. in [16], "in action research, the researcher wants to try out a theory with practitioners in real situations, gain feedback from this experience, modify the theory as a result of this feedback, and try it again."

Introduced by Kurt Lewin in 1946, action research is social research combining "generation of theory with changing the social system through the researcher acting on or in the social system" [20]. Patton [21] has categorized action research as "action-oriented, problem-solving research", with informal data collection and research publications different from those in basic and applied research. For example, there are a few academic publications on our collaboration with the VPSI [3–7], whereas we have produced many informal and internally circulated documents.

Conducting action research in the context of investigating information systems is not new. For example, Baskerville published a tutorial on an action research of information systems [22] and Checkland's soft systems methodology [15] is rooted in action research.

By using action research, we extend our knowledge by solving specific problems that we identify at the VPSI.

Design Research

In his book, "The Sciences of the Artificial" [23], Herbert A. Simon set the foundations for design methodologies relevant to various disciplines, including design science in IS. Looking through the design lens, the output of our research, the service cartography tool, is "a purposeful IT artifact created to address an important organizational problem" [17]. This puts our research efforts in Hevner et al.'s framework for IS design-science research.

But Sein et al. [18] point out that "traditional design science does not fully recognize the role of organizational context in shaping the design, as well as shaping the deployed artifact". However, they also mention a few researchers that have "a view of artifacts as emergent from organizational context", and they propose a design research method that does not separate the IT artifacts from the interaction with the organizational context. Their method is called action design research (ADR) and we find it best fits our approach for IS research.

Action Design Research

Action design research is a research method that has four stages (see Fig. 1), where each stage contains a set of principles. We briefly explain the stages, without focusing on the principles:

1. **Problem Formulation** is a stage in which researchers identify, articulate and scope a problem inspired by practitioners, researchers, end-users, technologies or prior research.
2. **Building, Intervention and Evaluation (BIE)** is a stage that is carried out as an iterative process interweaving "the *building* of an IT artifact, *intervention* in the organization and *evaluation*" [18].
3. **Reflection and Learning** occurs in parallel with the first two stages: researchers reflect on the problem formulated, and on the theories and tools chosen to develop a particular solution. The learning from this reflection leads to a refined problem formulation and solution, as both researchers and practitioners gain a better understanding of the emerging artifact.
4. **Formalization of Learning** is the most challenging stage, as the learning from the ADR project should result with generalized solution concepts for a class of field problems.

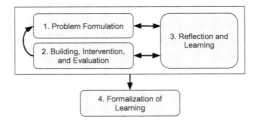

Fig. 1. Action design research method stages, adapted from [18].

In the following section, we illustrate how we applied ADR in our research project conducted at EPFL.

3   Service Cartography: Research Project in Collaboration with the EPFL Information Systems Department

The VPSI provides approximately 100 IT services to around 14,000 EPFL members, of which in 2015 [24] around: 3,000 are researchers and lecturers, 300 are professors, 10,000 are students, 5,000 are employees, including administrative staff, IT experts and others. Some of these members have dual roles (for example professors, researchers and lecturers are also employees). Besides EPFL members, the VPSI also serves many more visitors and the general public.

The marketing term "segment" is adopted to designate these separate groups of customers with different needs.

The VPSI's transformation into a service-oriented organization introduced a challenge to the internal organization. In parallel, to optimize the creation, management and operation of services, the VPSI developed a service strategy. The service strategy defines ways of collaborating in the service-oriented organization. This strategy also explains why services are needed and defines roles the VPSI members have in implementing these services.

The following subsections present the research collaboration with the VPSI through the ADR lens. Figure 2 depicts three major Building, Intervention and Evaluation (BIE) iterations, each lasting more than a year, and each having finely-grained, shorter iterations. As described in [18], "during BIE, the problem and the artifact are continually evaluated". Consequently, in each major iteration, the ADR team members gained better understanding of the problem and they tried to reformulate it. The artifact was updated in parallel to reflect the changing problem addressed.

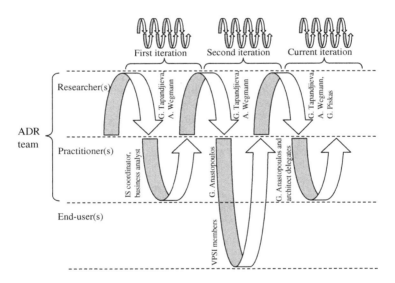

Fig. 2. Three major Building, Intervention and Evaluation (BIE) iterations for an IT-Dominant ADR project at EPFL's IT department, called the VPSI.

Due to the informal collaboration with the VPSI, in every major iteration, different practitioners belong to the ADR team. As a consequence, specific roles and responsibilities of the ADR team members are not defined. BIE iterations are conducted in a semi-structured manner where researchers rely on their observations and discussions with practitioners.

3.1 First Iteration

Problem Formulation. In the beginning of the collaboration with the VPSI, researchers identified that the VPSI management needed to *build and communicate a common view of the service-oriented enterprise architecture* in the context of the ongoing strategy formulation [3].

Building, Intervention and Evaluation. During the first iteration, researchers focused on modeling the architecture of several VPSI business processes. This iteration was used to understand and visualize the internal organization of resources and people in different roles. The idea was to build a map of IT resources, services, EPFL employees, users, external partners, protocols. As defined in [25], maps are "graphic representations that facilitate a spatial understanding of things, concepts, conditions, processes, or events in the human world". Cartography is the practice of making maps, therefore, the artifact was named an IT cartography. The design of the IT cartography was based on SEAM [26] conceptualization. SEAM is a service-oriented framework developed at LAMS [27].

To create the IT cartography artifact, the ADR team members chose a commercial tool called SOLU-QIQ [28] in which the researchers implemented the SEAM meta-model [5]. Figure 3 shows the cartography output during this iteration.

In this iteration, all decisions were made during frequent meetings and semi-structured interviews among researchers and two VPSI members (belonging to the ADR team), the EPFL's IS coordinator and a business analyst. The researchers also occasionally attended meetings with other VPSI members, to better relate to the problems VPSI was solving.

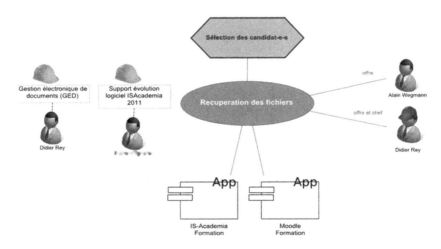

Fig. 3. IT cartography output, taken from [3], showing people and applications involved in one service implementation, with the process that consumes that service.

Reflection and Learning. As EPFL employees, the researchers used VPSI services on a daily basis, enabling themselves to experience first-hand the value services bring to the customer. The researchers also related their experience to the industry and academic service approaches. Moreover, the frequent discussions with the IS coordinator and the business analyst crystallized the understanding of services. In these discussions, all ADR team members made SEAM sketches to conceptualize their learning of services. The progress made was partly attributed to the SEAM sketches made.

At the end of the first iteration, the IT cartography showed only one service implementation per view, with a different notation from the usual SEAM notation. The different notation caused confusion among ADR team members, as the cartography did not represent the SEAM conceptualization from the discussions.

As a result of the reflection,

- the ADR team decided to apply the standard SEAM notation, and
- the IT cartography was spontaneously referred to as *service cartography*.

Finally, the service cartography built contained only a few example services, hence the ADR team believed it was not ready to be shared with all VPSI members.

3.2   Second Iteration

Problem Formulation. The ADR team gained a better understanding of the problem and they had ideas on how to improve the service cartography. The researchers found that there was no need to reformulate the problem after the first iteration.

Building, Intervention and Evaluation. At the start of this iteration, the second author of this paper, joined the VPSI, as the head of IS architecture. From the moment he joined, on top of his VPSI work, he became a member of the ADR team. In the first few months, the collaboration mainly involved knowledge transfer concerning SEAM, SOLU-QIQ and EPFL services in general. Afterwards, he became the main service cartography user, designer and developer. All ADR team members collaborated on designing the cartography overview page (see Fig. 4(a)) and the navigation between the detailed views. The second author was also the main advocate for having the standard SEAM notation in the service cartography. He succeeded in implementing a notation similar to the standard SEAM notation (see Fig. 4(b)). The first and second author conducted several interviews with other EPFL members, and they populated the service cartography with the information gathered. Subsequently, the head of IS architecture independently updated the service cartography, and after having the information for many services, he made the tool available to all VPSI members. In the following months, the ADR team observed that the VPSI members did not use the service cartography, despite having the tool at their disposal.

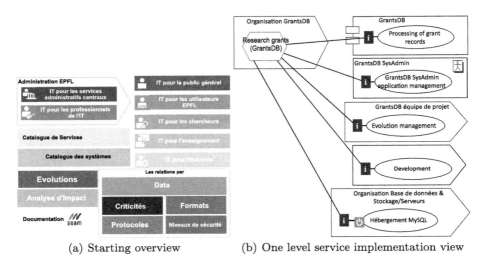

(a) Starting overview   (b) One level service implementation view

Fig. 4. The service cartography developed in the second iteration

The evaluation in this iteration was marked by a contextual inquiry [29] the first author conducted with a service manager. In this one-month contextual inquiry, the first author developed a master/apprentice relationship where the service manager was the master. The idea behind a contextual inquiry is to discover actions as they occur, to allow the service manager to talk about his work as it happens, and not to ask structured questions as in a traditional interview scenario [29]. The goal of this evaluation approach in the ADR project was to learn more about the daily work of a service manager and understand why the service managers did not consult the service cartography.

During the contextual inquiry, some observed activities of the service manager were (1) maintaining relationships with end-users and people involved in the service implementation, and (2) producing and updating the service documentation such as service description, service architecture, service change requests, and service knowledge base. The output views from the service cartography could be included in various service documentations, but navigating to the specific view of the service of interest was almost impossible.

Reflection and Learning. With the introduction of service orientation, the VPSI members are expected to separate the "service offering" and the "service implementation". The service offering is the end-customer's value-added level, whereas the service implementation is not visible to the end-customer and it represents the internal services and resources used to develop the offering. This separation is perceived as the major benefit of service-orientation and it focuses solely on the end-customer value. But what about the values of VPSI employees and other internal customers?

From the contextual inquiry, we learned that in the service documentation, service managers communicate their internal service-exchange. In this documentation, service managers describe the organization of the collaboration they have with other internal and external people, and the resources they use. The service cartography stores service-to-service, service-to-segment and service-to-person relationships within a defined collaboration context. The ADR team also became aware that the web pages with this information were difficult to find, were displayed in predefined views (see Fig. 4(b)) and could not be changed. By observing these pitfalls, the ADR team members realized they did not understand the needs of the users and the constraints imposed by the SOLU-QIQ tool. In addition, the maps generated with SOLU-QIQ were static, so the team decided

1. To stop using SOLU-QIQ and design a new employee-centric tool. The new tool should allow the VPSI employees to dynamically build their own service map, in order to fit exactly the needs of the employee at a given point in time for a specific purpose.
2. To initiate a frequent collaboration with one specific role, the architect delegate, in order to capture role-specific use cases.

3.3   Current Iteration

Problem Formulation. *How can an employee of a service-oriented organization visualize and communicate her work?* Based on ADR team's observations, an employee's work includes exchanging internal services and the value these services provide to her external stakeholders (customers, suppliers).

Building, Intervention and Evaluation. In the current iteration, the ADR team develops a new service cartography that enables VPSI members to communicate the internal service-exchange. The new service cartography is user centric. Instead of navigating between predefined views, the VPSI members search for the services or systems and interactively build service maps they need, starting from an empty canvas. Also, the VPSI members are independent from one another while dynamically building their map. The service maps they create can then be saved, exported and shared with other EPFL members. In addition, there is no restriction on the details shown: the map can show multiple service levels, starting from the lowest service level (e.g., network), all the way to the business services, and end-user level. For example, Fig. 5 shows two service levels, and it can be expanded. Furthermore, the service cartography has links to EPFL's service catalog and EPFL people's directory.

Additional features include a few predefined overviews:

– Aggregate overview relationships among all services with their context (not shown in this paper).
– The cropped overview in Fig. 6(a) shows collaboration for a specific service, where people are grouped around a service on which they worked at least once.

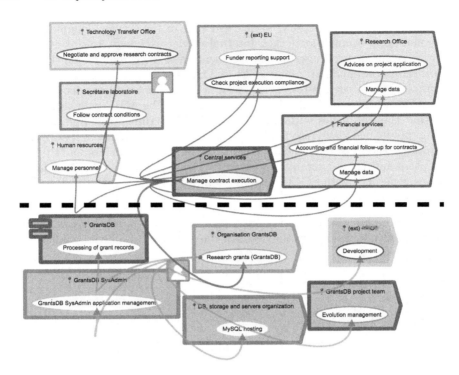

Fig. 5. The service cartography visualization tool in which users build their own map. The figure shows two service levels distinguished by the different colors of relationships. The service level of Fig. 4(b) is depicted below the dashed line.

– Figure 6(b) shows the opposite: the employee is visualized in the center of all services on which he worked at least once.

In this iteration, the development of the service cartography with all new features is done by a master's student, who is the third author. In addition, ADR researchers initiate a closer collaboration with the VPSI members who have an architect delegate role. An architect delegate is a specialist in one architectural domain, such as network, security and databases. The architect delegates and the head of IS architecture, form the architecture body that ensures "the coherence and efficiency of EPFLs information system" [30]. The researchers join the monthly architecture meetings to get regular feedback, ideas and requests for new features. They should develop a more intensive individual collaboration with each architect delegate, in order to better elicit all service cartography use cases in their context.

Reflection and Learning. In the previous two iterations, the researchers were focused on the design rather than on the action in the organization. By including the architect delegates in the early stages of the service cartography redesign,

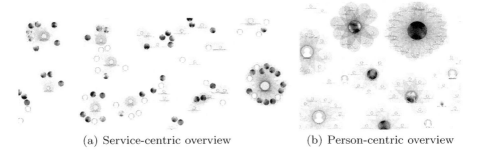

(a) Service-centric overview   (b) Person-centric overview

Fig. 6. Overviews in the service cartography

researchers hope to avoid some of the mistakes presented in [31], such as building a system that employees are reluctant to use.

After all project iterations, the researchers stabilized the formulation of the problem they are solving. They are currently reflecting on other approaches, both in academia and industry, that emphasize the internal service and value exchange. The importance of the internal service exchange is already indicated by Vargo and Lusch in [10]. Their second foundational premise states that without direct interaction with the end customer, employees lose sense of the internal service exchange among themselves, which leads to ignoring "quality and both internal and external customers" [10].

4   Limitations and Challenges

The emergent knowledge from an experience is "a knowledge which is contingent on the particular situation" [20] of a given moment (in this ADR project). All decisions taken "are subject to reexamination and reformulation upon entering every new research situation" [20]. The researchers cannot provide measurements for the improvements our service cartography brought to the VPSI and this paper only presents observations.

Most of the challenges faced in an ADR project are related to the dual role that the researchers and practitioners have. It is important for the practitioners to manage politics well, in order to "have a future in the organization when the research is completed" [32].

Involving VPSI members in this project requires managing relationships and negotiation with the employees and their supervisors. Sometimes the researchers encountered delays in securing the involvement of the VPSI members, thus making the time horizon of the ADR project unpredictable.

5   Related Work

The service cartography artifact is inspired by the field of enterprise architecture (EA). The objective of EA is to capture "the essentials of the business, IT and its evolution" in a holistic view [33]. Some EA authors use the metaphor of city planing and urbanization [34], and others use the metaphor of building a house [35]. The strong association with IT resources is a limitation of most EA approaches, such as the Zachmann framework [35], the IT4IT Reference Architecture [36] and TOGAF [37]. These approaches do not conceptualize the employees, employees' roles, and the contexts in which employees work; and they do not have a coherent representation of all organizational levels. For this reason the service cartography visualization is based on SEAM models, where people and their services can be modeled at any level of the hierarchy and seen in a concrete context.

The ADR team's goal is to visualize the relationships among services that exist in an organization. There are a few commercial tools [38,39] that automatically map service dependencies. These tools discover the technical details of services and application deployments, such as the communication endpoint, the address of the server or virtual machine where a service runs. They operate on the level of configuration items and address IT administrator's needs in finding a root-cause of a technical problem. Consequently, people's explicit responsibilities and collaborations for a specific service are not visualized, whereas we visualize all relationships that show collaboration among people, applications and technologies, all the way to the end user. These tools, however, can provide aggregate data for technical services, that can then be integrated in the service cartography visualizations.

Mega's HOPEX [40] and Link Consulting's EAMS [41] are commercial tools close to our service cartography. These tools are an Enterprise Architecture Management System (EAMS). Both of them are compatible with TOGAF and the second one has the feature to show the evolution of the models over time. The main difference with our tool is the focus on services and relationships among employees. Nevertheless, we get inspired by their features.

6   Conclusions

In this paper we have presented our experience of conducting a four-year ADR project of building a service cartography artifact, the directions taken and difficulties encountered. The researchers' activities and what they learned in the ADR project are presented in three major iterations. In the first iteration, the researchers tackled the architectural challenges, and the cartography tool was designed to communicate only the architectural perspective. In the second iteration, the ADR team populated the cartography tool with information gathered from interviews and made the output available to all VPSI members. Due to the static nature of the cartography tool, the VPSI members did not use it.

Thanks to the intensive collaboration and reflection on service orientation, all the ADR team members are able to see their work as an internal service-exchange. As the ADR team members did not need to rely on any tool, they concluded that it is tool agnostic to maintain a sense of service exchange in their work. But not all VPSI members can, or would like to, be included in an ADR project. Nevertheless, due to the lack, in both industry and academia, of concrete guidelines for managing internal collaboration in a service-oriented organization, the ADR team observed that the VPSI members still need a flexible and dynamic tool to visualize their services and to communicate about their collaborations. The researchers maintain their belief that by using the service cartography, VPSI members could (1) build a shared understanding of existing internal/external services, and (2) organize discussions in meetings, hence (3) adopt a service-oriented way of thinking.

References

1. Cartlidge, A., Hanna, A., Rudd, C., Macfarlane, I., Windebank, J., Rance, S.: An Introductory Overview of ITIL v3. The UK Chapter of the itSMF (2007)
2. Information Technology Infrastructure Library. https://www.axelos.com/best-practice-solutions/itil
3. Tapandjieva, G., Marchetti, D.R., Rychkova, I., Wegmann, A.: Towards the definition, implementation and communication of an IT strategy: the case of IT strategy at EPFL. In: Franch, X., Soffer, P. (eds.) CAiSE 2013. LNBIP, vol. 148, pp. 99–110. Springer, Heidelberg (2013). doi:10.1007/978-3-642-38490-5_8
4. Popescu, G., Tapandjieva, G., Wegmann, A.: Business and IT design with SEAM: an illustration with the PhD hiring process at école polytechnique Fédérale de lausanne. In: 2013 IEEE International Conference on Systems, Man, and Cybernetics, pp. 1938–1943. IEEE (2013)
5. Tapandjieva, G., Wegmann, A.: Specification and implementation of a meta-model for information systems cartography. In: Proceedings of the CAiSE2014 Forum at the 26th International Conference on Advanced Information Systems Engineering (CAiSE), vol. 1164, pp. 113–120. CEUR-WS. org (2014)
6. Tapandjieva, G., Gopal, A., Grossan, M., Wegmann, A.: Patterns for value-added services illustrated with SEAM. In: 5th International Workshop on Models and Model-driven Methods for Service Engineering (3M4SE 2014), as part of the Enterprise Computing Conference (EDOC) 2014 (2014)
7. Tapandjieva, G., Regev, G., Wegmann, A.: SLA: to sign or not to sign. In: Proceedings of the 2nd International Workshop on Socio-Technical Perspective in IS Development (STPIS), a CAiSE 2016 workshop, vol. 1604, pp. 15–24 (2016)
8. Zeithaml, V.A., Parasuraman, A., Berry, L.L.: Problems and strategies in services marketing. J. Mark. **49**, 33–46 (1985)
9. Demirkan, H., Kauffman, R.J., Vayghan, J.A., Fill, H.G., Karagiannis, D., Maglio, P.P.: Service-oriented technology and management: perspectives on research and practice for the coming decade. Electron. Commer. Res. Appl. **7**(4), 356–376 (2009)
10. Vargo, S.L., Lusch, R.F.: Evolving to a new dominant logic for marketing. J. Mark. **68**(1), 1–17 (2004)
11. Spohrer, J., Maglio, P.P., Bailey, J., Gruhl, D.: Steps toward a science of service systems. Computer **40**(1), 71–77 (2007)
12. Vandermerwe, S., Rada, J.: Servitization of business: adding value by adding services. Eur. Manag. J. **6**(4), 314–324 (1988)
13. Krafzig, D., Banke, K., Slama, D.: Enterprise SOA: Service-Oriented Architecture Best Practices. Prentice Hall Professional, Indianapolis (2005)
14. Lusch, R.F., Vargo, S.L., Wessels, G.: Toward a conceptual foundation for service science: contributions from service-dominant logic. IBM Syst. J. **47**(1), 5–14 (2008)

15. Checkland, P., Holwell, S.: Information, Systems and Information Systems: Making Sense of the Field. Wiley, Chichester, New York (1997)
16. Avison, D.E., Lau, F., Myers, M.D., Nielsen, P.A.: Action research. Commun. ACM **42**(1), 94–97 (1999)
17. Hevner, A., March, S., Park, J., Ram, S.: Design science in information systems research. MIS Q. **28**(1), 75–105 (2004)
18. Sein, M.K., Henfridsson, O., Purao, S., Rossi, M., Lindgren, R.: Action design research. MIS Q. **35**(1), 37–56 (2011)
19. Kolb, D.A.: Experiential Learning: Experience as the Source of Learning and Development. FT press, Upper Saddle River (2014)
20. Susman, G.I., Evered, R.D.: An assessment of the scientific merits of action research. Adm. Sci. Q. **23**, 582–603 (1978)
21. Patton, M.Q.: Qualitative Evaluation and Research Methods. SAGE Publications, inc, Newbury Park (1990)
22. Baskerville, R.L.: Investigating information systems with action research. Commun. AIS **2**(3es), 4 (1999)
23. Simon, H.A.: The Sciences of the Artificial. MIT Press, Cambridge (1996)
24. EPFL Annual Report 2015. http://information.epfl.ch/annual_report
25. Woodward, D., Harley, J.B.: The History of Cartography. Cartography in Prehistoric, Ancient, and Medieval Europe and the Mediterranean, vol. 1 (1987)
26. Wegmann, A.: On the systemic enterprise architecture methodology (SEAM). In: 5th International Conference on Enterprise Information Systems (ICEIS 2003) (2003)
27. SEAM. http://lams.epfl.ch/seam/
28. SOLU-QIQ - Accueil. https://www.soluqiq.fr/
29. Beyer, H.R., Holtzblatt, K.: Apprenticing with the customer. Commun. ACM **38**(5), 45–52 (1995)
30. Architecture—EPFL-SI. http://informationsystem.epfl.ch/architectureEN
31. Markus, M.L., Keil, M.: If we build it, they will come: designing information systems that people want to use. Sloan Manag. Rev. **35**(4), 11 (1994)
32. Coghlan, D., Brannick, T.: Doing Action Research in Your Own Organization. Sage, London (2014)
33. Lankhorst, M.: Enterprise Architecture at Work: Modelling, Communication and Analysis. Springer, Heidelberg (2009)
34. Longépé, C.: The Enterprise Architecture IT Project: the Urbanisation Paradigm. Butterworth-Heinemann, Oxford (2003)
35. Zachman, J.: A framework for information systems architecture. IBM Syst. J. **26**(3), 276–292 (1987)
36. IT4IT Reference Architecture, Version 2.0. http://pubs.opengroup.org/it4it/refarch20/index.html
37. Josey, A.: TOGAF® Version 9.1-A Pocket Guide. Van Haren (2011)
38. ServiceNow Wiki: Using a Next Generation BSM Map. http://wiki.servicenow.com/index.php?title=Using_a_Next_Generation_BSM_Map
39. Service Map in Operations Management Suite (OMS). https://docs.microsoft.com/en-us/azure/operations-management-suite/operations-management-suite-service-map
40. HOPEX—MEGA. http://www.mega.com/en/product/hopex
41. EAMS homepage. http://www.linkconsulting.com/eams/

Content Analysis of Customer Reviews to Identify Sources of Value Creation in the Hotel Environment

Elisabeta Molnar[1(✉)] and Remus Moraru[2]

[1] Partium Christian University, Primariei 36, 410029 Oradea, Romania
elis@partium.ro
[2] Doctoral School of Business Administration,
Bucharest University of Economic Studies,
Piața Romană 6, District 1, 010374 Bucharest, Romania
moraru_remus@yahoo.com

Abstract. A first goal of this paper is to determine what customers do value when choosing a hotel or having an experience with it and with what intensity they value these elements. Secondarily, the paper aims to identify sources of value creation in the hotel environment. For the purpose of this paper, the added value is understood as ability of the hotel to incorporate customers' feedback in its decisions that are affecting them and the choices they make. To achieve its goals, the paper employs content analysis. The paper provides valuable insights to hoteliers, which support their decisions to improve the business by incorporating proactively strategies of value creation.

Keywords: Hotel industry · Social value · Customers reviews · Content analysis

1 Introduction

Hospitality industry faces a continual evolution in the 20th century as a result of economic development and of the dynamic growth of tourism. Exactly in the periods of financial crises hotel investments did not stagnate, but found their ways of functioning all over the world bringing about a new concept of hotel infrastructure. Tourism has boasted virtually uninterrupted growth over time, despite occasional shocks, demonstrating the sector's strength and resilience. International tourist arrivals have increased from 25 million globally in 1950 to 278 million in 1980, 674 million in 2000, and 1186 million in 2015 [1]. International tourism now represents 7% of the world's exports in goods and services, up from 6% in 2014, as tourism has grown faster than world trade over the past four years [1].

Nowadays we find accommodation infrastructures ranging from simple mountain cabins to the most luxurious hotels, each with its particular specificities in order to please the clients who use them. The fierce competition in some regions called for specific quality standards and organizational methods [2]. Traditional restaurants, relaxing wellness departments, exclusive hotel lounges in luxurious hotels are at the

© Springer International Publishing AG 2017
S. Za et al. (Eds.): IESS 2017, LNBIP 279, pp. 251–260, 2017.
DOI: 10.1007/978-3-319-56925-3_20

service of customers who travel all over the world. Besides individual management one can find hotels with management contracts in franchise system, or hotel chains that bring about extraordinary advantages in the support of hotel industry [3].

World Tourism Organization UNWTO promotes and helps to implement different project in the hotel industry: Implementation of Hotel Classification Scheme; Training program focusing on the theoretical and practical implementation of the new classification criteria; Initiative the Hotel Energy Solution, and Collaboration Action for Sustainable Tourism [4].

The current paper discusses the role played by the customers' testimonials and reviews in developing strategies for value creation in the hotel environment. It aims to determine what customers do value when choosing a hotel or having an experience with it and with what intensity they value these elements. The paper starts with a review of how hotel environments have progressively changed and evolved. Then it explains the methodology used to identify sources of value creation in the hotel environments. Finally, analysis is conducted on selected hotels and conclusions are drawn based on results.

2 Insights into the Hotel Environment

Nowadays, no hotel can afford to operate in isolation, hoping to become successful or at least surviving on the market on its own. In Romania a significant development in hotel infrastructure took place between 1970–1980, in a period, when tourism visibly developed and when the activity of generating incomes was considered to be an essential priority. The political and economic situation before and shortly after 1990, influenced Romanian tourism in a negative way. This led to an imbalance in the hospitality industry as well. The state's infrastructure became privatized, and such a situation did not always benefit tourism. Thus, some time was needed in order to create an infrastructure that could meet the requests and standards already existing on the hotel market at European and global levels. If, from the point of view of infrastructure an acceptable level was reached, the level of services still needs a lot of investments, both in terms of technology and human resources.

For the tourism and hospitality industry, every visit, every request or inquiry, and every issue that hotel customer-relationship management and operations could address and successfully solve on-site represent opportunities to better understand guests and services [3, 5].

The labor market in tourism and hospitality industry shows a significant deficit regarding qualified workforce (both managers and staff) in the fields of tourism, hospitality industry, and restaurants. The educational system in this sense has major leaks that should be immediately taken care of; otherwise the deficit will affect negatively the newly realized investments in hospitality industry. This can be easily addressed, by learning everything necessary about guests, so their preferences and desires can be easily anticipated. Hotels who customize their communications to plan, guide and nurture their guests' experiences are able to turn those guests into loyal brand supporters and advocates [6].

Due to the change of consumer behavior and the desire of the guests for unique experience the question of competitiveness is a very important factor in the decision-making. Online platforms make it possible for the visitors to share their opinion about the hotel services and the destination as well [7].

Once the decision was made to focus the research on the benefits that would accrue from the implementation of social value parameters, the problem of developing an appropriate research question that would capture the essential objectives of the research came to the fore. A pertinent question that may be asked in this regard is "how do the anticipated or ex ante benefits of customer reviews are achieved ex post?" Such a question would enable sources of value creation in the hotel environment. It was therefore decided that the question would not be one of comparing the benefits of customer reviews achieved ex-post, but rather to concentrate on the question of what benefits were found by the hotel-managers, or at least perceived by them, ex-post. In this way the hotel-managers would be concerned about something that they had experienced directly and for which they had some knowledge on which to build their judgements or opinions.

The research questions in this paper derive from a review of the literature, from the practical process of attempting to undertake meaningful research, and from a desire on the part of the writers to explore the reality of hotel managers and their customers' attitudes. There was also, it must be admitted, a desire to provide a 'counterweight' to some of the attitudes found amongst the managers and some of academics, enforcement officers and officials, towards sources of value creation in the hotel environment.

It is noteworthy in Romania the Continental hotel group [8]. Established in 1991 the hotel chain has undergone a continuous development, and now it has properties in the country's main cities, in different regions, as Continental hotels chain and Ibis hotels chain, the least according with the Accor management agreement [9]. The Continental Hotels is the only Romanian chain of hotels that is represented in 2016 by the brands of success [8].

3 Methodology

The paper aims to answer two main questions: What customers value when choosing a hotel or during his/her stay at the hotel? With what intensity do customers value these elements? To answer the research questions we use a sample of eight hotels from Romania, from the chain "Continental Hotels". Continental Hotels is a hotel group, founded in 1991, consisting of eight hotels located in major cities of Romania, like Bucharest (two hotels), Sibiu, Arad, Oradea, Drobeta Turnu Severin, Suceava, and Targu-Mures (Table 1). The selection of the hotels has been made based on several criteria: experience on the Romanian market, number of rooms and capacity, and occupancy rate.

The paper employs as a method of research content analysis of customer reviews. Inductive content analysis has been performed over for all accessible reviews/ testimonials posted by customers on hotel website, for each one of the selected hotels. The choice for the method has been done based on its particular usability in the context of emerging fields of research, where no solid literature already exists [10].

Table 1. Sample of hotels researched.

No.	Hotel	Webpage	City
1	Grand Hotel Continental Bucuresti 5 stars	https://grand-hotel-continental-bucuresti.continentalhotels.ro	Bucharest
2	Continental Forum Sibiu 4 stars	https://continental-forum-sibiu.continentalhotels.ro	Sibiu
3	Continental Forum Arad 4 stars	https://continental-forum-arad.continentalhotels.ro	Arad
4	Continental Forum Oradea 4 stars	https://continental-forum-oradea.continentalhotels.ro	Oradea
5	Continental Targu-Mures 3 stars	https://continental-tirgu-mures.continentalhotels.ro/testimoniale	Targu-Mures
6	Continental Suceava 3 stars	https://continental-suceava.continentalhotels.ro	Suceava
7	Continental Drobeta Turnu Severin 3 stars	https://continental-drobeta-turnu-severin.continentalhotels.ro	Drobeta Turnu Severin
8	Hello Hotels Bucuresti 3 stars	https://hello-hotels-bucuresti.continentalhotels.ro/	Bucharest

The data collected include all customer testimonials and reviews, written and visual, available on the official websites of the selected hotels. All valid reviews per single website (for each hotel) have been considered for analysis, so as to eliminate potential differences in interpretation due to technical factors and the complexity of the data provided. Main advantages of considering these reviews for analysis include the certainty that the reviewer has experienced the stay at the hotel and benefited from the services offered [11].

The content extracted has been posted online during the period of 10 months (January 2016–October 2016). After extracting the reviews in Excel and evaluating their content qualitatively, the data has been formatted and coded with the purpose to highlight main trends and perceptions of the customers.

4   Analysis and Discussion

To answer the research questions settled in the paper we started first with analysis of the traveler rating and hotel ranking for each one of the hotels from our sample, as they are provided by TripAdvisor. TripAdvisor® is the world's largest travel site, which offers advice from millions of travelers and a wide variety of travel choices [12]. As it is shown in Table 2, we observe that all hotels from the Continental Hotels group receive a good to very good rating from TripAdvisor travelers with the highest score for Grand Hotel Continental Bucuresti (4.5 of 5 bubbles) followed by Continental Forum Sibiu and Continental Forum Oradea (4.0 of 5 bubbles) and the lowest score for Continental Targu-Mures (3.0 of 5 bubbles). Also, according to TripAdvisor Ranking, two hotels in the group are ranked in top 10% in their city, five in top 25% and 1 in top 50%.

Table 2. TripAdvisor ranking for each of the selected hotels.

No.	Hotel	TripAdvisor Traveler Rating	TripAdvisor Ranking
1	Grand Hotel Continental Bucuresti 5 stars	4.5 of 5 bubbles Based on 489 traveler reviews	#10 of 169 hotels in Bucharest
2	Continental Forum Sibiu 4 stars	4.0 of 5 bubbles Based on 317 traveler reviews	#7 of 30 hotels in Sibiu
3	Continental Forum Arad 4 stars	3.5 of 5 bubbles Based on 222 traveler reviews	#2 of 13 hotels in Arad
4	Continental Forum Oradea 4 stars	4.0 of 5 bubbles Based on 146 traveler reviews	#4 of 24 hotels in Oradea
5	Continental Targu-Mures 3 stars	3.0 of 5 bubbles Based on 96 traveler reviews	#6 of 13 hotels in Tirgu Mures
6	Continental Suceava 3 stars	3.5 of 5 bubbles Based on 55 traveler reviews	#3 of 12 hotels in Suceava
7	Continental Drobeta Turnu Severin 3 stars	3.5 of 5 bubbles Based on 45 traveler reviews	#1 of 7 hotels in Drobeta-Turnu Severin
8	Hello Hotels Bucuresti 3 stars	3.5 of 5 bubbles Based on 380 traveler reviews	#36 of 169 hotels in Bucharest

Next, we proceeded to content analysis of the customers' testimonials and reviews available on the hotels websites. These are extracted from the testimonials posted by customers on online review platforms like Booking, Expedia, Google, Facebook, etc. Analysis was conducted by taking into consideration several indicators: frequency in time of reviews, percentage of positive vs. negative feedback, key words repeating, how many customers recommending the hotel, and other indicators. A synthesis of data analyzed function of selective indicators are presented in Tables 3 and 4.

The current research is limited to analysis and interpretation of data available on the hotels official websites. Therefore, generalization of our findings relies heavily on the consistency and accuracy of information researched, posted on the hotels websites [13].

Our research findings suggest that, for all of the eight hotels researched, information about location and surroundings is valued on the first place by customers who experienced the stay at the hotel, followed by information about room and information about staff. Other information valued by customers regards breakfast and restaurant, price or hotel facilities.

Table 3. Key elements of reviews and their intensity.

No.	Hotel	Key elements of reviews and description	Frequency in time
1	Grand Hotel Continental Bucuresti 5 stars	Location and surroundings (great, close to everything, beautiful, close to city center)	11
		Room (lovely, luxury, clean, wonderful, comfortable)	9
		Staff (outstanding, very helpful, knowledgeable, courteous, friendly, efficient)	7
		Breakfast (very good selection, delicious, excellent)	4
		Facilities (SPA)	2
		Building, furniture (like a palace)	2
		Price (good, affordable)	2
		Restaurant (small, fancy)	1
		Housekeeping (twice a day)	1
		Overall experience (great, very nice, perfect, top category, 5 star treatment, extremely fancy and amazingly friendly, fantastic, super, excellent, amazing, truly wonderful, lovely)	12
	Total no. of reviews and their provenience: 20 (1 Expedia, 2 Google, 14 Booking, 1 Hotel.de, 2 Hotels.com)		
2	Continental Forum Sibiu 4 stars	Location and surroundings (close to old city, walking distance, central, near everything, pedestrian traffic, quiet, excellent)	11
		Room (clean, comfy, design, Wi-Fi, large)	6
		Breakfast (good variety, amazing, nice, complete)	5
		Staff (very good, friendly)	4
		Parking, accessibility for cars	2
		Price (convenience, good value for money)	2
		Facilities (SPA)	1
		Overall experience (comfort, traditional, relaxing, great, clean, best upper-class hotel, excellent, super)	7
	Total no. of reviews and their provenience: 18 (15 Booking, 1 Expedia, 2 h.de)		
3	Continental Forum Arad 4 stars	Location (perfect, heart of the city, in the center, excellent, super)	10
		Room (large, clean, extra-large comfortable wide bed, strong Wi-Fi, AC)	5
		Staff (ok, nice, very good)	5
		Breakfast (good service, best selection, quality)	5

(continued)

Table 3. (*continued*)

No.	Hotel	Key elements of reviews and description	Frequency in time
		Panorama and balcony (amazing view, wonderful)	4
		Restaurant (wonderful for dinner)	3
		Bar	1
		Price (fair)	1
		Parking lot (secure)	1
		Overall experience (perfect, delightful surprise, pleasant, wonderful, surprise, design, good taste)	7
		Total no. of reviews and their provenience: 14 (13 Booking, 1 Facebook)	
4	Continental Forum Oradea 4 stars	Location and surrounding (close to Aqua Park, close to downtown, great)	4
		Staff (very friendly, kind)	4
		Room (inclusive smoking rooms)	2
		Breakfast (rich)	2
		Spa (swimming pool)	2
		Balcony	1
		Price	1
		Overall experience (relaxing, nice, super)	5
		Total no. of reviews and their provenience: 10 (10 Booking)	
5	Continental Targu-Mures 3 stars	Location (central, walking distance, excellent)	10
		Staff (very kind, helpful, friendly)	7
		Room (heating, TV, Wi-Fi, large rooms, renewed, clean AC)	5
		Price (good quality/price ratio)	3
		Panorama (city view)	2
		Breakfast (good, complete)	1
		Overall experience (it is worth, worth the stay, very good quality, clean, quiet, perfect location for business travelers)	9
		Total no. of reviews and their provenience: 15 (15 Booking)	
6	Continental Suceava 3 stars	Location (good, convenient in the city center, walking distance from everything, great)	5
		Staff (kind, friendly, great, very welcoming, nice)	5
		Room (recently renovated, clean, large)	3
		Breakfast (okay, very good)	2
		Price (good price/quality ratio, good value for money)	2
		Panorama (wonderful view)	1

(*continued*)

Table 3. (*continued*)

No.	Hotel	Key elements of reviews and description	Frequency in time
		Restaurant (good quality food)	1
		Overall experience (quiet, very pleasant, full pleasure, real business, good stay)	4
		Total no. of reviews and their provenience: 9 (7 Booking, 1 Facebook, 1 Continental Survey)	
7	Continental Drobeta Turnu Severin 3 stars	Location and surroundings (excellent, close to the city center, close to the river port, railway station, a beautiful park)	5
		Staff (very friendly, excellent, warm)	4
		Room (comfortable, clean)	3
		Panorama (nice view of Danube river)	3
		Breakfast (nice)	1
		Price (best value for money)	1
		Overall experience (nice, comfort, best hotel in the region, old communist hotel recently renovated, ancient style very well conserved, wonderful)	6
		Total no. of reviews and their provenience: 9 (9 Booking)	
8	Hello Hotels Bucuresti 3 stars	Location (close to railway station and subway, bus stops, very good)	7
		Room (clean, simply and comfortable, good shower, comfortable bed)	7
		Price (good value for price, excellent value for money)	6
		Staff (friendly, kind)	4
		Breakfast (good)	3
		Overall experience (nice place, enjoyable stay, modern, good taste, for youth or single persons, clean, perfect for one night)	4
		Total no. of reviews and their provenience: 14 (12 Booking, 1 Agoda, 1 Facebook)	

In what regards the overall experience shared by the customers, this can be summarized through the following main key words: great/very good stay, (very) nice place, perfect location, pleasant and super. These words are repeating for majority of the researched hotels. Other experiences shared by the customers, from which hoteliers can learn and build upon, are qualified as: friendly, traditional, relaxing, delightful, clean and comfort [14].

Although the positive reviews are dominant in the total reviews analysed, negative testimonials are present as well (Table 4). These offers valuable lessons and insights to hotel managers in terms of the additional sources of value creation for their customers.

Table 4. Negative feedback.

No.	Hotel	Negative feedback
1	Grand Hotel Continental Bucuresti 5 stars	A more traditional type bar with atmosphere
2	Continental Forum Sibiu 4 stars	Air conditioning couldn't be changed to warm (in September); Rooms two small; Breakfast was a bit poor
		Air conditioning not great; Parking cost; Wellness area not open on Mondays; Bathroom needs improvements
3	Continental Forum Arad 4 stars	The dum-dum from the night club upstairs; Breakfast needs improvements; Bed comfort needs improvements
		Parking cost; The loud music from the restaurant, the base was vibrating the room even if it was on 9th floor
4	Continental Forum Oradea 4 stars	It needs renovation; Preparing beef is not a favorite for the kitchen crew; Bathroom, swimming pool need improvements; Restaurant and food and staff needs improvements; Parking cost
5	Continental Targu-Mures 3 stars	Cleaning is an issue; Parking cost; Elevator too slow
		Furniture needs renovation; Breakfast could be better, not much to choose from; Breakfast restaurant very busy; Breakfast was average; No closed parking; No isolation of rooms against noise; The room could have been a bit warmer; No cable to plug the mini-bar; Old bed sheets; Parking cost; Wi-Fi connection was poor and the bed was a little too soft
6	Continental Suceava 3 stars	Small parking; A strange smell on corridors; There is no hair dryer in bathroom
7	Continental Drobeta Turnu Severin 3 stars	Restaurant needs improvements (no fish); The place should be renovated; Breakfast too expensive relative to content; AC needs improvements
8	Hello Hotels Bucuresti 3 stars	A glass window between shower and room – no intimacy; Breakfast outside, in another building

As it can be observed, there are not dominant elements that need immediate action or improvement for the entire Intercontinental Hotels group, but particular issues, which require separate treatment for each hotel. However, the cost of parking, as a distinctive cost (instead of free parking) seems to be a common concern for most of the hotels researched.

5   Conclusions

The research results offer good insights to hoteliers by helping them to identify what customers appreciate during their experience with the hotel and what is the value added to them. Also, hoteliers learn how to incorporate customer feedback in the hotel environment to add more value in both online and offline.

As the current research interprets only the customers' testimonials and reviews available on the hotels websites, which are limited and may be biased in what regards selection of them, further research is needed in order to compare those reviews with customer reviews available on other specialized online platforms, like Booking, TripAdvisor of Continental Survey. The future steps for this research should respond to the following subsidiary questions: How do the perceived benefits of customer reviews vary from one hotel to another hotel? How do hotel managers regard the avoidance of the negative feedback? How do the benefits of elements of reviews taken into consideration differ from other benefits that are not related to social value creation?

References

1. UNWTO Tourism Highlights, 2nd edn. (2016)
2. Poater, A., Garriga, A.: Tourism in European cities: insights into the dynamics of weekend hotel accommodation. Tour. Econ. **15**(1), 41–86 (2009)
3. Vogt, C.A.: Customer relationship management in tourism management needs and research applications. J. Travel Res. **50**(4), 356–364 (2011)
4. UNWTO Annual Report 2015. http://www2.unwto.org/en
5. Skift Report 2015. Hospitality Analytics: How Data Can Make Hotels Smarter. http://www.wikitourism.co.za/images/a/a3/Skift-SnapShot-Hospitality-Analytics.pdf
6. CENDYN: Hospitality Cloud, Journey to loyalty: Personalizing the guest experience, Plan, Guide and Nurture the Guest Experience. http://www.cendynone.com/resources/library/journey-to-loyalty/
7. Juhasz, K.D.: Hotel competitiveness measurement methods. Knowl. Horiz. Econ. **7**(3), 15 (2015)
8. Continental Hotels. https://www.continentalhotels.ro
9. Koushik, D., Higbie, J.A., Eister, C.: Retail price optimization at intercontinental hotels group. Interfaces **42**(1), 45–57 (2012). Special Issue
10. Ali, H.: The value of trust in service sector marketing. In: Armistead, C. (ed.) The Future of Service Management. Kogan Page, London (1994)
11. Carey, M.A.: The group effect in focus groups. Planning, implementing, and interpreting focus group research. In: Critical Issues in Qualitative Research Methods, pp. 225–241 (1994)
12. Gallouj, C.: Asymmetry of information and the service relationship: selection and evaluation of the service provider. Int. J. Serv. Ind. Manag. **8**(1), 42–64 (1997)
13. Maxwell, J., Lyon, T., Hackett, S.C.: Self-regulation and social welfare, the political economy of corporate environmentalism. J. Law Econ. **53**, 583–617 (2000)
14. Yusof, S.M., Aspinwall, E.: TQM implementation issues: review and case study. Intl. J. Oper. Prod. Manage. **20**(6), 634–655 (2000)

Beyond Quality of Service:
Exploring What Tourists Really Value

Jesús Alcoba[1(✉)], Susan Mostajo[2], Rowell Paras[2], and Romano Angelico Ebron[2]

[1] Centro Superior de Estudios Universitarios La Salle, Madrid, Spain
jesus@lasallecampus.es
[2] De La Salle University-Dasmariñas, Dasmariñas, Philippines
{stmostajo,rrparas,rtebron}@dlsud.edu.ph

Abstract. Experience economy is the last segment in the growth of economic value. After the appearance of services and the quality assurance culture, experience design and management point to experience quality as the new way to assess what the customer really values. Focusing on tourism industry, the present article compares a wide sample of tourists' numerical scores and verbal assessments, analyzed with a sentiment analysis engine. The objective is to acquire a deeper knowledge of the concept of experience quality to find out what the tourist really values.

Keywords: Quality of service · Quality of experience · Tourism · Experience · Design · Service science · Meaning · Narrative · Sentiment analysis

1 Introduction

One of the most significant aspects of economy is the analysis of the continuous evolution of what customers consider valuable. Customers exchange their money for that which provides them with value, and this exchange is the basis of economy. Recently, the evolution of economic value has experienced two major changes.

The first is nested in the approach that analyzes the impact caused by the switch from an exchange of goods market into an economy that is mainly based on services. This approach has had two parallel and interrelated pathways [1]. The first, called service-dominant logic, was originally conceived as a marketing issue [2] but it has undergone a progressive evolution [3] that has expanded its descriptive value [4]. The second pathway, originally termed as science of service systems [5] is now named service science [6], it is understood as the interaction between people, technology and shared information [7] and considers service systems as the basic abstraction that explains the co-creation of value [8].

The second concept is related to Pine and Gilmore's schema of the progression of economic value called experience economy [9]. Under this approach experiences are personal and memorable, customers can add them to their biographies, and so they can help them to construct their identities [10]. Generally, a customer tends to use products or services that relate with his/her own conception of the world and to reject those that are unrelated or opposed to such conception [11]. If the experience, through its narrative,

S. Za et al. (Eds.): IESS 2017, LNBIP 279, pp. 261–271, 2017.
DOI: 10.1007/978-3-319-56925-3_21

generates a meaning for the customer, it will then become part of his/her biography. Finally, depending on the extent to which this biographical element becomes emotionally charged, it will end up occupying a significant place in the customer's identity. In sum, the key elements are: experience, narrative, meaning, biography, and identity [12].

Within this conceptual framework, one of the key responsibilities of companies is finding out what the customer really values in order to design services, otherwise the value proposal could hardly be successful. In such endeavor, evaluation of the experience plays a paramount role.

The present study aims to go deeper into the concept of quality of experience with a sentiment analysis focus, in order to approach what the tourist really values, thus trying to overcome classic approaches to the assessment of the quality of services.

2   Quality of Service vs Quality of Experience

Quality reflects the value of a product or service, and was recognized as a strategic tool in attaining operational efficiency and business performance [13] to maintain organizational competitiveness. Consequently, it became a focus of many organizations in service delivery and had interestingly led them to create processes and strategies to ensure that a desired quality standard is achieved. Few of these processes are Philip Crosby's "zero defect management" – that is, to eliminate faulty products before it reaches the customer [14] and Japanese' principle on "Kaizen management" which is directed towards keeping processes, products and services quality, and their continuous improvement [15].

Service is an action in which one person produces to assist another directly and personally [16]. It involves the human factor, wherein both the service provider and the customer has participation in the quality of service rendered [17]. The customers' involvement in the process provides them the opportunity to have a hands-on experience, including modification, in the service delivery which has a relationship on how the service is perceived. The company's quality of service is the foundation of its worth or image in the market because the latter equates how the company is perceived according to the meaning created from the customers' experience. This, consequently, became one of the hallmarks in the creation of product designs which may involve emotions, interaction, sustainability, service or transformation [11]. Thus, new approaches in designing more meaningful services came out because it is from the quality of service experienced by the customers where emotional bond with the company is developed [10].

Experience in a psycho-physiological perspective is an individual's interaction with the physical and social environment involving processes of sensory awareness and perception towards interpretation and creation of meaning. In business, experience is the customer's engagement to the services of the company which in a way creates a memorable event [18] and when combined with its surrounding experiences goes beyond itself to bring value to a customer's life [19]. It is inherently subjective because it is about the individual's own thoughts and emotions; and the memories and understanding of the experience exist in the mind only of the person engaged to it who eventually generates meaning of that experience.

Quality of experience is attached to customer experience, as experience is the consequence of the interaction between the user's inner state, the features of the service, and the context in which the interaction takes place [20]. According to the International Telecommunications Union, quality of experience refers to the overall acceptability of a service, as perceived subjectively by the end-users influenced by their expectations and context [21]. It is difficult to translate in an objective manner because the rating may not be the real measurement of the experience. Hence, quality of experience is not always numerically quantifiable but qualitatively described.

From the above arguments, the link between quality of service and quality of experience is noticeable, however the degree of relationship between them is yet to be established. Although services are external to the customer and experiences are inherently personal [9], it cannot be absolutely claimed that there is no subjectivity on how quality of service is assessed by a customer. To some extent, assessment of both quality of service and quality of experience are customer-dependent because the value of a service and experience is typically based on customer's satisfaction including individual differences in perception, needs, and expectations; and that some customers maybe innately easier to please than others.

Taking into consideration the nature of each service and experience, the very lean difference leads professionals in the field to think of metrics to as much as possible objectively measure both quality of service and quality of experience. As argued, customer satisfaction has been adopted frequently as key performance indicator by those that aim to achieve and maintain competitive advantages over their competitors [22]. Thus, companies develop feedback mechanisms to measure the quality of service they offer. The Net Promoter Score (NPS), for instance, was designed to measure customer satisfaction and loyalty to predict company growth [23]. Correspondingly, customers' enthusiast to recommend the company is directly correlated with growth rates, and an indicator that customers have received good economic value from the company. However, critics challenged the predictive value of NPS for customer loyalty because for them loyalty cannot be measured by a single number and requires further assessment. Generally, feedback quality models are done through survey questionnaires to estimate the gap between customer's perception and expectations of a service. Models like SERVQUAL and SERVPERF to quantitatively measure quality of service have been conceptualized. Congruently, qualitative analysis is usually done to measure quality of experience wherein the instrument's format allows customers to express their opinions in words. Calculation of a *semantic mean* of customers' opinions could be one of the approaches in the assessment of the quality of experience [10, 12, 16]. Moreover, it has been noticed that recent quality assessment tools are a combination of both quantitative and qualitative platforms wherein customers use both numbers and words to express what they think and feel about the service and/or experience. Given these perspectives, the necessity to continuously explore more profound approaches to measure quality of service and quality of experience in order to understand what customers really value for the improvement of processes on service design is highly respected, hence the direction of this study.

3 Quality of Service vs Quality of Experience in the Tourism Sector

Quality of service is considered as a standard used to assess the effectiveness of a particular leisure service agency, including the tourism sector [24], thus the quality of service involved with tourism plays an important role in the delivery process [25]. In some earlier studies, service quality has been defined to the extent where the service fulfills the needs or expectation of the customers [26] and conceptualized as the overall impression of customers towards the service weakness or supremacy [27]. Therefore, service quality has been viewed as the difference between the perceived services expected performance and perceived service actual performance [28].

As mentioned in the previous section, SERVPERF and SERVQUAL models have been widely utilized to measure the perceived service quality across service sectors including tourism [29]. Researchers have adjusted the SERVQUAL model to fit with their research agenda or identified alternative metrics to evaluate quality of service in tourism at large. For instance, Pawitra and Tan used SERVQUAL in order to analyze the image of Singapore from the perspective of tourists from Indonesia [30]. The authors noted that the relationship between customer satisfaction and service attribute performance is linear.

Tourism service providers aim to improve the quality of services and the level of tourist satisfaction with the belief that this initiative will create not only a meaningful customer experience but satisfied customers as well. Satisfied customers will lead to loyal visitors who continuously repurchase the product or service and will further recommend it to others [31]. Conversely, when the service provider fails to meet the customer's expectations, customers often switch to a different provider [32]. Empirical studies recommend that customer satisfaction mediates the association between quality of service and company performance [33, 34]. Accordingly, perceptions of quality of service result from incidents of customer satisfaction [35]. Moreover, some research findings suggest that satisfaction is an antecedent to service quality [36]. However, it was revealed that service quality is related but not exactly the same with satisfaction because perceived service quality is a global judgment or attitude relating to the superiority of service [31] while satisfaction is related to a specific transaction [35].

There are factors affecting service quality in tourism industry and destination is considered as highly important. Accordingly, the destination facilities and accessibility, and attraction directly influenced tourist satisfaction [37]. The geographical location where the event takes place should be highly accessible, whether traveling by plane, train, bus or automobile. Appropriate signage should also be displayed at various transportation modes to provide direction to points of interest [38]. Another factor is accommodation – the place where tourists stay, whose location can be a source of satisfaction or dissatisfaction for the tourists. For example, proximity from the hotel to the sporting venue is an important factor for many travelers [39]. Hence, hotels that are not within walking distance from the sporting venue often provide transportation services. Furthermore, there are various accommodation options. One may choose from simple to a more complex and luxurious accommodations. Regardless of the classification, service quality will always be the providers' priority.

Moreover, tourists may judge accommodation based on several considerations, including the evaluation of interactions, hotel environment and the value associated with staying at the place. The interaction which takes place between the guest, the accommodation provider's personnel and/or other guests during the stay is considered necessary [40]. The accommodation's environment such as facilities, surroundings and landscapes are contributors of either good or bad comments from guests. Literature suggests that physical evidences like noise level, odors, temperature, colors, textures and comfort of furnishings may influence perceived performance in service. Such variations in physical environment can affect perceptions of an experience independently of the actual outcome [41]. Ambient conditions, facility design, and social conditions directly influence the physical environment [40]. The guests' perception of value associated with accommodation is likewise given high importance in ensuring quality service. Guests judging the value of the accommodation to be worth the cost are more likely to stay until the end of the reservation.

Conventionally, experience refers to the sum of all interactions with people, place, product, services, organizations, governments and culture. This construct has become increasingly popular within tourism, hospitality and leisure sectors as well as marketing [42]. Tourism industry has been a developing experience – based products through which experiences become the product and provide visitors with distinctive, meaningful experience as a unique attribute of the destination [43]. Tourism experience is essentially sequential, which means that performance quality prompts experience quality. Experience quality has a significant impact on customers' loyalty, advocacy and satisfaction [44]. Tourists usually participate in numerous activities and interactions, causing them to feel, react and decide in different ways. Interaction and connection with local community further develops individual's sense of place, which promotes quality of life and sustainable tourism development, hence, high levels of quality tourism experience [43]. Tourist trips are life experiences, highly memorable for travelers both during and after the service. In tourism, emotional reactions are particularly important because they influence tourists' evaluation of the service, and therefore their satisfaction [45]. The tourists' quality of experience is characterized by the tour as a whole which encompasses various tourism components such as transportation, accommodation and most importantly the destination. Tourist satisfaction comes to the end of the process as a goal of the service provider. It is essential to ensure that the guest will consider repeat of business therefore, the constructs of both satisfactions with trip experience and leisure life domain would influence one's sense of well-being and or revisit intention [46].

Quality of experience and service quality in tourism are undeniably associated yet different from each other. The former is more complex than the latter, so one cannot be directly derived from the other. This study aims to explore the relationship between both concepts in order to obtain information that can help a better design of the tourist experience.

4   Methodology

As in a previous work [12], reviews from TripAdvisor.com of the top eight frequently visited provinces where the tourism destinations in the Philippines are located, based on Department of Tourism website 2015, were used. Namely: Aklan, Bohol, Cebu, Davao, Ilocos, Ifugao, Manila, and Palawan were considered. Only foreign tourists' online reviews were gathered from 2012 to 2015, with a total of 6371 reviews.

In each review the overall rating numerical score was registered (1 to 5). Also, the verbal evaluation was downloaded, and it was analyzed with Bitext, a proprietary deep linguistic sentiment analysis engine [47]. The sentiment analysis produced 44123 coded lines in total.

The culture of the quality of services has mainly based upon numerical scores garnered from a question or a statement. This study has understood that the scores that customers give on web sites belong to this type (in this case, the single score that has been used is the Tripadvisor.com "overall rating", a simple score that the tourist performs before writing a review). In contrast, it has been understood that verbal assessment, given that it is more descriptive and that it is written using those terms that are part of the customer's sematic constellation, is closer to an assessment of the customer's experience.

The analysis strategy attempted to find out if variables were related, that is, if the overall perception of the service, measured as a number on a 1 to 5 scale, related to the experience assessment, expressed as a verbal assessment. To do so, the sentiment analysis engine was pivotal, given that this technology can transform subjects' statements into numerical scores.

5   Results

Table 1 provides the statistical descriptions of this study. As can be observed, in contrast with the 1 to 5 range in the numerical variable, the verbal variable goes from −18.3 to 32 in the category of hotels, from −10.8 to 14 in the category of restaurants, and from −18.8 to 20 in the case of attractions. Considering these ranges, it must be pointed out that means are higher in the case of numerical than verbal scores, which seems to suggest a higher positivity on the part of customers when it comes to assess the different elements. In this way, in the case of hotels the numerical mean is 4.39 as opposed to the 2.087 verbal mean, in the case of restaurants, it is 4.189 in contrast with 2.001, and in the case of attractions, it is 4.17 versus 1.862.

In order to test the relationship between the two mean groups, an ANOVA test was carried out, and the results are shown in Table 2 (a previous test with a correlation matrix was discarded because the excessive size of the sample caused false positives - significant correlations in close to zero values). As can be noticed, in the three cases one can observe a linear effect of the numerical score on the verbal score.

Table 1. Descriptive statistics

	Hotels		Restaurants		Attractions	
	N	V	N	V	N	V
Valid N[a]	30093	30093	3512	3512	8862	8862
Mean	4.39	2.087	4.189	2.001	4.17	1.862
Median	5	2	4	2	4	2
Mode	5	2	5	2	5	2
Standard deviation	.891	2.475	1.00	2.594	.925	2.602
Skewness	−1.686	−.205	−1.247	−.646	−.990	−.244
S. Standard Error	.014	.014	.041	.041	.026	.026
Kurtosis	2.747	2.957	1.031	1.508	.515	3.198
K. Standard Error	.028	.028	.083	.083	.052	.052
Minimum	1	−18.3	1	−10.8	1	−18.8
Maximum	5	32	5	14	5	20

[a]Only cases with both a numerical and verbal score were considered.

Table 2. ANOVA

		Sum of squares	gl.	F	Welch
Hotels	Inter	11430.47	4	496.94***	420.58***
	Intra	173018.91	30088		
Restaurants	Inter	2437.66	4	100.84***	80.60***
	Intra	21194.34	3507		
Attractions	Inter	2688.21	4	103.84***	88.14***
	Intra	57321.97	8857		

***$p < .000$, **$p < .010$, *$p < .050$

Finally, in order to test the effect size between the means of the numerical and verbal scores, a Cohen test was carried out; and the result is shown in Table 3. As can be observed, although there is indeed a linear relation, the differences in the means show a low magnitude. A moderate magnitude can only be observed between the means of the verbal score in Level 3–4 of the numerical score in the assessment of restaurants (.59).

Table 3. Cohen's d effect size between numerical and verbal scores

	Numerical score variable levels			
	1–2	2–3	3–4	5–6
Hotels	.16	.22	.39	.22
Restaurants	.29	.16	.59	.09
Attractions	.21	.21	.29	.19

6 Conclusions

The analysis of the results shows that, as expected, there is a linear relation between numerical scores and verbal assessments. In other words, the positive assessments that a customer makes of a hotel, a restaurant or an attraction, lead the customer to give a high numerical score and also to make a positive verbal assessment. Obviously, it must be pointed out that the opposite would not have made sense (a customer who is positive when giving a numerical score and negative when verbalizing the assessment).

However, what is significant is that the differences in the means show a low magnitude, underlining that the aforementioned linear relation is lower than expected. If we accept that the numerical score is a means to study the assessment of the quality of a service, and that the verbal assessment is a means to study the quality of the experience, the result would certainly be that the variables are different and that the assessment of one of them cannot be directly drawn from the other one, or vice versa. And without accepting this assumption, yet it can be observed that customers are more positive in their numerical scores than in their verbal assessment. A fact that is relevant because the opposite could also have been true, that is, that the tourist used the verbal assessment to be more positive than in the numeric score. More research is needed to clarify why the direction of the assessment is of a greater positivity in the numerical scores and not the other way round.

The natural question that arises next is which of them really responds to what the customer values. The narratives reflect how the customer psycho-cognitively processes the entire experience, thus a description of a deeper and internal state, which may result difficulty to quantify for the human being. Given that in order to design experiences the narrative is pivotal, using the verbal assessments would be more useful, because the customer talks about the experience using his own words. Within this context, wondering then what the use of the numerical score could be is interesting, above all if this does not coincide exactly with the verbal assessment and if it is also more positive.

Another interesting aspect of this result is how to value the effect that a customer's numerical scores has on other customers, if these scores do not really reflect his assessment of the quality of the experience. In this sense, it might be useful to go deeper into the concept of *semantic mean*, thus producing a verbal index that would be more reliable when a customer comes to make use of other customers' opinions.

In any case, an evident practical conclusion for a better design of the tourist experience is to incorporate sentiment analysis technologies, since they do not only complete the quantitative vision of the quality of the experience, but also provide vivid and precise descriptions of what the customer really values.

Although the display of assessments is wide, this study is limited insofar as it only includes opinions about one country. Verifying this same comparison in other destinies would be very interesting. Besides, quite surely more accurate results could be reached refining the sentiment analysis engine until it is completely linked to the analyzed context and country. Although it was not the goal of this paper, it is also important to emphasize that the use of qualitative or mixed analysis techniques (IPA, QCA) could shed more light on this phenomenon.

In any case, and taking into account the analysis of the literature mentioned earlier and the analyzed data, the key conclusion is that the growth of the economic value evolves continually, and that we should transcend the classic assessment of the quality of service in order to create models that would allow us to know more about what the authentic epicenter of economy is, and this is knowing what the customer really values.

Acknowledgement. The authors of this paper would like to express their gratitude to Bitext for providing the sentiment analysis technology that made possible the data exploration.

References

1. Spohrer, J., Anderson, L., Pass, N., Ager, T.: Service science and service-dominant logic. In: Otago Forum, vol. 2. Otago (2008)
2. Vargo, S.L., Lusch, R.F.: Evolving to a new dominant logic for marketing. J. Mark. **68**(1), 1–17 (2004)
3. Vargo, S.L., Lusch, R.F.: Service-dominant logic: continuing the evolution. J. Acad. Mark. Sci. **36**(1), 1–10 (2007)
4. Vargo, S.L., Lusch, R.F.: Institutions and axioms: an extension and update of service-dominant logic. J. Acad. Mark. Sci. **44**(1), 5–23 (2015)
5. Spohrer, J., Maglio, P.P., Bailey, J., Gruhl, D.: Steps toward a science of service systems. Computer **40**(1), 71–77 (2007)
6. Spohrer, J., Maglio, P.P.: The emergence of service science: toward systematic service innovations to accelerate co-creation of value. Prod. Oper. Manag. **17**(3), 238–246 (2009)
7. IfM & IBM: Succeeding through service innovation: a service perspective for education, research, business and government. In: Cambridge Service Science, Management and Engineering Symposium, University of Cambridge Institute for Manufacturing, Cambridge (2008)
8. Maglio, P.P., Vargo, S.L., Caswell, N., Spohrer, J.: The service system is the basic abstraction of service science. Inf. Syst. e-Bus. Manag. **4**(7), 395–406 (2009)
9. Pine, J., Gilmore, J.H.: Welcome to the experience economy. Harv. Bus. Rev. **76**(4), 97–105 (1998)
10. Alcoba, J., Mostajo, S., Clores, R., Paras, R., Mejia, G.C., Ebron, R.A.: Tourism as a life experience: a service science approach. In: Nóvoa, H., Drăgoicea, M. (eds.) IESS 2015. LNBIP, vol. 201, pp. 190–203. Springer, Cham (2015). doi:10.1007/978-3-319-14980-6_15
11. Alcoba, J.: Beyond the paradox of service industrialization: approaches to design meaningful services. In: Wang, J. (ed.) Management Science, Logistics and Operations Research. IGI Global, Hershey (2014)

12. Alcoba, J., Mostajo, S., Paras, R., Mejia, G.C., Ebron, R.A.: Framing meaningful experiences toward a service science-based tourism experience design. In: Borangiu, T., Drăgoicea, M., Nóvoa, H. (eds.) IESS 2016. LNBIP, vol. 247, pp. 129–140. Springer, Cham (2016). doi: 10.1007/978-3-319-32689-4_10
13. Jain, S., Gupta, G.: Measuring service quality: SERVQUAL vs SERVPERF scales. VIKALPA **29**(2), 25–37 (2004)
14. Krishnan, C.: Zero defect management – a study on the relevance in modern days. Int. Res. J. Eng. Technol. **02**(05), 578–581 (2015)
15. Aurel, T., Simina, R., Stefan, T.: Continuous quality improvement in modern organizations through Kaizen management. In: 9th Research/Expert Conference with International Participations "QUALITY 2015", pp. 27–32. B&H, Neum (2015)
16. Alcoba, J.: The paradox of service industrialization and the creation of meaning. Int. J. Serv. Sci. Manag. Eng. Technol. **3**(2), 50–62 (2012)
17. Wolak, R., Kalafatis, S., Harris, P.: An investigation into four characteristics of services. J. Emp. Gen. Mark. Sci. **3**, 22–43 (1998)
18. Pine, J., Gilmore, J.: The Experience Economy: Work is Theatre and Every Business is a Stage. Harvard Business School Press, Boston (1999)
19. Lasalle, D., Britton, T.: Priceless: Turning Ordinary Products into Extraordinary Experiences. Harvard Business School Press, Boston (2003)
20. Hassenzahl, M., Tractinsky, N.: User experience – a research agenda. Behav. Inf. Technol. **25**(2), 91–97 (2006)
21. Kuipers, F., Kooij, R., Vleeschauwer, D., Brunnström, K.: Techniques for measuring quality of experience. In: Osipov, E., Kassler, A., Bohnert, T.M., Masip-Bruin, X. (eds.) WWIC 2010. LNCS, vol. 6074, pp. 216–227. Springer, Heidelberg (2010). doi:10.1007/978-3-642-13315-2_18
22. Migueis, V., Novoa, H.: Using user-generated content to explore hotel service quality. In: Borangiu, T., Drăgoicea, M., Nóvoa, H. (eds.) Exploring Services Science. LNBIP, vol. 247, pp. 155–169. Springer, Heidelberg (2016)
23. Reichheld, F.: The one number you need to grow. Harv. Bus. Rev. **82**(6), 133 (2004)
24. Godbey, G.: Leisure and Leisure Services in the 21st Century. Venture Publishing, State College (1997)
25. Wyllie, R.W.: Tourism and Society. Venture Publishing, State College (2000)
26. Dotchin, J.A., Oakland, J.S.: Total quality management in services Part 2: service quality. Int. J. Qual. Reliab. Manag. **11**(3), 27–42 (1994)
27. Zeithaml, V.A., Berry, L.L., Parasuraman, A.: The behavioral consequences of service quality. J. Mark. **60**, 31–46 (1996)
28. Kara, A., Lonial, S., Tarim, M., Zaim, S.: A paradox of service quality in turkey: the seemingly contradictory relative importance of tangible and intangible determinants of service quality. Eur. Bus. Rev. **17**(1), 5–20 (2005)
29. Albacete-Saez, C.A., Fuentes-Fuentes, M.M., Llorens-Montes, F.J.: Service quality measurement in rural accommodation. Ann. Tour. Res. **34**(1), 45–65 (2007)
30. Pawitra, T.A., Tan, K.C.: Tourist satisfaction in Singapore – a perspective from Indonesian tourists. Manag. Serv. Qual. **13**(5), 399–411 (2003)
31. Tian-Cole, S., Cromption, J.L.: A conceptualization of the relationships between service quality and visitor satisfaction, and their links to destination selection. Leis. Stud. **22**(1), 65–80 (2003)
32. Sparks, R., Westgate, M.: Broad-based and targeted sponsorship strategies in Canadian women's ice-hockey. Int. J. Sports Mark. Sponsors. **4**(1), 59–84 (2002)
33. Babikas, E., Bienstock, C.C., Van Scotter, J.R.: Linking perceived quality and customer satisfaction to store traffic and revenue growth. Decis. Sci. **35**(4), 713–737 (2004)

34. Gotlieb, J.B., Grewal, D., Brown, S.W.: Consumer satisfaction and perceived quality: complementary or divergent constructs? J. Appl. Psychol. **79**(6), 875–885 (1994)
35. Parasuraman, A., Zeithaml, V., Berry, L.: SERVQUAL: a multi-item scale for measuring customer perceptions of service quality. J. Retail. **64**(1), 12–40 (1988)
36. Bolton, R.N., Drew, J.H.: A multistage model of customers' assessments of service quality and value. J. Consum. Res. **17**, 375–384 (1991)
37. Abu Ali, J., Howaidee, M.: The impact of service quality on tourist satisfaction in Jerash. Interdiscip. J. Contemp. Res. Bus. **3**(12), 164–187 (2012)
38. Ebrahimpour, A., Haghkhah, A.: The role of service quality in development of tourism industry (2010). https://ssrn.com/abstract
39. Bernthal, M.J., Sawyer, L.L.: The importance of expectations on participatory sport event satisfaction: an exploration into the effect of athlete skill level on service expectations. Sport J. **7**(3) (2004). http://www.thesportjournal.org/. Accessed 6 June 2012
40. Brady, M.K., Cronin, J.J.: Some new thoughts on conceptualizing perceived service quality: a hierarchical approach. J. Mark. **65**, 34–49 (2001)
41. Bitner, M.J.: Evaluating service encounters: the effects of physical surroundings and employee responses. J. Mark. **54**, 69–82 (1990)
42. Jennings, G., Lee, Y., Ayling, A., Lunny, B., Cater, C., Ollenburg, C.: Quality tourism experiences: reviews, reflections, research agendas. J. Hosp. Mark. Manag. **18**(2–3), 294–310 (2009)
43. Haven-Tang, C., Jones, C.: Delivering quality experiences for sustainable tourism development: harnessing a sense of place in Monmouthshire. In: Morgan, M., Lugosi, P., Ritchie, J.R.B. (eds.) Tourism and Leisure Experience: Consumer and Managerial Perspectives, vol. 44, pp. 163–181. Channel View Publications, Bristol (2010)
44. Fernandes, T., Cruz, M.: Dimentions and outcomes of experience quality in tourism. The case of port wine cellars. J. Retail. Consum. Serv. **31**, 371–379 (2016)
45. Otto, J., Ritchie, B.: The service experience in tourism. Tour. Manag. **17**(3), 165–174 (1996)
46. Kim, H., Woo, E., Uysal, M.: Tourism experience and quality of life among elderly tourists. Tour. Manag. **46**, 465–476 (2015)
47. Benjamins, V.R., Cadenas, D., Alonso, P., Valderrabanos, A., Gomez, J.: The voice of the customer for Digital Telcos. In: 13th International Semantic Web Conference. ISWC, Trentino (2014)

Multivariate Analysis of EU Convergence in Higher Education Services

Alina Mihaela Dima[(✉)], Simona Vasilache, and Shahrazad Hadad

The Bucharest University of Economic Studies, Calea Grivitei 2-2A,
010731 Bucharest, Romania
{Alina.Dima,Simona.Vasilache,
Shahrazad.Hadad}@fabiz.ase.ro

Abstract. The paper analyses the evolution of convergence in higher education during 2002–2013, based on previously used macroeconomic and transition indicators: Gross Domestic Product (GDP) per capita, Research and Development (R&D) investment level, government expenditure on education (% of GDP), share of exports of high-tech in total manufactured exports, costs of exploiting intellectual property. We resort to cluster analysis to identify and characterize main groups of higher education systems in Europe. Using forecasting techniques for the 2014–2020 period, we foresee the dynamics of the previously outlined clusters, and the perspectives of convergence in the near future. We analyse the factors facilitating or inhibiting convergence, and the necessary exchanges between groups of countries which are mostly prone to harmonization. The paper integrates previous studies of the authors regarding European level convergence, bringing together best practices and strategies to achieve convergence not only regionally, but also within the European higher education sector.

Keywords: European Union · Economic convergence · Higher education · Multivariate analysis · Forecasting techniques

1 Introduction

The debates on European level convergence have not left apart, in the last years, the field of higher education. On the contrary, in our opinion, they should stem from here, from analysing the convergence perspectives, at European level, in a sector which is genuinely future-oriented. We have focused, thus, on outlining the mechanisms through which economic convergence and higher education systems convergence, in terms of inputs, as well as outputs, influence each other, and we have forecasted the perspectives of convergence in EU, in the horizon of year 2020. The analysis we have performed is based on clusters perspective, starting from the assumption that, in various sectors such as education and R&D, as well as at a macroeconomic level, countries have common behaviour or converge in groups. We have analyzed the evolution of the clusters, between the moment of the Lisbon strategy adoption and the 2020 milestone, in order to interpret the effects of these evolutions, from both perspectives: economic convergence and convergence in higher education. Our analysis revealed that the two

© Springer International Publishing AG 2017
S. Za et al. (Eds.): IESS 2017, LNBIP 279, pp. 272–282, 2017.
DOI: 10.1007/978-3-319-56925-3_22

types of convergence influence each other, in the sense that, for instance, investments in R&D depend on economic growth and, on its turn, economic development depends on patents and licenses, which are R&D products. Regardless of the level of convergence, drawing especially on the education stream, higher education institutions need to get thoroughly involved in service innovation, which has arisen from the growth of service activities across various industries, in order to improve their position in an ever competing market place [1].

2 Literature Review

Several concerns mark the future higher education landscape and among them are the need for increased financial support, the shrinking population of young students, the decline in the number of non-traditional students, and changes in the educational process itself due to increased digital literacy of the students [2]. While the reshaping of education in universities as public institutions is highly dependent on the evolution of the forms of the mind [3] what happens with the economic systems is significant in configuring the future of education. Traditionally, universities were repositories of the past, perpetuating its cultural patterns and lessons learned [4]. Education, in this paradigm, was transmitting images of yore. A major shift was Toffler's approach [5] "All education springs from images of the future and all education creates images of the future. Thus all education, whether so intended or not, is a preparation for the future. Unless we understand the future for which we are preparing, we may do tragic damage to those we teach. Unless we understand the powerful psychological role played by images of the future in motivating – or de-motivating – the learner, we cannot effectively overhaul our schools, colleges or universities, no matter what innovations we introduce." [6]. Although economic forecasts may not prove sufficient for volatile matters as digging into the future of mentalities, they are, however, necessary. The link becomes more obvious in the case of universities of applied sciences, which are shaping the future labour market, and are also being shaped by it, in the bi-univocal relationship identified by Toffler.

The relationship between education and growth is a highly debated topic in the literature, mostly analyzing "catch-up" or disparity reduction models and trying to estimate factors and predict timing when CEE countries will get the level of most advanced countries of the EU. Literature abounds in theories where education is a critical factor of growth and economic development [7]. Thus, starting from the neo-classical model of Solow [8], continuing with [9], and [10] who demonstrated that a higher education level makes the labour force more able to deal with technological innovations, education was seen as a motor for productivity [11], a directly productive investment [12], a catalyst for regional cooperation and innovation [13] or simply a condition for development, and, finally, a driving force for growth.

[14] reinforces the triple role of the education mechanism in the economy: education stimulates the effectiveness of the individual in production, facilitates the spread of innovations, and stimulates research and technological progress. However, the issue of causality is a main concern among the empiricists: is higher education a driver/ influencing factor for GDP growth or is it the other way around? The same concern

applies to the relationship between education and income inequality. Trying to overcome these limits, the Walrasian model used by [15] tested various relations between educational attainment, GDP and income inequality and suggested that this mechanism is quantitatively more important than a causal relationship between inequality and growth.

However, a considerable approach on the variation in countries' experience of growth remains unexplained by education. [16] remarked that countries where educational allocation is relatively unequal across individuals, they tend to perform relatively poorly in terms of economic growth. Later on, [14] concludes that the quality and distribution of education offer us some extra insight into the relationship between education and growth.

The education impact is to raise the personal competences or to improve the performances of the individuals [17]. Another effect of education is to increase the flexibility of labour and its capacity to absorb new ideas, to adapt it to the new technologies or to apply them at the working place. In the literature, human capital is an essential element of innovation and research and development activities [18].

Many papers develop the idea that creativity contributes to innovation and other organizational outcomes, principle applicable also to higher education institutions [19]. Moreover, on a broader sense, education for service innovation ensures the filling of the current gap between the demand for and the availability of skilled workers while placing enormous pressure on higher education institutions to deliver consistent results, otherwise being discarded as irrelevant [20].

[14] suggested that it is important to understand how groups of countries get formed into convergence clusters, and how different policy (investment in education, R&D, etc.) can change the structure of these clusters, in order to stimulate the growth and reduce economic disparities in time.

Researchers [21–25] analysed the convergence between de old EU member states and new EU member states in terms of convergence of gross domestic product per labour-unit. In reality, these studies confirm the theoretical expectations since the integration process should bring transition economies some important advantages aimed at accelerating the growth of real gross domestic product per labour-unit.

Underlying this disparities reduction was the recognition that higher education institutions and systems were central to the achievement of Europe's economic and social goals [26]. For the less developed countries, education will be a key factor to rapid absorption of technology and to catch up with the level of the most advanced countries since in the last decades, the European Union has made remarkable progresses in designing and implementing measures for a greater European unity [27].

Ultimately, the added value of the convergence will bring additional investments in infrastructure, human resources, modernization and diversification of regional economies. The social and economic convergence contributes finally to the improvement of the institutional capacity and to the modernization of public administration, increasing transparency and encouraging good governance [28].

3   Research Methodology

We have selected five indicators relevant for higher education convergence, which link economic conditions to outcomes in research and innovation: growth of GDP (annual %), R&D investment level, government expenditure on education (% of GDP), charges for the use of intellectual property, and high technology exports (% of manufactured exports). These indicators are based on World Bank data, updated as of 2014. The first indicator characterizes, broadly, the economic background against which future convergence in education is assessed, while the following two indicators reflect economic inputs into the educational system, and the last two economic outputs of the higher education system.

Thus, we achieve a balance allowing us to analyse convergence in the input, as well as in the output quadrant, and spot possible disequilibria. Our hypothesis is that convergent outputs depend on convergent inputs, that is, economic convergence influences and determines higher education convergence. However, as intellectual capital accumulates, this relationship becomes non-linear, and achieved outputs will, in their turn, impact economic convergence, speeding up its pace.

We hypothesise that the relation between input and output convergence may be expressed as follows, in an inverse variation formula:

$$OutC = k/InC \tag{1}$$

In other terms, the more the intellectual capital accumulates, the less output convergence is dependent on input variables, and becomes self-sustaining.

To prove this, we have used a dynamic multiple correspondence analysis, outlining, for the inputs, as well as for the outputs quadrants, on the background of the general GDP evolution, the clusters of countries which are closer together. We see, on the 2002–2013 timeframe, which are the modifications and in which way inputs influence outputs, and vice-versa. We then use forecasting techniques to assess the future evolution of the selected indicators until 2020, and redo the multiple correspondence analysis in order to diagnose further trends and their influence on the economic-educational convergence binomial. Thus, our main purpose in this analysis is to combine a highly visual tool, offering insights into how inputs, outputs and the relationship between them modifies in real time, with a future-mapping tool, which outlines trends of evolution, allowing a prospective analysis as well as the elaboration of strategies.

The forecasting analysis is an essential tool for national governments, which would offer them valuable insights in real time, information that could be further used for the correction of certain impact factors [29]. Our forecasting analysis is based on a dynamic regression, assessing the future evolution of a variable as a consequence of past and present evolutions of that variable and of other variables related to it, that is, on a multivariate data set. We have used a single-equation model, taking output indicators as relevant phenomena in the analysis, and the parameters influencing their future evolution as explanatory variables.

While multiple correspondence analysis techniques are highly visual, allowing us to see the disposition of countries in the bi-dimensional space, forecasting techniques offer us a more systematic perspective on how the EU system is likely to evolve, from a convergence perspective. In the context in which the concept of convergence being time-dependent, but also linked to entities coming together, we believe that the complementary nature of the two sides of the analysis is beneficial to our goals which are directed towards describing and visualising, as well as towards systematically understanding how convergence is likely to take place, in the future.

However, although we consider exports of R&D outcomes among our indicators, the countries we analyse are not closed systems, especially in the higher education field. Thus, mutual influences should be included in the model, in order to give a more accurate overview of the co-evolution of the considered countries. Another limitation consists of the lack of information on private flows of capital in the higher education field, which would have given a more complete image of the factors driving convergence.

4   Analysis and Interpretation

We have plotted the input variables – share of GDP allocated to education, share of GDP allocated to R&D and GDP annual growth (%), on two dimensions, using multiple correspondence analysis, at various moments in time. Dimension D1 is the general financial allocation, representing R&D expenses and GDP growth, while dimension D2 is the specific financial allocation, reflecting GDP share allocated to education. The results are presented below (Fig. 1):

Figure 1 accounts for the visual representation of the multiple correspondence analysis varies, over the 10 years considered, from a relative concentration of the European countries, accounting for a more convergent state of their educational systems, to a dissipation of the educational systems, which is obvious especially as a preamble of the economic crisis of 2008.

It may be seen that, at the initial moment of the analysis, there are two distinct clusters, one represented by advanced European economies (Northern Europe, UK, France, Belgium, Austria), the other one represented by aspiring, developing economies (Romania and Bulgaria). The first cluster is characterized by larger annual increases in GDP and by larger investments in R&D, contrasting to their investments in education. Consequently, developed countries become less interested in education as such, and more oriented towards research achievements that will, on their turn, influence their productivity and economic growth. Our hypothesis is, at this point, confirmed. On the other hand, the countries aspiring, at the moment, to join the EU invest very little in R&D, and not so much in education, thus not being able to achieve performance in either of the sectors. At this instant, education and R&D are regarded as 'luxury' sectors, which are most likely to attract massive investments after other urgent issues are solved, in other words, education and R&D are not perceived, by developing countries, as sectors situated in the first line of convergence.

In 2005, we may see a reorientation of some of the developed countries, while the developing countries' investments in education incur a decrease. Some developing countries, as UK, Austria, Germany, increase their investments in education and

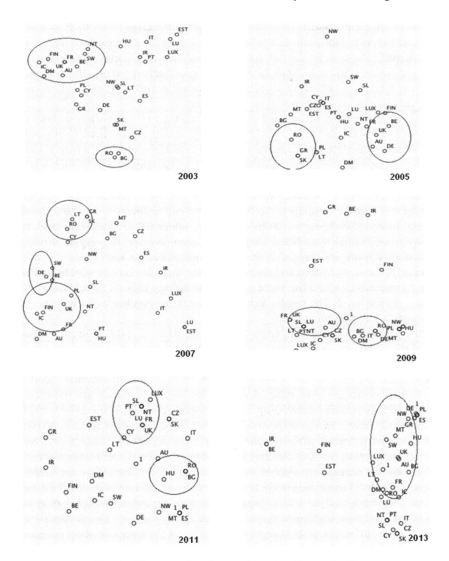

Fig. 1. The countries clusters dynamics (inputs level)

decrease their investments in research, on a trend of slowing down of their increase in GDP. According to OECD data, as GDPs begin to fall, investments in education increase, in developed European countries. This somewhat paradoxical situation can be explained by a justified belief in the regeneration of the economy through a well-developed and financed educational and research system.

In 2007, at the moment of their accession to the European Union, developing countries exhibit an increase of their investments in research, to bridge the gap with advanced industries, on the background of a moderate investment in education.

Developed European countries, on the background of the crisis, limit their investments in R&D, and invest moderately in education. Thus, possible convergence, at this point, is mismatched, the two groups of countries being as far apart as in 2003, after a potential of coming closer, in 2005. In 2009, the crisis again reorients all countries from research to education, where investments are easier to sustain. Under these circumstances, an apparent convergence may be foreseen.

However, the trend is maintained in 2011, when the two groups of countries begin to reinvest in both education and R&D, following similar tendencies. An alignment is visible, between the developed and the developing countries clusters, which is likely to perpetuate in the future. The increased absorption of European funds, which are, then, used in the national systems, may explain the path towards convergence in the R&D investment field. Also, the requirements of the Bologna process may trigger an equalizing of funding initiatives in the higher education fields. Thus, countries becoming more convergent in terms of inputs allocated to education and R&D, according to our initial hypothesis, they should become more convergent in terms of outputs, that is, more convergent from an economic standpoint.

Similarly, we have plotted the output variables, namely the costs with intellectual property (dimension 1), and the share of high technology exports (% of manufactured exports) – dimension 2. The results, using multiple correspondence analysis, at various time moments, are presented in Fig. 2.

Figure 2 displays patterns of aggregation from 2003 to 2009, based on the availability of data. We may see that the patterns are roughly similar, in the case of outputs, compared with previously analysed inputs. Although the funding initiatives are rather heterogeneous, the outputs are somewhat stable, with groups of countries tending to maintain their positions, over time. However, developed countries are more convergent, and the gap between developing and developed countries is increasing, the latter being oriented more towards exporting high-tech products, than towards exploiting their intellectual property. *Although there isn't a noticeable misbalance between the inputs and the outputs level, in the long run more convergent investments in education and R&D do not necessarily lead to more convergent outcomes, thus correcting our initial hypothesis.* A conjunction of legal and political factors, besides the financial aspect that we have analysed here, leads to a proper or improper exploitation of the intangible goods, thus leading to their adequate selling, or not. The economic crisis, however, leaves its marks on the outputs, as well as on the inputs, leading to countries, even the developed ones, being less able either to sell their high tech products, or to exploit their intellectual property. Contrarily, developing countries try to foster innovation, and begin to take the fruits of investing in applied research, as their relation to intellectual property improves, as of 2007. But it may be, as well, a relative improvement, as compared to the slowdown of the developed nations, as 2009 features a different picture, with developing countries coming back to exporting their intellectual products, rather than exploiting them for their own economic good. The interference of the economic crisis leads to an inconsistent pattern of evolution, for both developed and developing clusters, which mismatches the initial separation between the two axes: developed countries – exploitation of their intellectual property, developing countries – export of their high-tech products.

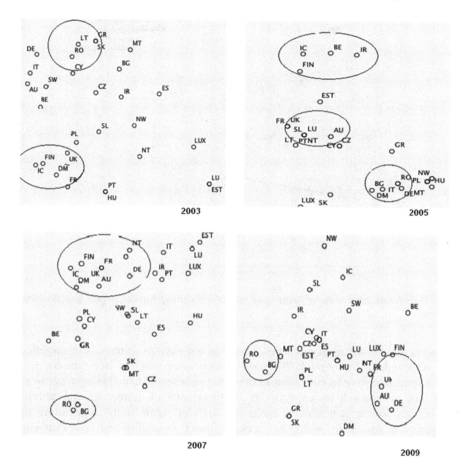

Fig. 2. The countries clusters dynamics (outputs level)

We have further employed economic forecasting techniques, for both inputs and outputs, for the 2014–2020 time frame, in order to assess potential convergence in the near future. The results are presented in Fig. 3, where the horizontal axis represents the time frame, while the vertical axis is the economic growth of the considered groups of countries:

Figure 3 It may be seen that developed countries gradually slow down, in both input and output indicators, a trend which was outlined also by the cluster analysis, while developing countries are slightly catching up. In the absence of any crisis, the two sides of the balance, the input and the output indicators, evolve similarly, with a more pronounced gap in the case of developing countries, due to the previously discussed reasons. However, due to economic exchanges and competition pressure, as well as legislative harmonization in the field of IP, convergence is faster in the field of outputs, because of legislative tension, and European-wide competition between educational and research systems, that push to convergence, as in the field of inputs, where

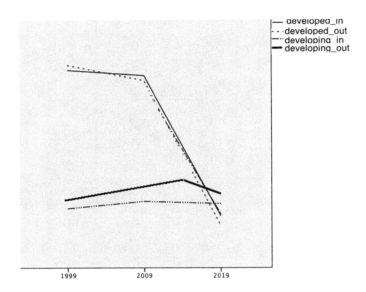

Fig. 3. Forecasted evolutions of input and output indicators, for developed and developing countries

it involves similar financial resources, which is not easy to achieve. Convergence is rather achieved, in the horizon 2020, on a divestment trend of rich nations, which reorient form investing in education to investing in research and finding ways to gain money out of research by engaging in different sorts of entrepreneurial activities, allowing developing countries to catch up, without significantly augmenting their investments in education, which has a clear impact on quality and competitiveness, quality being also improvable through service innovation.

5   Conclusions

Our research aimed at linking inputs to outputs of higher education and research in the EU, outlining similar or dissimilar future patterns of evolution. Also, we have checked for the influence of economic convergence on education convergence, and vice-versa. Our analysis showed that, although eclectic, due to various external factors, among the most prominent is the economic crisis, patterns of evolution of financial inputs and intangible outputs seem to be similar, across the clusters which are formed. Also, economic conditions significantly impact the dynamics of future convergence in the higher education area. At the times achieved, convergence is not a static phenomenon, exhibiting sometimes significant variations every two years, as our mapping shows. Further investigations should, then, be dedicated to ways to consolidate convergence, or to avoid fake convergence, which appears due to economic conditions, making developed countries to invest less in education and research, thus giving the *impression* that developing countries are catching-up. However, our present analysis, by combining visual patterns and more analytical, forecasting techniques, gives a fair overview of the links between the desirable convergence in higher education, and other related economic and social moves.

Convergence within the European Union will enhance the solidarity principle between EU member states. Also, it will contribute to the achievement of its fundamental objectives, namely, to support the aims of economic growth (higher education R&D investments in certain regions of the EU are positively correlated to innovation and ultimately to economic growth [30]), the improvement of the social and territorial cohesion, disseminating the benefits of the single market within the European Union, while enhancing the overall competitiveness of the EU economy [28].

References

1. Danjum, I., Rasli, A.: Imperatives of service innovation and service quality for customer satisfaction: Perspective on higher education. Procedia-Social Behav. Sci. **40**, 347–352 (2012)
2. Blass, E., Jasman, A., Shelley, S.: Visioning 2035: the future of the higher education sector in the UK. Futures **42**(5), 445–453 (2010)
3. Gidley, J.M.: Evolution of education: from weak signals to rich imaginaries of educational futures. Futures **44**, 46–54 (2012)
4. Readings, B.: The University in Ruins. Harvard University Press, Boston (1996)
5. Toffler, A.: Learning for Tomorrow: The Role of the Future in Education. Random House, USA (1974)
6. Hicks, D.: The future only arrives when things look dangerous: reflections on futures education in the UK. Futures **44**, 4–13 (2012)
7. Levine, R., Renelt, D.: The sensitivity analysis of cross-country growth regressions. Am. Econ. Rev. **82**, 942–963 (1992)
8. Solow, R.M.: A contribution to the theory of economic growth. Q. J. Econ. **70**, 65–94 (1956)
9. Benhabib, J., Spiegel, M.: The role of human capital in economic development evidence from aggregate cross-country data. J. Monetary Econ. **34**, 143–173 (1994)
10. Barro, R.J., Sala-i-Martin, X.: Disparities reduction. J. Political Econ. **100**, 223–251 (1999)
11. Cuaresma, C.J.: Convergence of educational attainment levels in the OECD: more data, more problems? Econ. Educ. Rev. **25**, 173–178 (2006)
12. Diebolt, C., Jaoul-Gramare, M.: Convergence of higher education and economic growth during the european construction: a contribution to the cliometrics of growth (EU-15). Res. Comp. Intl. Educ. **1**, 14–30 (2006)
13. Santamarta, J.C., Mora-Guanche, A.: Impact of erasmus master programmes on regional innovation and higher education: the case of The Canary Islands. Procedia-Social Behav. Sci. **191**, 1255–1260 (2015)
14. Johnes, J.: Efficiency and productivity change in the English higher education sector from 1996/97 to 2002/03. Working Papers 575369, Lancaster University Management School, Economics Department (2006)
15. Teulings, C.N., van Rens, T.: Education, growth and income inequality. Tinbergen Institute, Discussion Papers 02-001/3, Tinbergen Institute (2002)
16. Castello, A., Domenech, R.: Human capital inequality and economic growth: some new evidence. Econ. J. R. Econ. Soc. **112**, C187–C200 (2010)

17. Carneiro, P., Heckman, J.J.: The evidence on credit constraints in post-secondary schooling. Econ. J. **112**, 705–734 (2002)
18. Winston, T.H.K., Leung, K.M.: Education, Technological Progress and Economic Growth. Research Collection School of Economics (2003). http://ink.library.smu.edu.sg/soe_research/1184
19. Ahmad, I., Zafar, M.A., Shahzad, K.: Authentic leadership style and academia's creativity in higher education institutions: intrinsic motivation and mood as mediators. Transylvanian Rev. Adm. Sci. **11**, 5–19 (2015)
20. Baumann, T., Mantay, K., Swanger, A., Saganski, G., Stepke, S.: Education and innovation management: a contradiction? How to manage educational projects if innovation is crucial for success and innovation management is mostly unknown. Procedia-Social Behav. Sci. **226**, 243–251 (2016)
21. Verblane, U., Vahter, P.: An analysis of the economic convergence process in the transition countries. FEBA Working Paper No. 37 (2005)
22. Estrin, S., Urga, G.: Convergence in output in transition economies: Central and Eastern Europe 1970–1995. Centre for Economic Policy Research (1997)
23. Lenain, P., Rawdanowicz, Ł.: Enhancing Income Convergence in Central Europe after EU Accession. OECD Economics Department Working Papers 392. OECD Publishing (2004)
24. Sohinger, J.: Growth and convergence in European transition economies. East. Eur. Econ. **43**, 73–94 (2005)
25. Alho, K.E.O., Kaitila, V., Widgrén, M.: Speed of convergence and reallocation: new EU member countries catching up with the old. ETLA Discussion Papers. No. 963 (2004)
26. Middlehurst, R., Teixeira, P.: Governance within the EHEA: dynamic trends, common challenges and national particularities. In: Scott, P., Curaj, A., Vlasceanu, L., Wilson, L. (eds.) European Higher Education at the Crossroads: Between the Bologna Process and National Reforms, pp. 573–598. Springer, Dordrecht (2012)
27. Dan, H.: The euro zone – between fiscal heterogeneity and monetary unity. Transylvanian Rev. Adm. Sci. 68–84 (2014)
28. Szilard, L., Lazăr, I.: The European Union Cohesion Policy in Romania - strategic approach to the implementation process. Transylvanian Rev. Adm. Sci. **8**, 114–142 (2012)
29. Rădulescu, M., Banica, L.: Neural networks-based forecasting regarding the convergence process of CEE countries to the eurozone. Transylvanian Rev. Adm. Sci. **10**, 225–246 (2014)
30. Bilbao-Osorio, B., Rodríguez-Pose, A.: From R&D to innovation and economic growth in the EU. Growth Change **35**, 434–455 (2004)

Does Community Service Make Any Difference in University Rankings?

Carmen Păunescu[✉]

UNESCO Department for Business Administration, Bucharest University of Economic Studies,
Piaţa Romană 6, District 1, 010374 Bucharest, Romania
carmen.paunescu@ase.ro

Abstract. The paper analyzes the way in which different attributes of the community service are taken into consideration among the indicators used in international university rankings. Also, the paper discusses the extent to which these factors influence positioning of the university in the overall ranking. For this purpose, eleven international rankings of world universities have been examined to identify those indicators that address community service issues. These are meant to measure how seriously a university takes its obligations to society by investing in community. Furthermore, two main criteria form the basis of analysis, namely social responsibility and regional engagement, as these include explicit measures for community service. The results suggest that there is a positive correlation between ranking in social responsibility and the overall ranking for universities investigated, and between ranking in regional engagement and overall ranking, but of different intensity.

Keywords: University · Community service · International rankings · Social responsibility · Regional engagement

1 Introduction

There is a growing popularity of university rankings and the criteria used to assess universities worldwide. International rankings of universities gain more and more interest and priority from students, university administrators, academics, government officials and businesses as well. A number of rankings are published internationally, most of them aiming to identify the top universities in the world. The literature in the field includes various discussions around what attributes form a world-class university [1–5]. Being placed in one of these international rankings provides a strong publicity for a university and offers guarantees for retention of high-quality students and academics from all over the world.

As part of the higher education landscape, international rankings have a double impact on university strategy. On one hand, there is a positive impact quantified through the increased public awareness about the quality of education and research reputation, transparency of the institution, degree of internationalization or overall competitiveness [6]. On the other hand, the impact is negative because of the somehow exclusive nature

© Springer International Publishing AG 2017
S. Za et al. (Eds.): IESS 2017, LNBIP 279, pp. 283–294, 2017.
DOI: 10.1007/978-3-319-56925-3_23

of the methodologies used, bias in the number, choice and weighting of indicators, or lack of consideration of the whole complexity of the higher education system [7].

Universities in developing countries are also interested in relevant rankings that could assist them with developing appropriate policies to become more competitive worldwide. Much criticism has been directed to the criteria used in rankings as the universities are different to one another, fulfil various roles and have distinct strengths [8, 9]. Complexity and multiple functions of higher education systems, the diversity of their missions, the lack of a consensus on what constitutes the quality of an educational system, and the lack of universally available, measurable indicators to reflect these functions and missions explain criticism on university rankings [3]. Thus, more rankings for higher education institutions are being developed to compete with the established ones.

The current paper analyzes the role played by the university's community service in international rankings. It is an attempt to distinguish those indicators, which measure the community service role of a university, that are taken into consideration in university rankings. The extent to which different aspects of the community service are reflected among the criteria and indicators used for universities' rankings is discussed. Also, the paper analyzes the extent to which these factors affect positioning of the university in the overall ranking. The paper starts with a review of how university community service have progressively evolved. Then it explains the methodology used to assess the community service role of university in international rankings. Finally, analysis is conducted on several university rankings and conclusions are drawn based on results.

2   The Community Service Role of University

Universities have always been engines for the economy, in their triple role they play, trying to adapt continuously to the environmental changes that are constantly evolving. Not only they try to align their offerings with the current expectations of the students, employers and society as a whole, but universities should also assume an active role in the external community where they are present [10]. Therefore, they have to act as a binder between academic and community actors, necessary to create new skills and jobs required at the local or regional level. Moreover, universities have to discover the way in which they will engage in community service, by either being part of it or starting it from the scratch [11].

Currently, modern universities incorporate, as an integral part of their mission, the contribution to the economic and social development of local and regional communities, considering it a main function of their existence, besides the educational and research purposes.

The existing literature defines community service as a process in which people from all levels of the society unite their forces to improve living conditions and the level of development in the community [12]. Community service resides in the idea that people share common goals and values that help them build and develop in the same direction and on common grounds. The values can be either humanistic values like care for others and environment accountability or a sense of identity and belongingness based on social, cultural or ideological aspects.

Nowadays, universities can bring an enormous power in the changing process of the society, as well as they play a leading role in helping communities develop a way of sustainable growth. Capacity building of the professors and students seems to be one of the main ways of working towards enhancing community service [13]. However, building capacity becomes very hard and critical in current globalization era, where resources are getting scarce and technology is ever changing [14].

Formal collaborations between community groups and academic institutions to promote economic and social development have increased substantially over the past 10 years. The bulk of research on community-university partnerships has focused more on the experiences of institutions of higher education, leaving a gap in the understanding of community experiences [15]. There has also been a strong emphasis on partnerships or relations that the universities can develop with the outside world, that lead to knowledge creation and a beneficial exchange [16, 17]. No real social or environmental problem can be solved or even simply understood without these two groups working together.

Community service, as the third function of a university besides education and research, has been central in any higher education discourse for many years [12, 15, 17, 18]. However, the concept is defined differently by various scholars. Some view it as a university's service to the community by knowledge dissemination and good practices implementation [19], while others perceive it as part of the entrepreneurial culture of the universities that can bring extra income necessary to support education and research [20]. Also, the term is used simultaneously with other various terminologies for the same scope, like for example community engagement, outreach or extension [21].

With the new competition, universities themselves had to create better connections with the external communities in their areas, promote a more caring social image, and supply constantly added value in order to attract students, academics, donors and employers [22]. By doing so, many universities have developed partnerships and networks with inside and outside communities, have invested in R&D departments that promote social responsibility and have developed worldwide projects on social innovation. Certainly, universities have developed themselves as a more powerful actor on the social map.

Sanderson and Watters (2006) argue that this recent change in the higher education system refers to the "corporatisation" of higher education [23]. Thus, universities take part of the environment as part of a loop. They are generators of skilled human capital, they are relying on qualified human capital and afterwards, they improve the value of the human capital generated. Since universities are part of many loops, not only this one, they play a significant role on the scene of community service and its development, as they have to be involved in educating people in this regard, in creating the responsibility, as well as in sustaining businesses and communities to do so or they do it themselves.

3   Methodology

The main research questions to which the paper aims to answer are two-fold: How is the community service role of university regarded among the indicators used for international university rankings? To what extent does university's community service influence its positioning in the overall ranking? The paper attempts to prove that the university's community service plays a key role in positioning of universities in international rankings. Other scholars have also researched how the social role of universities is taken into consideration in rankings to determine the quality and characteristics of a university. For example, Lukman et al. (2010) developed a model, which enables a comparison between universities with respect to their three missions: research, educational and environmental [7]. The model enables inclusion of additional indicators (like environmental performances) and allows for a better understanding of university development and for fostering improvement. Also, by referring to two European initiatives (U-Map and U-Multirank), Boulton (2011) comments on how different roles fulfilled by universities can be compared in meaningful ways [24].

Table 1. World university rankings investigated and their community service-related indicators

No.	University international ranking	Indicators linked to community service
1	Academic Ranking of World Universities/ ARWU	To some extent: per capita academic performance of an institution
2	Center for World University Rankings/ CWUR	To some extent: broad impact
3	CHE University Ranking	To some extent: town and university, job market and career orientation, academic studies and teaching
4	Leiden Ranking/CWTS	N/A
5	Performance Ranking of Scientific Papers of World Universities/PRSPWU	N/A
6	QS-Top Universities Ranking/QS	To some extent: academic reputation
6a	QS Stars University Ratings/QS Stars	Social responsibility To some extent: inclusiveness
7	Scimago Institutions Ranking/Scimago	Societal impact (web size, domain's inbound links)
8	Times Higher Education–World University Rankings/THE	To some extent: teaching reputation and research reputation
9	University Ranking by Academic Performance/URAP	N/A
10	U-Multirank	Regional engagement
11	World's Best Universities Rankings/US-News	To some extent: global and regional research reputation

For the purpose of this paper eleven international rankings of world universities have been examined to identify those criteria/indicators that include community service aspects (Table 1). These indicators are meant to measure how seriously a university takes its obligations to society by investing in community service as well as to assess the overall performance of higher education institutions. The university rankings selected for investigation have been scrutinized to identify the extent to which different aspects of university's community service are considered in their methodologies and successively measured.

The rankings have been selected based on their relevance for the current research and, mainly, based on their popularization and recognition worldwide.

Both deductive and inductive content analysis has been performed over all accessible methodologies and list of criteria/indicators for each one of the university international ranking researched. The choice for the method has been done based on its particular usability in the context of emerging fields of research, where only some literature already exists. The results of the analysis led to the list of indicators presented in Table 2, which relate, to different extents, with community service aspects. Following this list of indicators and given the complexity of the data extracted, two university rankings have been selected for further research, namely QS Stars University Ratings and U-Multirank, as they include explicitly indicators which measure the university's community service. These indicators are further introduced in the coming section of the paper. As such, the following criteria formed the basis of the current analysis: social responsibility (for QS Stars) and regional engagement (for U-Multirank). However, future research should analyze the impact of aligned indicators which measure community service in several university rankings and compare their influence in those rankings.

To answer the two-fold questions, the paper relies on the data available on the online platforms of the two university rankings. Data processed in the paper refers to QS Stars University Ratings 2016 and U-Multirank 2016.

4   Discussion

Analysis of data was performed independently for each one of the two university rankings investigated. Comparisons are following and lessons are drawn successively.

4.1   Data Analysis for QS Stars Universities Ranked in Social Responsibility

The social responsibility role is often central to the university's community service mission. According to QS Stars methodology, social responsibility of a university is measured against four criteria: community investment and development, charity work and disaster relief, regional human capital development, and environmental impact. As a result of this evaluation, universities are categorized in five groups by the number of stars, starting from 1 to maximum 5 stars.

Nowadays, universities are key part of modern communities, aim to directly improve and develop the areas in which they are present, and inevitably gain from the regions in which they are based, both financially and in more intangible ways. Also, universities invest in community by supporting charities and different national and international social causes and by getting their students involved in the local or global environment for the benefit of society [25]. Universities are interested in investing in their regional human capital development as well. Additionally, impacting positively on the environment is another aim of educational institutions focused on developing future generations of professionals and leaders.

Our analysis conducted on QS Stars universities ranked in social responsibility reveals that 88 universities across the globe are ranked in this category, out of which 41 are 5-stars universities, 9 are 4-stars, 20 are 3-stars, 7 are 2-stars and 11 are 1-star universities (Table 2). QS Stars ratings offer a good opportunity for universities to compare themselves in different categories, like social responsibility, and learn from good practices for other universities better ranked in the category.

Table 2. Distribution of QS Stars universities ranked in social responsibility, by country (Total: 88 universities ranked).

No. of stars	5 stars	4 stars	3 stars	2 stars	1 star
Total	41 out of which:	9 out of which:	20 out of which:	7 out of which:	11 out of which:
Social responsibility	Mexico 6	Indonesia 2	Malaysia 3	Malaysia 2	India 2
	Brazil 4	Egypt 1	Columbia 2	Chile 1	Kazakhstan 2
	Australia 3	Jordan 1	Lithuania 2	Colombia 1	Netherlands 2
	Indonesia 3	Malaysia 1	Mexico 2	Indonesia 1	Australia 1
	UK 3	Peru 1	UK 2	Kazakhstan 1	Peru 1
	USA 3	Russia 1	China 1	Russia 1	UK 1
	Ireland 2	Spain 1	Costa Rica 1		United Arab Emirates 1
	Spain 2	UK 1	Ecuador 1		
	Chile 1		Netherlands 1		
	Columbia 1		Russia 1		
	Cyprus 1		Saudi Arabia 1		
	Ecuador 1		South Africa 1		
	Honduras 1		United Arab Emirates 1		
	Iraq 1		Vietnam 1		
	Jordan 1				
	New Zealand 1				
	Pakistan 1				
	Peru 1				
	Philippines 1				
	Thailand 1				
	Turkey 1				
	Ukraine 1				
	Vietnam 1				

A comparison of the extents to which the universities ranked by QS Stars in social responsibility are found in other QS Stars ratings reveals interesting results. Thus, out of the 41 5-stars universities ranked in social responsibility, 37 universities are found in QS Stars overall ranking, 36 in ranking by employment, 9 in ranking by innovation and 31 in ranking by inclusiveness (Table 3). These findings demonstrate that there is a strong positive correlation between rankings in social responsibility and employment, on one hand, and rankings in social responsibility and inclusiveness, on the other hand. Also, a better ranking in social responsibility corroborates with a higher overall ranking.

Table 3. Frequency with which QS Stars universities ranked in social responsibility are found in other QS Stars rankings.

Social responsibility rating	Criteria	No. of universities in each star category					
		Total	5 stars	4 stars	3 stars	2 stars	1 star
5 stars (n = 41)	Overall ranking	39	10	9	19	1	–
	Employment	39	27	8	3	1	–
	Innovation	9	8	–	1	–	–
	Inclusiveness	31	26	4	1	–	–
4 stars (n = 9)	Overall ranking	9	1	1	4	3	–
	Employment	9	4	2	2	1	–
	Innovation	2	–	1	1	–	–
	Inclusiveness	6	4	1	1	–	–
3 stars (n = 20)	Overall ranking	20	–	6	9	5	–
	Employment	19	8	3	2	5	1
	Innovation	3	3	–	–	–	–
	Inclusiveness	14	11	2	1	–	–
2 stars (n = 7)	Overall ranking	7	–	–	3	4	–
	Employment	6	1	1	2	1	1
	Innovation	2	2	–	–	–	–
	Inclusiveness	5	4	1	–	–	–
1 star (n = 11)	Overall ranking	11	–	3	5	3	–
	Employment	11	3	2	3	2	1
	Innovation	3	1	–	2	–	–
	Inclusiveness	7	2	2	2	1	–

Similarly, out of the 9 4-stars universities ranked in social responsibility, 8 universities are found in QS Stars overall ranking, 9 in ranking by employment, 2 in ranking by innovation and 6 in ranking by inclusiveness. Also, out of the 20 3-stars universities ranked in social responsibility, 20 universities are found in QS Stars overall ranking, 19 in ranking by employment, 3 in ranking by innovation and 14 universities in ranking by inclusiveness (Table 3). Finally, out of the 7 2-stars universities, 6 are found in ranking by employment and 5 in ranking by inclusiveness and out of the 11 1-star universities, 11 are found in ranking by employment and 7 in ranking by inclusiveness. Therefore, the correlations found above are also confirmed for 4-stars, 3-stars, 2-stars and 1-star universities ranked in social responsibility.

A comparison for each criterion by number of stars reveals also interesting results. Thus, out of the 41 5-stars universities ranked in social responsibility, 24 are 5-stars universities ranked in employment and 26 are 5-stars universities ranked in inclusiveness. Therefore, we might conclude again that there is a positive strong correlation between ranking in social responsibility and ranking in employment and inclusiveness. Also, out of the 9 4-stars universities ranked in social responsibility, 6 are 5- and 4-stars universities ranked in employment and 5 are 5- and 4-stars universities ranked in inclusiveness. And, out of the 20 3-stars universities ranked in social responsibility, 13 are

5-, 4- and 3-stars universities ranked in employment and 14 are 5-, 4- and 3-stars universities ranked in inclusiveness.

These results suggest that the better the ranking of a university in social responsibility, the higher the likelihood of university is to be better ranked by employment and/or inclusiveness. Also, a better ranking in social responsibility will increase the likelihood of the university to be better positioned in the overall ranking.

4.2   Data Analysis for U-Multirank Universities Ranked in Regional Engagement

U-Multirank introduces a multi-dimensional model of assessment of universities' performances that takes into consideration diversity of higher education's mission, including the community service role. According to the U-Multirank methodology, regional engagement is used to measure the university's presence in the community by taking into account indicators like: student internships in the region, bachelor and master theses with regional organizations, regional joint publications, graduates working in region, and income from regional sources. As result of this evaluation, a university is given scores from E (weak), to D (below average), C (average), B (good), up to A (very good).

There are 1195 universities included in the ranking by regional engagement. The distribution of universities by country and by top scores (from 5A up to 1A) is shown in Table 4. The universities investigated cover all broad subject areas (from education, humanities, arts and social sciences, to agriculture, health and engineering), offer bachelor and master level of study programs and take consideration of all indicators for regional engagement listed by U-Multirank.

The analysis reveals that out of 1195 universities included in the ranking, 2 universities are ranked at the highest level of performance, which is 5A, 8 at 4A level, 54 at 3A level, 80 at 2A level and 332 at 1A level of performance (Table 4). The other levels of performance that are following are ranging from 5B, 4B, 3B, 2B, 1B, where B stands for good, to C categories (average), further to D categories (below average) and up to 5E, 4E, 3E, 2E, 1E, where E stands for weak. Data summarized in Table 4 suggest that 476 out of 1195 universities, which represent 39.83% manage to achieve a very good level of performance in regional engagement, at least for one of the indicators measured in this category. Also, it is worth to mention that about 35% of the universities included in the ranking rank below average in regional engagement.

Our analysis of the extents to which university ranking in regional engagement affects its positioning in overall ranking reveals interesting results. Thus, out of the top 30 universities ranked in regional engagement, 6 universities (20%) are found in top 100 overall ranking, 14 (46,6%) in top 200 overall ranking, 19 (63,3%) in top 300 overall ranking and 24 (80%) in top 400 overall ranking. The left 6 universities are not included in top 400 overall ranking (Table 5). These findings suggest that there is a positive correlation between ranking in regional engagement and the overall ranking, but of low intensity. Further research should investigate possible correlations and synergies between rankings in regional engagement and other dimensions used in U-Multirank (e.g., teaching and learning, research, knowledge transfer, international orientation) to determine the extent to which they might affect positioning of a university in overall ranking.

Table 4. Distribution of U-Multirank universities ranked in regional engagement, by country (Total: 1195 universities ranked).

Top scores	5A	4A	3A	2A	1A
Total	2 out of which:	8 out of which:	54 out of which:	80 out of which:	332 out of which:
Regional engagement	Australia 1	Russia 3	Russia 6	Spain 7	France ≥ 18
	Lithuania 1	Spain 2	Ireland 4	Latvia 6	Italy ≥ 18
		Latvia 1	Lithuania 4	Poland 6	Japan ≥ 17
		Romania 1	Poland 4 Romania 4	Lithuania 5	China ≥ 16
		Switzerland 1	Australia 3	Portugal 5	Korea ≥ 17
			Italy 3 Portugal 3	Belgium 4	Spain ≥ 12
			Belgium 2	Russia 4	Japan ≥ 11
			Norway 2	Finland 3	Australia ≥ 10
			Ukraine 2	France 3	Israel ≥ 10
			Belarus 1	Italy 3	Russia ≥ 10
			Brasilia 1	Australia 2	Taiwan ≥ 10
			Bulgaria 1	Ireland 2	US ≥ 10
			Canada 1	Romania 2	Poland ≥ 9
			Ciudad de Mexico 1	Slovenia 2	Germany ≥ 9
			Cyprus 1 Egypt 1	Austria 1	Netherland ≥ 9
			Germany 1	Belarus 1	Russia ≥ 8
			India 1	Canada 1	Ukraine ≥ 8
			Malaysia 1	Chile 1	Ireland ≥ 6
			Panama 1	Columbia 1	Portugal ≥ 6
			Spain 1	Croatia 1	Romania ≥ 6
			Saud Arabica 1	Cyprus 1	Turkey ≥ 5
			Slovenia 1	Czech 1	Belgium ≥ 4
			Turkey 1	Denmark 1	UK ≥ 4
			Uruguay 1	Egypt 1	Greece ≥ 3
			Vietnam 1	Estonia 1	India ≥ 3
				Hungary 1	Latvia ≥ 3
				India 1	Lithuania ≥ 3
				Iraq 1	Norway ≥ 3
				Israel 1	Brasilia ≥ 2
				Lebanon 1	Bulgaria ≥ 2
				Luxembour 1	Canada ≥ 2
				Malaysia 1	India ≥ 2
				Moldova 1	Mexico≥ 2
				Namibia 1	etc.
				Netherland 1	
				South Korea 1	
				Slovakia 1	
				Ukraine 1	
				UK 1	
				US 1	

Table 5. Overall ranking of top 30 universities ranked in regional engagement.

U-Multirank		
Top 30 universities ranked in regional engagement	Country	Overall ranking of the university (in top 400 universities)
1 Edith Cowan U	Australia	81
2 Lithuanian U Education Sci	Lithuania	137
3 Aurel Vlaicu U Arad	Romania	237
4 Autonomous U Barcelona	Spain	213
5 Higher School Economics	Russia	325
6 U AS & Arts Western Switzerland	Switzerland	94
7 Don State Tech U	Russia	199
8 Moscow Inst. Physics & Tech	Russia	162
9 Riga Stradins U	Latvia	346
10 U Barcelona	Spain	104
11 Queensland U Tech	Australia	Not in top 400
12 U Ramon Llull	Spain	49
13 U Salento	Italy	Not in top 400
14 Siberian Federal U	Russia	260
15 South Ural State U	Russia	Not in top 400
16 Vilnius Gediminas Tech U	Lithuania	78
17 Kaunas U Tech	Lithuania	123
18 Lithuanian U Health Sci	Lithuania	Not in top 400
19 Lomonosow Moscow State U	Russia	2
20 Sumy State U	Ukraine	60
21 Polytech. Inst. Braganca	Portugal	324
22 Indian Inst. Tech Kharagpur	India	127
23 King Andulaziz U	Saud Arabia	110
24 Polytech. Inst. Lisbon	Portugal	Not in top 400
25 Lobachevsky State U	Russia	337
26 Nizhny Novgorod Tech U	Russia	Not in top 400
27 U Sherbrooke	Canada	217
28 Swinburne U Tech	Australia	159
29 Belarusian State U	Belarus	274
30 U Ljubliana	Slovenia	326

5   Conclusions

The paper brings new perspectives on the role played by the university's community service in international rankings and the extent to which indicators that measure the university's service in community should be defined and quantified in those rankings. Our analysis revealed that there are several indicators already defined to a certain extent in the ranking methodologies, as well as measured in practice, which are meant to assess the university's community service.

The results of our investigation on QS Stars University Ratings and U-Multirank show that the degree to which universities manifest involvement in community service is strongly correlated with the university's commitment to graduates' employability and to accessibility of the university to students. Also, the higher the university commitment to community service, the better the positioning in the overall ranking. A number of implications can be derived from this study, particularly for policy makers in developing countries, who design policies and methodologies to assist universities becoming more competitive worldwide and to fulfil various roles and distinct strengths the universities have.

Our study, like most studies, is not without limitations. Thus, the research does not compare all the rankings as there are not aligned indicators for community service to do so. A future research should try to find indicators in the different university rankings that could be aligned with the community service variables, which will allow for comparisons and could prove the influence of those variables in different university rankings. Also, further research could test possible correlations and synergies between rankings in social responsibility and regional engagement, on one hand, and other dimensions used in university rankings, like knowledge transfer, innovation, research, quality of teaching, on the other hand.

Despite its limitations, this paper offers valuable insights for those academics and government institutions who may want to review the effectiveness of current university ranking methodologies of their country and make changes in order to foster community service and to improve the university's positioning in the ranking.

References

1. Daraio, C., Bonaccorsi, A., Simar, L.: Rankings and university performance: a conditional multidimensional approach. Eur. J. Oper. Res. **244**(3), 918–930 (2015)
2. Marconi, G., Ritzen, J.: Determinants of international university rankings scores. Appl. Econ. **47**(57), 6211–6227 (2015)
3. Millot, B.: International rankings: universities vs. higher education systems. Int. J. Educ. Develop. **40**, 156–165 (2015)
4. Ordorika, I., Lloyd, M.: International rankings and the contest for university hegemony. J. Educ. Policy **30**(3), 385–405 (2015)
5. Olcay, G.A., Bulu, M.: Is measuring the knowledge creation of universities possible? A review of university rankings. Technological Forecasting and Social Change (2016). http://dx.doi.org/10.1016/j.techfore.2016.03.029
6. Serhiivna, S.Y., Fedorivna, G.O., Oleksandrivna, B.V.: The international and national university rankings as a constituent of university's competitiveness. Market. Manage. Innovations. **1**, 183–195 (2016)
7. Lukman, R., Krajnc, D., Glavič, P.: University ranking using research, educational and environmental indicators. J. Clean. Prod. **18**(7), 619–628 (2010)
8. Kroth, A., Daniel, H.D.: International university rankings - a critical review of the methodology. Zeitschrift fur Erziehungswissenschaft **11**(4), 542–558 (2008)
9. Buela-Casal, G., Gutierrez-Martinez, O., Bermudez-Sanchez, M.P., et al.: Comparative study of international academic rankings of universities. Scientometrics **71**(3), 349–365 (2007)

10. Mtawa, N.N., Fongwa, S.N., Wangenge-Ouma, G.: The scholarship of university-community engagement: interrogating Boyer's model. Intl. J. Educ. Develop. **49**, 126–133 (2016)
11. Dlouhá, J., Huisingh, D., Barton, A.: Learning networks in higher education: universities in search of making effective regional impacts. J. Clean. Prod. **49**, 5–10 (2013)
12. Checkoway, B.: Community development, social diversity, and the new metropolis. Community Develop. J. **46**(suppl 2), ii5–ii14 (2011)
13. O'Rafferty, S., Curtis, H., O'Connor, F.: Mainstreaming sustainability in design education – a capacity building framework. Int. J. Sustain. High. Educ. **15**(2), 169–187 (2014)
14. Shiel, C., Leal Filho, W., do Paço, A., Brandli, L.: Evaluating the engagement of universities in capacity building for sustainable development in local communities. Eval. Program Plann. **54**, 123–134 (2016)
15. Afshar, A.: Community-campus partnerships for economic development: community perspectives. http://www.bos.frb.org/commdev/pcadp/2005/pcadp0502.pdf
16. Bruning, S.D., McGrew, S., Cooper, M.: Town-gown relationships: exploring university-community engagement from the perspective of community members. Public Relat. Rev. **32**(2), 125–130 (2006)
17. Pacheco, P., Motloch, J., Vann, J.: Second chance game: local (university-community) partnerships for global awareness and responsibility. J. Clean. Prod. **14**(9–11), 848–854 (2006)
18. Acworth, E.B.: University-industry engagement: the formation of the knowledge integration community (KIC) model at the Cambridge-MIT Institute. Res. Policy **37**, 1241–1254 (2008)
19. Weerts, D.J., Sandmann, L.R.: Building a two-way street: challenges and opportunities for community engagement at research universities. Rev. High. Educ. **32**(1), 73–106 (2008)
20. Guan, J., Zhao, Q.: The impact of university-industry collaboration networks on innovation in nanobiopharmaceuticals. Technol. Forecast. Soc. Chang. **80**(7), 1271–1286 (2013)
21. Bender, G.: Exploring conceptual models for community engagement at higher education institutions in South Africa. Perspect. Educ. **26**(1), 81–95 (2008)
22. Boyle, M.E., Silver, I.: Poverty, partnerships, and privilege: elite institutions and community empowerment. City Community **4**(3), 233–253 (2005)
23. Sanderson, D.M., Watters, J.J.: The corporatisation of higher education: a question of balance. In: Debowski, S. (ed.) Proceedings Higher Education Research and Development Society of Australia Annual Conference, Perth, Western Australia, pp. 316–323 (2006)
24. Boulton, G.: University rankings: diversity, excellence and the european initiative. Procedia Soc. Behav. Sci. **13**, 74–82 (2011)
25. Hammersley, L.: Community-based service-learning: partnerships of reciprocal exchange? Asia Pac. J. Cooperat. Educ. **14**(3), 171–184 (2013)

Digital Transformation at the University of Porto

José António Faria[1]([✉]) and Henriqueta Nóvoa[2]

[1] INESC TEC and Faculty of Engineering, University of Porto,
Rua Dr. Roberto Frias, 4200-465 Porto, Portugal
`jfaria@fe.up.pt`
[2] Faculty of Engineering, University of Porto,
Rua Dr. Roberto Frias, 4200-465 Porto, Portugal
`hnovoa@fe.up.pt`

Abstract. Digital transformation has become a priority for any organization and Universities are not an exception. This paper deals with the digital transformation that is being undertaken at the University of Porto (U. Porto). In order to understand its major guidelines, it is important to clarify the term "digital transformation" and "digital business strategy", comparing these new concepts with prevailing views of aligning the business strategy with the IT strategy. A newly created Shared Services Center (UPSSC) is at the core of the digital business strategy of U. Porto. This endeavor triggered a profound university wide re-engineering, that was enabled by a new work management system, hereafter referred to as UPWMS. The discussion of the distinctive characteristics of this work management system and how it copes with the diversity and heterogeneity of the requests received by the Shared Services, is the main objective of the paper. A discussion of the main cornerstones of the project in the context of the digital transformation strategy finishes the paper.

Keywords: Digital transformation · Digital business strategy · Business process management · Knowledge based services

1 Introduction

Digital transformation (DX) has become a priority for any organization. Nevertheless, exploiting efficiently all the opportunities and potentialities open up by the wealth of digital technologies available, redefining completely business models across the entire value chain is not straightforward and, for sure, is a challenging task. This challenge is more pressing for organizations that permanently try to assure its competitive positioning in a global market, but the same concern is becoming pertinent for universities, as competition to select the best students and researchers is increasing. This natural quest for efficiency and quality improvement powered by technologies is a natural trend, and institutions have to be able to rise above new challenges.

This paper deals with the digital business strategy of U. Porto, a large public University in Portugal. This strategy encompasses multiple dimensions and actions, but this paper will mainly focus in one of its core elements, called UPWMS – the U. Porto work management system, designed to support the operation of the recently created Shared Service Center (UPSSC).

© Springer International Publishing AG 2017
S. Za et al. (Eds.): IESS 2017, LNBIP 279, pp. 295–308, 2017.
DOI: 10.1007/978-3-319-56925-3_24

First, the terms "digital transformation" and "digital business strategy" are introduced and compared with the prevailing views of aligning the business strategy with the IT strategy. Framing the concept according to literature is the focus of the first section of this paper. After a brief introduction to U. Porto, the main directions of the digital business strategy will be depicted in Sect. 3.

Section 4 will focus on the work management system UPWMS, as a key piece of the digital business strategy, and Sect. 5 introduces the conceptual framework relating the work taxonomy and the work lifecycle considered in UPWMS.

Finally, the paper finishes with a discussion of the main cornerstones of the project, the implementation problems faced in this particular facet of the digital transformation strategy, as well as prospective developments.

2   Conceptual Background

Over the past decades, strategic planning of information systems shed a light on the importance of having an IT and IS strategy aligned with the business strategy, as IT technologies and new applications boomed. The strategic information systems era that started in the 80's, after the so-called DP (Data Processing) and MIS (Management Information Systems) eras, had the goal of improving competitiveness by changing the nature or conduct of business, clearly stating that IS/IT investments were a source of competitive advantage [1].

Digital transformation transposed the prior approach to an all-new level. In fact, according to [2], a digital transformation strategy impacts a company more comprehensively than an IT strategy, by addressing potential effects on interactions across company borders with clients, competitors and suppliers. As clearly stated by [3], the role of IT strategy has passed from a functional-level strategy, to one that reflects a fusion between IT strategy and business strategy. This powerful idea of fusion between IT and business strategy, considering IT as essential to the overall business strategy, is also present in [4].

The key factors that characterize the strategic information systems approach are also present in a digital transformation strategy, such as [1]: (1) an external, not internal focus, (2) adding value, not cost reduction, (3) sharing the benefits, (4) understanding customers and what they do with the product or service, (5) a business-driven innovation, not technology-driven, (6) incremental development and (7) using the information gained from the systems to develop the business.

In fact, a recent framework published by [5] to formulate a digital business strategy recovered and redefined four of these dimensions, namely: (1) the use of technologies, reflecting the capability of the company to explore new technologies, (2) changes in value creation, reflecting the influence of digital transformation on a firm's value creation, (3) structural changes, referring to the modifications in organizational structures necessary to exploit the new technologies and, finally (4) the financial aspects, namely the ability to finance a digital transformation endeavor.

The transformational ability of technologies to re-define the company's business model is at the core of a digital business strategy [5].

In summary, according to [3], the idea behind digital transformation and a digital business strategy is simply that of an organizational strategy formulated and executed by leveraging digital resources to create additional and differentiated value. The creation of value is at the core, coming from elevating the performance implications of IT strategy beyond efficiency and productivity metrics to those that drive competitive advantage and strategic differentiation [3].

In the next section, the major guidelines of the digital transformation approach adopted by U. Porto will be presented.

3   The Digital Transformation at U. Porto

The University of Porto is a public university founded in 1911, with a number of students close to 30 thousand. The University integrates fourteen schools and fifty research centers spread in three campus [6]. The schools were founded at different moments of the long history of the university, therefore having different cultures and procedures. Thus, we can say that within the University there is a recognized heterogeneity of processes and practices, that if, in one hand, is a manifestation of its richness, is also very challenging to manage. The Rectory of U. Porto oversees this federate organization, balancing at every moment the righteous aspiration of autonomy of each of the schools, with the need to improve the overall efficiency of the University as a whole [7].

The University has a consolidated management information system developed in the past twenty years, a system that manages all the scientific and academic activities of the University [8]. The information and management eco-system also integrates an ERP that deals with the economic and financial processes, and several other modules for human resources management, performance assessment of academic staff and library management.

In summary, and for the sake of the work ahead, it is important to highlight the following distinctive features of the University: (1) being geographically spread by three different poles, in different locations of the city, (2) having several schools with heterogeneous procedures, due to different backgrounds and cultures and, finally, (3) having a shared and consolidated management information system.

As U. Porto is a very active organization, always striving for a permanent quality improvement of its processes and services, the principles of digital transformation have been intensely discussed indoors – how to create that additional differential value in our products and services, being recognized by our peers? In fact, public politics for modernization and streamlining of administrative processes are already underway at a national level, being a key concern of the central government; but, these generic guidelines have to be consubstantiated in innovative and well thought approaches, that take into account the capabilities of each organization. Due to the size of the University and the hundreds of processes managed daily, re-engineering and certification of all the processes and the de-materialization of document management, keeping at the same time the agility of the technologic infrastructures, is a very exigent process, requiring an innovative approach to succeed.

A synthesis of the key digital transformation guidelines that resulted from this internal discussion and from the main cornerstones of a DX is depicted in Fig. 1.

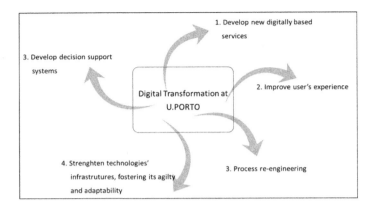

Fig. 1. Principal guidelines of the digital transformation at U. Porto

Digital transformation at U. Porto had a turning point in 2013 with the creation of a new Shared Services Center (UPSSC), with the objective of improving the quality of services delivered to the academy and reducing the overall costs. At that time, it was decided that instead of having administrative units in each school, running similar, but slightly different processes, it would be more efficient to centralize these services in a Shared Service Center. In order to be successful, this new approach required a profound re-engineering of all the supporting processes, a task that had to be dealt with profound sensitivity and attention, in order to overcome the natural resistance to change of the different schools. The main focus of this SSC is directly linked to guideline 3 of the digital business strategy (Fig. 1), as the initial priority of this center was precisely to re-engineer all current processes performed at an university level. Plus, the new center had to quickly demonstrate its ability of providing superior quality services adapted to the specific needs of each school. It was soon recognized that to accommodate this challenge it was crucial to develop a management system that could streamline transversal processes, supplying at any point in time all the required information to anyone involved in a particular process, irrespectively of its physical location [7].

It has to be noted that, although the vast majority of processes were already digitally processed at the time of the creation of the UPSSC, units such as Legal Support and Human Resource often used email in its internal communications. With the UPSSC, the process de-materialization became imperative, as process requests are now dealt either at the central unit, or at the local schools.

Anyway, as clearly stated in Sect. 2, digital transformation goes well beyond de-materialization of processes, encompassing an innovative use of new technologies (cloud, social, mobile and analytics) as a way to promote new services, re-define business models and innovative interactions with its users. Thus, in summary, a digital business strategy at the university should:

- Facilitate the communication, collaboration and co-creation of value in all the stakeholders (management, technicians, teachers, students and external community);
- Optimize process management within organic units and throughout all the University;
- Speed up the development and adaptation of processes and services according to new societal requirements and legislative and regulatory changes, as well as organizational alterations;
- Improve data and information usage in all the decision support processes, either at an operational or strategic level, allowing decisions to be taken based in data and real data;
- Separate management and process execution from the physical place where the process is carried out;
- Widen service's offering at an University level;
- Allow a more effective knowledge sharing, fostering learning and quality improvement.

It was quickly perceived from the intended digital business strategy, that traditional workflow engines couldn't cope with the heterogeneity of processes and cultures of the different schools [7]. The lack of flexibility of standard BPM suites is widely documented in the literature [9, 10]. This fact lead the University management to decide to develop the UPWMS platform, an agile platform that could handle adaptive and emergent processes. The objective was two-fold: (1) in a first phase, achieve in a short-period, a successful de-materialization of processes with all the associated documentation and integrating its multiple variants, and (2) in a second phase, the developed system had to possess the agility to evolve, harmonizing and consolidating procedures as they naturally arise.

The main features of the UPWMS platform as a critical component of an overarching Digital transformation of U. Porto will be presented in the next section.

4   The UPWMS Work Management System: A Key Piece in the DX

As highlighted in the previous section, U. Porto engaged a profound organizational transformation in 2013 with the creation of a Shared Services Centre UPSSC. This movement made the full process dematerialization a mandatory and urgent issue as the processes' workflow travels across the different Schools of the University and the central units.

Given the heterogeneity of the processes, it was assumed that conventional workflow applications would not be an effective solution to support the operation of UPSSC, and the Dean Office decided to engage on the development of an agile work management platform. In the first stage, the platform should allow to dematerialize the full range of business processes of the University (several hundreds) in a relatively short time and accommodate the multiple variants in their procedures. In a second stage, the work management system was expected to promote the harmonization, consolidation and optimization of the working procedures.

This commitment resulted in the UPWMS - Administrative Process Management System, a system that handles and manages all the requests directed by the academic community to the UPSSC, as well as all the internal processes triggered subsequently. With the UPWMS, everyone involved in the fulfillment of the business processes works together to ensure that requests are handled expeditiously, being in accordance with the best interests of all the parties and in conformance with the legal framework. In summary, the UPWMS assures that:

- using the front office virtual helpdesk, the members of the academic community register their service requests to UPSSC and monitor their progress;
- in the back office, the area managers of the different units of UPSSC (Human Resources, Accountability, Funding, Procurement, Law Office Units, etc.) plan, assign and monitor the work to be done;
- the staff members manage the processes in "their hands", recording the dates of the tasks, archiving the documents produced, exchanging messages with other stake-holders in the processes, i.e. with colleagues and managers to inform and get informed about what is required at every step of the process;
- the different school's boards monitor the activity developed by UPSSC through activity reports and service level indicators integrated with business intelligence mechanisms, providing a comprehensive vision of the on-going business processes, a critical view for effective decision making.

4.1   Work Taxonomy

The work to be done in organizations may be of a very different nature, but, for the sake of simplicity, we will classify it in three major types: processes, projects and tickets. Figure 2 shows a high level work taxonomy as described hereafter.

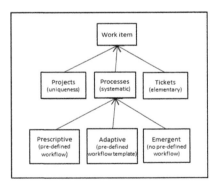

Fig. 2.   Enterprise work taxonomy

Much like a project, a process also involves a sequence of tasks but executed according to pre-defined standard procedure. Each time a trigger event occurs, a set of tasks is executed according to that predefined procedure. Tickets consist of simpler work items that normally do not justify to be further being decomposed into elementary tasks.

As far as business processes are concerned, several classification may be considered ranging from structured workflows, to semi-structured and ad-hoc business processes [11].

Hereafter, workflow processes, adaptive processes and emergent processes will be considered. Workflow (or prescriptive) processes applies when there is a prescriptive workflow that is fully known in advance, prior to execution start. The main limitation of these processes is their low flexibility, since they are only able to deal with situations that have been predicted a priori and that are contemplated in the process model. However, as remarked by many authors [12, 13], many processes do not possess such a deterministic behavior.

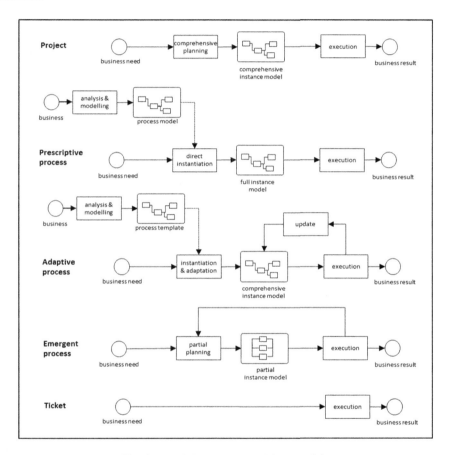

Fig. 3. Work item types workflow models

When process execution is subject to factors that are difficult to anticipate (such as errors, delays, new data that questions earlier decisions), it is not feasible to represent all possible flow paths in the process model. In such cases, only a template workflow providing a reference for the course of action of the processes can be specified. This template will then be further adapted at instance level, according to the events that may occur at run time. These processes are labelled as adaptive processes as they are characterized by a continuous adaptation of the workflow [14, 15].

In an emergent process [16] not even a reference exists for the process workflow. The main stages of the process may be a-priori known, but the detailed work plan has to be designed on the fly, case by case, as new information and data is being collected during process execution. This is the case of knowledge intensive processes, whose management highly depends on the judgement of a specialist, such as a doctor, a lawyer, a design engineer or a top manager facing a complex problem.

Figure 3 summarizes all the work item types introduced and the workflow models that apply to each of them and to their instances.

Adaptive and emergent processes are also known as knowledge based processes, because the course of action depends on the judgement of the situation of the actors involved in the process, and therefore, on their tacit knowledge. All these constraints can hardly be externalized in a formal process model, as required in a standard workflow application.

A wide selection of very specialized work management tools is available in the market, typically targeted to a particular type of work: project management systems suites, business process management suites, case and issues management systems and collaborative platforms (Fig. 4).

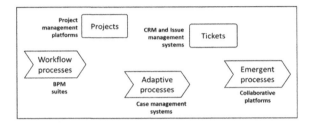

Fig. 4. Widely available work management solutions

Very often, an organization may not find a solution addressing all its needs and has to implement several disconnected work management tools. For sure, this is detrimental to the overall business performance, because the resources (namely people) involved in the different work items are the same, causing overload and inefficiency. UPWMS was designed to overcome this difficulty, as it intends to provide an all-encompassing work management environment, allowing to manage within the same platform the full range of work items of the organization, from projects to knowledge based processes and tickets.

4.2   Work Lifecycle Management

Irrespectively of its type, the lifecycle of a work management item involves a number of standard stages, as sketched in Fig. 5.

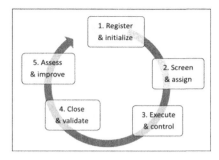

Fig. 5. Work item lifecycle

In order to be managed, any single piece of work should be registered and initialized in the work management system by the requestor. Then, the back-office area manager has to screen incoming requests and assign them for execution to a member of his team. The next stage is task execution (orchestrated by the work manager) and control (by the area manager). The close and validation stage deals with work completion and validation by the interested parties (typically the area manager and the requestor). Finally, relevant data produced along the lifecycle is logged in the performance management systems for further calculation of key performance indicators (KPI's) and assessment of compliance to the service level agreements (SLA's).

This lifecycle applies to each work instance. In a typical work management system, multiple instances will co-exist in a given moment and dispute the set of shared resources for execution. In order to plan, schedule and prioritize effectively the set of concurrent wok items, an overall work management cockpit has to be put in place.

The main management tools belonging to the UPWMS's cockpit are represented in Fig. 6. Globally, these tools allow area managers to plan and control the set of work items within his area, and individual staff to manage the set of tasks in their hands.

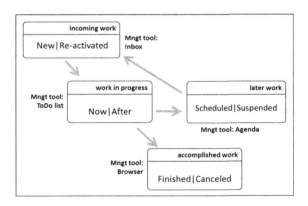

Fig. 6. Work management tools

The design of the cockpit in UPWMS got inspiration in the "Getting Things Done Methodology" [17, 18], namely on the fact that there should be a clear separation between the incoming work that was not yet screened, the work in progress to be done in the short term, the work to be done later that does not require immediate attention, and the work that was already completed.

4.3   From Tasks to Documents and Messages

Looking at a typical process model as the one represented in Fig. 7, we see that besides the tasks of the main workflow, the models contains two other main elements: messages and documents.

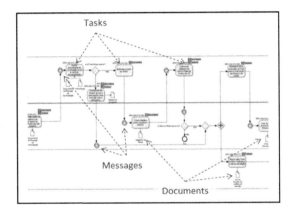

Fig. 7. Typical process model representing tasks, documents and messages

This is particularly true in intensive knowledge based processes once these processes always involve human interaction and collaboration (messages) and deal with semi-structured information (documents) such as reports, contracts, proposals, drawings, etc.

Therefore, a comprehensive work management platform such as UPWMS should provide tools addressing three main dimensions:

- work management: dealing with plans, deadlines and tasks;
- information management: dealing both with structured and unstructured data (documents);
- communication management: dealing with messages, notifications and alerts.

Figure 8 shows a high-level domain model representing the main entities involved in the configuration of the work items, both at type and instance level required to support the three management dimensions introduced above.

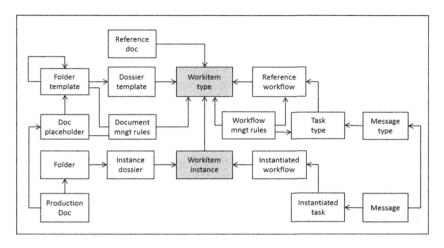

Fig. 8. Work items configuration entities

For each work item type, there is a reference workflow specifying the main phases (for processes) or work packages (for projects), and the set of tasks within each phase or work package. When a new work item instance is created, the corresponding workflow is also instantiated, thus allowing workflow at the instance level to be modified on the fly during the execution. To each work item type there are also assigned typed messages, a structured dossier to hold the documents produced along the process and a set of reference documents, such as procedures, work instructions and documents' templates.

The dossier template is instantiated for each work instance so that it can be modified at instance level. A set of management rules assigned to the work item type specify the kind of changes that are allowed, both in the workflow and in the documents folders. In a prescriptive workflow process, no changes are allowed, whereas in an emergent process both the workflow and the dossier may be fully re-designed for each instance by authorized users granted with administrator permissions level, without needing any support from the IT service.

This way, the set-up of a new work item type can be accomplished by non-technical users that belong to the business units and are fully aware of their needs. Admin users may also fine tune the permissions assigned to each staff member both in terms of access to the content of the work items (status and documents), as well as in terms of the allowed operations (sending messages, uploading and updating documents, changing the workflow, etc.). They can also define milestones and reference delays for task execution. All the data collected during the execution of the work is made available via web services to external business intelligence applications, providing area and top managers with the information they need to manage effectively their units.

5   Discussion and Conclusions

Nowadays, UPWMS is an unavoidable reality in the daily operation of the University and a key enabler of the on-going digital transformation. More than one thousand processes are active and running, processes that are very heterogeneous. In fact, the catalogue of the services provided by the UPWMS to the academic community includes more than one hundred entries, each one corresponding to a different process type. Some of these processes involve just a few tasks and may take just a couple of hours to be solved, while other process types involve several tens of tasks and take months to be concluded. Irrespectively of this, with UPWMS it is possible to know at any moment in time the exact situation of each process, the expected milestones, as well as to access all the data, documents and interactions that took place in its context.

As discussed in the previous section, UPWMS is highly flexible and has the ability to evolve smoothly on time according to the pace of the digital transformation of the University. In an organization such as the Service Shared Center of U. Porto, composed of several business units quite different in terms of size, internal organization, nature of the work to be done, this flexibility is a key factor for the success of the digital transformation of the University. In fact, a digital transformation is always a complex and long lasting project that takes time (a lot of time, in fact). All along this transformation process, the organization is constantly evolving, constantly seeking to improve the services provided. Thus, if the management systems do not have the required agility to adapt to this transformation, instead of being an asset, it becomes a liability and a hurdle. This fact was recognized from the start: the objective was always to have a system that

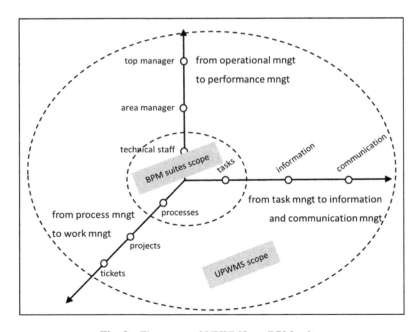

Fig. 9.   The scope of UPWMS vs. BPM suites

naturally evolves as the maturity of the organization evolves, instead of automatizing existent processes, thus compromising immediate and small scale improvements.

It is important to fully understand this fundamental difference between the traditional work management solutions, namely standard BPM suites widely available in the market, and the more comprehensive UPWMS (Fig. 9). In fact, whereas the main focus of the BPM suites is mostly on process automation at an operational level (task enactment and routing), UPWMS has a much broader scope, once it goes from (i) task management to information and communication management, (ii) prescriptive workflow processes to knowledge based processes and tickets and, finally (iii) operational task level to performance management.

We believe that the UPWMS at its current state is clearly aligned with the major guidelines of a digital business strategy: the key concern of constantly adding value to its stakeholders and sharing the benefits, the preoccupation of fully understanding the evolving requirements of very exigent customers, triggering incremental and small-scale developments and, in the end, the cyclic improvement of the system fed by all its users with the final objective of providing better services.

Regardless of all the features and functionalities of the system, we strongly believe that the real DX occurs inside people's minds, when they overcome the natural resistance to change and manage to start working collaboratively, in the end continuously improving work effectiveness. UPWMS has provided the right tools to foster such a collaborative state of mind, which is the essence of a learning organization. As Claudio Ciborra stated [19], the constant "trying out" and experimentation is the true hallmark of organizational change. An idea that definitely applies here: in fact, there is not a single right way of doing a process or an ideal process model, but truly an incremental improvement process.

Although there is still a long way ahead, UPWMS has proved to be a determinant factor for the success of the ongoing digital transformation at U. Porto.

Acknowledgments. The authors would like to thank N. Almeida and M. Fernandes for their invaluable work in developing UPWMS.

References

1. Ward, J., Peppard, J.: Strategic Planning for Information Systems. Wiley, Chichester (2002)
2. Matt, C., Hess, T., Benlian, A.: Digital transformation strategies. Bus. Inf. Syst. Eng. **57**(5), 339–343 (2015)
3. Bharadwaj, A., El Sawy, O., Pavlou, P., Venkatraman, N.: Digital business strategy: toward a next generation of insights. MIS Q. **37**(2), 471–482 (2013)
4. Mithas, S., Tafti, A., Mitchell, W.: How a firm's competitive environment and digital strategy posture influence digital business strategy. MIS Q. **37**(2), 511–536 (2013)
5. Hess, T., Benlian, A., Matt, C., Wiesböck, F.: Options for formulating a digital transformation strategy. MIS Q. Exec. **15**(2), 123–139 (2016)
6. University of Porto (2014). http://www.up.pt
7. Faria, J.A., Nóvoa, H.: An agile BPM system for knowledge-based service organizations. In: Nóvoa, H., Drăgoicea, M. (eds.) IESS 2015. LNBIP, vol. 201, pp. 65–79. Springer, Cham (2015). doi:10.1007/978-3-319-14980-6_6

8. TIC: SIGARRA - information system for the aggregated management of resources and academic records. https://sigarra.up.pt/up/en/WEB_BASE.GERA_PAGINA?P_pagina=2418

9. Mangan, P., Sadiq, S.: On building workflow models for flexible processes. J. Aust. Comput. Sci. Commun. **24**(2), 103–109 (2002)

10. Stavenko, Y., Kazantsev, N., Gromoff, A.: Business process model reasoning: from workflow to case management. Procedia Technol. **9**, 806–811 (2013)

11. Škrinjar, R., Trkman, P.: Increasing process orientation with business process management: critical practices. Int. J. Inf. Manag. **33**(1), 48–60 (2013)

12. Hill, C., Yates, R., Jones, C., Kogan, S.L.: Beyond predictable workflows: enhancing productivity in artful business processes. IBM Syst. J. **45**(4), 663–682 (2006)

13. Alter, S.: Service system fundamentals: work system, value chain, and life cycle. IBM Syst. J. **47**(1), 74 (2008)

14. Sadiq, S.W., Orlowska, M.E., Sadiq, W.: Specification and validation of process constraints for flexible workflows. Inf. Syst. **30**(5), 349–378 (2005)

15. Pesic, M., Schonenberg, H., van der Aalst, W.M.P.: DECLARE: full support for loosely-structured processes. In: 11th IEEE Enterprise Distributed Object Computing Conference (EDOC 2007), p. 287 (2007)

16. Markus, M., Majchrzak, A., Gasser, L.: A design theory for systems that support emergent knowledge processes. MIS Q. Exec. **26**(3), 179–212 (2002)

17. Heylighen, F., Vidal, C.: Getting things done: the science behind stress-free productivity. Long Range Plan. J. **41**(6), 585–605 (2008)

18. Allen, D.: Getting Things Done. Penguin, London (2001)

19. Ciborra, C.: The Labyrinths of Information: Challenging the Wisdom of Systems. Oxford University Press, Oxford (2002)

The Role of Digital Tools and Platforms
for Training Programmes Developed by the Organisations
of the Banking Sector

Maria Menshikova[✉], Alberto Romolini, Illa Sabbatelli, and Marco De Marco

International Telematic University Uninettuno,
Corso Vittorio Emanuele II, 39, 00186 Rome, Italy
{m.menshikova,a.romolini,illa.sabbatelli,
marco.demarco}@uninettunouniveristy.net

Abstract. E-learning technologies represent important instruments for teaching. New technologies play a big role in the financial education of citizens and employees' training activities. In recent years in the banking sector digital tools for education programmes are available in growing numbers and play a significant role. In fact, after the financial crisis, many banks have offered not only online training to their employees but also to their clients, both real and potential. The objective of this approach is to help banks' customers to understand some fundamental financial concepts and the risks of some financial operations. Given this framework, the aim of this paper is to provide an exploratory research of the online training courses and initiatives developed by the banking sector institutions and addressed to both their employees and their real and potential clients. The paper provides a preliminary analysis of the programmes based on the experience of an Italian bank.

Keywords: Digital tools and platforms · Training programmes · E-learning · Financial education · Banking sector

1 Introduction

In recent years e-learning tools are being employed by companies of different sectors and industries. A special demand for such technologies and solutions is noted in the enterprises of the service sphere in general, and in the organisations of the financial and banking sector in particular.

From the one side, we can speak about financial education as an emerging issue after the recent financial crisis that has involved many countries. The lack of financial knowledge has been seen as one of the main reasons for the wrong allocation of individual investments by citizens. Since then, many international institutions have defined principles for financial education in order to strengthen the knowledge of stakeholders' financial issues. From the other side, many companies operating in the financial and banking sector began to perceive e-learning technologies for the continuous training of employees as a fundamental instrument for transformation and innovation of business, as well as its sustainability. Thus, e-learning technologies and digital tools for training

© Springer International Publishing AG 2017
S. Za et al. (Eds.): IESS 2017, LNBIP 279, pp. 309–322, 2017.
DOI: 10.1007/978-3-319-56925-3_25

are an essential element to achieve two main goals: (1) to develop financial education for external users, which is for real and potential clients; and (2) to increase knowledge and skills of internal users, namely employees of an organisation.

Given this approach, the objectives of this paper are to systematise already existing information about the use of digital and web-based tools for internal and external learning by financial institutions, as well as to explore the online training initiatives developed by bank organisations and dedicated to the two particular clusters of users described above (customers and employees). Despite that the need for e-learning and training innovation in the banking sector is relevant in the business communities of different geographical and economic contexts and many countries throughout the world, in this study the focus is on the Italian experience where an emerging need for general financial education can be observed.

The paper provides a description of some digital and web-based initiatives and actions for training addressed to internal and external users in the banking sector, based on the analysis of a single case study related to a big Italian bank, UniCredit. Due to the need of financial education in the Italian context, and the importance of online training for the professional development of banking workers requested by the new business environment in a digital economy, UniCredit has developed a specific approach to new e-learning technologies, exploring new perspectives and creating new tools and platforms to perform the training.

The paper is organised as follows. Section 2 delineates the need for a new financial education in the global context in general, and with a particular attention in Italy, starting from the assumption that financial education can be described as a way to ensure stability of the national and international financial markets. Section 3 analyses the use of e-learning and web-based technologies for internal areas of a banking institution such as staff development and knowledge management. Section 4 explores the initiatives in the field of information and communication technology (ICT) and web-based training through the presentation of the case of an Italian banking institution. The analysis is based on the case study approach [1–3], taking into consideration the experience of UniCredit Bank, one of the biggest European players in the financial market. Finally, Sect. 5 presents the conclusion of the paper, highlighting the contribution of the authors in this research area, as well as outlining managerial implications, limitations and some directions for further research.

2 The Need for a New Financial Education

After the crisis that occurred in the financial markets a very big challenge is represented by "financial education" that could be described as an instrument for providing effective consumer protection, which can impact on the macroeconomic situation [4]. In recent years, the governments and institutions of many countries have understood the importance of financial literacy of their citizens: the lack of financial literacy is one of the main causes of wrong investments [5], as well as over-confidence in financial literacy [6, 7].

As noted by some scholars [8–10] the population has aged and the new generation has the mental attitudes such as "want it now" and "think, buy now, pay later". For these reasons, financial education is considered a "pillar to ensure the stability of the financial markets" [11] and an "essential component to enjoy the benefits of the European Single Market" [12]. As demonstrated by the previously cited documents [11, 12], the European Commission assumes that governments could help national financial markets by developing a national strategy that integrates a methodology for evaluating the level of financial literacy and promotes online educational programmes to increase low financial literacy and reduce lack of transparency that affect enterprises' and citizens' access to financial services.

According to the Policy Handbook created by the Organisation for Economic Cooperation and Development (OECD) [13], the International Network on Financial Education (INFE) adopted the "High-level Principles on National Strategies for Financial Education" in 2012 in order to solve the abovementioned problems.

Similar principles were also approved by the Group of Twenty (G20) in the prior year [11]. Since financial literacy levels are quite low worldwide, the summit of the G20 has insisted on the elaboration of new policies to support general financial literacy through the development of new models.

In April 2016 the OECD published a report [14] that explains that this requirement is particularly pertinent to Europe, given the sophistication of the retail financial markets, impact of the population ageing and reforms of public pension systems. European citizens lack the financial awareness and skills to face these changes. According to the OECD, 21 of the 48 European economies covered in the report use the OECD/INFE Principles, which are planning, developing, implementing or revising national strategies for financial education. Five countries analysed in the report have already revised or are revising their first national strategy; twelve countries are implementing their first national strategy, while four European countries described by the OECD are in the process of designing/developing their first national strategy. It is important to note that some countries have not yet developed a national strategy.

Regarding Italy, financial education is very low and little has been done until last year to develop a national strategy and promote financial education programmes.

Data from the Global Financial Literacy Excellence Center (GFLEC) [15] shows Italy is the country with the lowest level of financial education among major advanced economies. Only 37% of Italian people know the main financial principles. In BRICS countries such as Brazil the level reaches 35% and in South Africa, 42%. The scenario changes depending on age, and some data [16] explain that Italian teenagers are the least prepared among peers in other OECD countries.

Moreover, the Italian Government began planning the implementation of a law (DDL 1196) on financial education and "economic citizenship" only in 2016 [17]. All the financial institutions have been involved in this project: the Bank of Italy, the Italian Securities and Exchange Commission, the Federation of Co-operative Banks and the Ministry of Education. Schools also have undertaken programmes to improve financial literacy as stated in the document of the Ministry of Education signed on 10 June 2015 [18]. Considering the particular emerging situation in Italy described above, financial education in this country may present an interesting case through which to conduct an exploratory research to understand the ways and effects of technological innovation of training in the banking sector. The main elements of this innovation could be e-learning platforms and learning management systems (LMS), Web 2.0 learning environments, as well as Massive Open Online Courses (MOOCs).

The use of MOOCs for financial education programmes can be studied as one of the elements for training innovation. Open education in banking and finance is fundamental and will have an increasing role in ensuing years. Many banks, after the recent crisis, offer free e-learning courses to their clients. The Bank of Italy, in particular, offers an e-learning course "The knowledge of Euro" [19].

Regarding the e-learning platforms and LMS, it should be noted that in most cases we are talking about a closed internal system developed or applied within a single organisation in order to perform some training programmes. Ellis's Field Guide to Learning Management Systems [20], describes LMS as a software application that automates the administration, tracking and reporting of training events. The author highlights that a robust LMS should integrate with other enterprise application solutions, enabling management to measure the impact, effectiveness and overall cost of training initiatives.

With regard to the Web 2.0 learning environment, according to Sigala [21], it can be defined as an open learning environment or "...a social software or interactive collaborative software that affords the development of learning environments with user-centred activities and socially engaging tasks that require users to take an active role in applying and enhancing knowledge by using such tools as weblogs; Wiki; RSS; podcasts; tagging; social search; social gaming or metaverse environments..."

Thus, it can be concluded that the e-learning technologies, platforms, systems and environments are fundamental not only for a general financial education but also for the training of employees operating in a company or in a specific business unit.

3   E-learning and Web-Based Technologies for Staff Development and Knowledge Management

E-learning technologies are not only fundamental modernisation instruments for European universities, but also for enterprises and institutions. New technologies have a big role in the ongoing training of employees and in business development and sustainability.

In particular, banks are increasingly using LMS for educational courses within the organisation. The Interbanking Convention For Problems of Automation (CIPA) in collaboration with the Italian Banking Association (ABI) provides annually an updated report on the use of ICT in the Italian banking system.

In general, when studying the adoption and use of new forms of staff training/learning, it is important to consider some specific features of the banking industry.

The banking industry, which a few years ago was highly protected, recently faced significant changes relating to the competitive environment and technological innovation, as well as institutional changes that have driven companies to a high rate of strategic and organisational transformation. The new competitive scenario requires highly educated professionals within the banking institutions, especially those who have direct contact with customers.

As stated by Zimkova [22] the success of any bank in today's business environment depends upon the satisfaction of its customers, and therefore employees are requested to be able to provide quality financial services. The technological changes (including discussion forums and social networks, chats, e-mails, multimedia tools and Internet-based applications) push the banking institutions to constantly update the service delivery process, as well as to significantly diversify their business portfolios [23]. The institutional changes, however, lead to a reconsideration of corporate strategies related to the organisational structure and professional composition of a bank, as well as pressure on the recovery of efficiency and productivity [24].

One of the transformations, which occurred recently, is undoubtedly related to the ways of implementing training courses aimed at skills development and professional improvement of bank employees, as well as knowledge management within a banking institution or network.

In fact, basing on the speech of Giuseppe Zadra [25], the former General Director of ABI at the European Bank Training Network (EBTN) Conference, the e-learning system in the banking system can be defined not only as a tool aimed at delivering training courses but also and especially as a management tool of the company's knowledge and an access tool to necessary information.

Comacchio and Scapolan [24] expound that the decision-making process related to the adoption of e-learning among the banks can be impacted by various factors. Some of them are as follows:

- Rationalisation and optimisation that comprises the fact that online courses and modules are more flexible in delivery and use (anytime and anywhere) and lead to sustainability of training related to economies of scale, and subsequently to cost reduction [23, 27];
- Limitative thrusts [24];
- Cultural and institutional factors [23, 24];
- Connectivity, i.e. the ability to facilitate communication and dissemination of knowledge in a company [24].

In a recent article on e-front portal, Andriotis [27] argues that e-learning in a fast-paced, knowledge-driven industry like finance is very important for various reasons: cost factor; privacy and control; staying up to date; new employees on-board (training for newly hired personnel); training insight (statistics about a single course's progress and a single employee performance).

Moreover, Comacchio and Scapolan [24] claim that e-learning can bring added value to both the banking institutions and their employees. According to the authors, in the former case, it derives from the improvement of the learning processes and the quality of training integrated with work or networking, as well as from the saving of training costs (travel, accommodation, training, rent, instructor costs, opportunity costs, etc.). The added value for employees, however, comes from factors such as "self-service" of training courses or personalisation of training programmes in accordance with individual learning models. Karaaslan's [26] study supports this statement, defining web-based training as a new opportunity to create a harmonious labour force with new technology and to increase the efficiency of business productivity.

With regard to the level of e-learning adoption, Comacchio and Scapolan [24] verified different types of pressure that can have a significant impact on the decisions for introduction and development of an e-learning strategy in banking institutions. In this regard, the authors highlight the following types of isomorphism that may emerge:

- Coercive isomorphism (e.g. policies of ABI, which, through ABIFormazione, promotes the dissemination of e-learning culture within the banking industry, as well as indications from public institutions dealing with training and education);
- Mimetic isomorphism (e.g. suggestions from colleagues working in the Human Resources Management (HRM) area of other banks);
- Regulatory isomorphism (e.g. pressure exerted by legislation in the field of training developed by ABI).

A recent research conducted by ABI, demonstrates that, in addition to the associations' initiatives, the majority of banks in Italy have also started to introduce and implement e-learning experiences inside the organisations. The findings of the study highlight that the main target groups studying in banks through online courses are employees of the sales departments, bank operators, bank consultants, and, in some cases, branch managers and staff of the Head office.

With regard to the content of the didactic modules for bank employees, we can claim that, according to the EBTN conference proceedings, an extensive training process is focused on the practical and operational aspects with a very strong commercial value. This type of content allows credit companies to meet the specific needs of their clients, obtaining personnel that are more qualified and strengthening their relationship with customers. Some authors [23, 24, 28] argue that the content of the training modules in banks are mainly aimed at the following areas:

- Introducing new products, and new banking services;
- Creating basic knowledge, as well as specialist (technical and operational) skills (e.g. legal aspects of the Stock Market, Banking Quality, Civil Law, Distraint);
- Training on internal systems and business processes (e.g. Customer Service and Customer Relationship Management);
- Training on informatics and new technologies (e.g. Microsoft Office, Business-related Internet service, Basic Internet, Advanced Internet);
- Language and communication skills training.

4   Use of E-learning by UniCredit Bank

4.1   Case Study Description

The aim of this part of the paper is to explore the initiatives of an Italian Bank Institution developed in the field of ICT and web-based training. The bank described in this study is UniCredit.

UniCredit Group is an Italian global banking and financial services company. Its network spans 50 markets in 17 countries, with more than 8,500 branches and over

147,000 employees. Its strategic position in Western and Eastern Europe gives the Group one of the region's highest market shares.

As discussed in previous sections, all the initiatives realised by the Banking Group have been divided into two particular areas:

- Internal training dedicated to staff and delivered through the most innovative learning modes (ICT and web-based); and
- Educational programmes designed by the UniCredit Banking Group in the area of financial education and addressed to external stakeholders (citizens, customers, enterprises).

In order to examine the abovementioned training activities delivered by UniCredit, data from secondary sources has been collected (corporate website, official documentation of the Banking Group, information published in third-party sources) and analysed by means of the qualitative manual content analysis technique [29]. The content analysis has started from analysis of the texts according to two content (subject) categories: internal training and external educational programmes. Next, all the text units located in these two groups have been analysed by the researchers involved in the study; the definition of the research problem at the beginning of the study has allowed us to define the main directions to data analysis, and the patterns to be examined (such as goals of the internal and external education, content of the educational courses, main digital tools and platforms used to achieve the educational goals, main effects and consequences of the application of these initiatives).

4.2   Internal Education - Staff Training at UniCredit

The strategy of the HRM in UniCredit Bank is structured in such a way that it can allow the organisation to adapt to the current regulatory and economic environment, increased competition and technological innovation that have a significant impact on the activities of the Banking Group.

Recently, UniCredit has seen the implementation of several initiatives aimed at improving competitiveness and efficiency of the company. These initiatives have significantly strengthened the Banking Group in the current market context, contributing to the organisation's sustainability.

One of the key areas in the process of adapting the company to new external conditions is definitely related to the development of HRM initiatives, including employee development. Official data of the UniCredit reveals the initiatives in this area are divided in three directions: (1) Talent attraction and development; (2) Increase of professional skills and competencies; and (3) Investments in communication and dialogue.

In the last 3 years, the UniCredit Banking Group has continued to invest in training and development with the aim of increasing the responsibilities of managers regarding the development of their subordinates and promoting essential behaviours for the achievement of business objectives, with a special attention to cooperation, synergy and risk management.

In this part of the paper, in fact, we are analysing the training initiatives for the organisation's staff, related to some of the areas mentioned above but carried out only through the modalities based on new technologies, digital platforms and the Internet.

With regard to the development of professional skills, the priority of the Group over the last few years has been to ensure the necessary balance between costs and quality in defining the training approach. The attention to cost containment has led to an overall decrease in training hours, providing, however, a mix of excellent and effective training solutions, among which the key role is played by e-learning.

A recent agreement between the Banking Group and the Organization of Trade Unions specifies the online training at UniCredit represents a large part of the training provided by the company, including all mandatory training courses for the personnel of the organisation.

In the Integrated Report [30], there are references to more than 76.800 h of online training carried out from October 2014 to December 2015 for the employees of the Group in the retail and corporate areas.

As stated by the Sustainability Report [31], the courses provided to the staff of the bank are represented by three main macro-areas – technical training, management skills training and language training (Fig. 1):

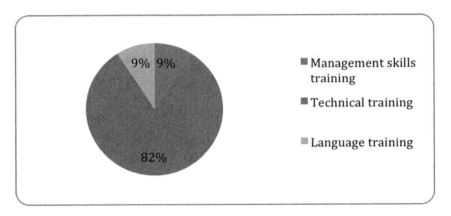

Fig. 1. Distribution of training hours by training type, % (Source: Sustainability Report [31])

The official documents of the Banking Group (Sustainability Reports, Integrated Reports) indicate that one of the important areas of the technical training is to develop the skills of the personnel to give necessary support to the customers – improving knowledge on the products, understanding customers' needs and strengthening sales and relationship-building skills.

Another area of the technical training aims to improve risk management skills. Thanks to the UniCredit Risk Academy and UniCredit Business Integrated Solutions, an online training course aimed at providing all employees with basic knowledge on concepts, goals and tools of operational risk management has been launched by means of new methods of distance education.

In addition, in the context of specific technical skills development, starting from 2012 the Banking Group created several workshops aimed at the workers from different competence lines where they can learn also through new technologies and digital tools.

In view of the growing importance of social media and their potential influence on business activities and reputation, in 2015, some of the employees participated in specialised workshops to explain to the company's managers about their use of social media. Based on their considerations, from 2016 an online mandatory training programme should be designed to help the bank staff to understand how to use social media in relation to UniCredit [30].

As regards managerial training, the Sustainability Report [31] states that in accordance with the company's commitment to increase the accountability of its managers and encourage them to use feedback in order to benefit their performance and their personal development, UniCredit has invested heavily in programmes aimed at improving essential skills for managers.

In 2015, special attention was paid to the implementation of the Leadership Programmes for the development of key competencies associated with each banding for current and future managers [29].

In addition to these global initiatives, the Group also invests in specific training in the different geographical areas in correspondence to the activation of specific programmes. Indeed, in recent years, training courses for the managers of the different countries have been initiated:

- In Croatia, when a performance management process was launched, special programmes for managers have been created. Training courses for managers through e-learning in this area are aimed at improving the ability of assessment of teams in order to promote professional development, also the strengthening of individual responsibility in the phase of development needs definition.
- In Hungary, a programme of transformation for managers of the retail network has provided them with guidance on how to manage employees' development in the best way. A particular emphasis has been given to the building of a culture of feedback and promotion of collaboration among different divisions.

With regard to the development of knowledge and skills through internal communication and dialogue, the UniCredit Group can boast of its new Intranet "OneGate" in seven languages, which allows all employees to access all the necessary internal information. This tool is considered a key element for improving the effectiveness of internal communication, facilitating access to information and managing knowledge within the organisation. This digital platform is a centralised access point online to all the information of the Group. The Intranet offers different easy-to-use content, tools and widgets to support the employees in their daily work. Some examples are as follows:

- Sharing of the Strategic Plan content with the staff to ensure commitment and involvement;
- Publication of informative articles, video messages realised by the CEO, as well as internal presentations aimed at illustrating updates related to the Strategic Plan 2015–2018;
- Creation of a fruitful social dialogue between the bank and the employees;

- Description of the examples and best practices to outline the best attitudes to adopt in different situations.

4.3   External Education - Financial Education Delivered by UniCredit

In addition to investments in internal training programmes addressed to staff, UniCredit has been investing for several years in training for actual and potential customers. These initiatives are aimed at supporting the competitiveness of enterprises and the ability of citizens to make economic choices sustainable over time, which ultimately will have a positive impact on the sustainability of the Banking Group itself.

To strengthen the financial inclusion of individuals (young people, families, the elderly and immigrants), UniCredit is implementing the training courses in different modes (traditional, e-learning, blended), also thanks to collaboration with different national and international stakeholders (consumer associations, schools, universities, and trade associations). The aim of this training is to create courses on topics related to banking and finance, designed for specific needs of participants, and delivered by highly qualified staff of the bank.

With regard to the enterprises, UniCredit is engaged in the development of particular programmes on various issues, including export and digitisation, or specific topics regarding the hospitality sector, realised and taught by both recognised external partners and highly qualified staff of the Banking Group. These training activities addressed to the external environment are provided, at least partially, through the Internet and/or by means of digital technologies.

The first initiative has been designed in collaboration with ADOC, Federconsumatori and the National Union of Consumers for the project "La Banca è anche per me". In the framework of this initiative the educational video pills on various topics such as Multi-channel Bank, Current Account, Payment Cards, Money Remittance, etc. that have been realised and subsequently distributed through the official channels of the partners.

The video, structured through info-graphics and text, with commentary in five languages (Italian, English, French, Romanian and Chinese), aims to provide information on some precautions to be taken in consideration for a conscious and sustainable choice, as well as promote access to the bank for citizens and immigrants and contribute to their financial education.

Another initiative, developed by UniCredit and addressed to both citizens and enterprises, is called "In-formati". This initiative represents a financial education programme that offers free courses to customers and non-customers over all Italian territory. The programme is aimed at those who want to increase their knowledge on the topics of banking and finance to make their financial activities more conscious and responsible.

In particular, the educational offer of this programme is aimed at three specific target groups: (1) individuals (young people, families, the elderly, immigrants), (2) small and medium enterprises (SMEs), (3) non-profit associations.

The trainers involved are the specialists of UniCredit who have decided to participate in the programme as volunteers, "donating" to the Community in which they live and work as part of their own free time, knowledge, desire of research and innovation, as well as their own professional experience. The internal training school of the Group supports about 1,200 trainers through provision of an adequate training and a constant update on the topics covered by the programme. Even if the training courses mainly take place in traditional classrooms, the trainers in collaboration with the Oil Project – the largest MOOC in Italy – have realised the educational videos on the main topics of the course. These videos are freely available on the Internet for anyone who wants to acquire basic knowledge in the field of banking and finance [32].

Another interesting project to be considered in this paper is Go International! This project is the educational offer of UniCredit developed to allow companies that have started or intend to start a process of opening up towards foreign markets to deepen knowledge about international trade systematically. The training in the framework of this initiative has been developed by the internal training school of UniCredit in synergy with the structures of international business of the Banking Group, and with the best professionals in the industry, to help the participants trigger two important levers [32]:

- Knowledge – a clear and complete framework regarding a specific international context (standards, techniques, tools) to face the process of opening to foreign markets in a better way, and to assess in advance the potential risks to be addressed;
- Action – analysis of operational and practical issues on the context and tools necessary to achieve the growth objectives.

The idea of UniCredit is to make available for SMEs the experience and know-how of the whole Banking Group to help the companies to adequately face the opportunities offered by international markets in a context that requires an in-depth and updated knowledge of some aspects, points, critical issues and risks.

The Go International! training courses can be used by enterprises through five different formats. This allows participants to access the format closest to their own learning needs and available time. One of the most popular formats is that related to distance education, entitled "Video Seminar". The Video Seminar is a format delivered throughout Italy that offers the possibility to follow live video seminars at a bank's headquarters or at one of the offices of local stakeholders. Moreover, during the live broadcast, participants will be able to interact with the speakers through their own devices – smartphones or tablets – by means of an online platform for instant messaging [33].

#Cashlessgeneration is the last initiative of financial education that we are describing in this study. This initiative, developed by UniCredit, involves 9,000 students from 130 high schools and 5 universities. The information on the corporate website of the Banking Group shows that this initiative is part of the path traced by the Law on the Reform of the national education system that has, among others, the aim of enhancement of knowledge regarding legal, economic and financial issues, as well as of education aimed at entrepreneurship development. According to the Programme for International Student Assessment (PISA) 2012 survey, Italian students are at the bottom of the rating among OECD countries that assesses the level of financial knowledge and expertise. To overcome this gap UniCredit regularly organises events aimed at dissemination of information about certain topics. Among these is #Cashlessgeneration – a project dedicated to the new non-cash payment instruments.

After the 2015 edition of #Cashlessgeneration, realised by UniCredit in collaboration with the Association of Consumer Protection and Orientation, it was decided to extend the 2016 edition through cooperation with WeSchool (the Italian MOOC platform) to improve the educational proposal and reach a wider audience of participants by using the digital platform of the new partner [34].

5   Conclusion

Nowadays digital tools and platforms are being employed for various types of training by different businesses and non-profit organizations. The contribution of the authors to the scientific research in the field of e-learning application by the bank institutions can be defined as the systematisation of information about the learning/training initiatives performed in the banking sector for different goals and objectives. Moreover, the authors have attempted to explore the significance and role of e-learning initiatives (both internal and external) for a bank or other financial organisations, by describing one of the "best practices" of online training implementation in the example of UniCredit Bank. Based on the case analysed in this paper we can confirm the applicability and usefulness of e-learning technologies and digital tools for training realised by banking organisations in the two particular areas: (1) financial education for external users (real and potential clients), and (2) knowledge and skills development of internal users (employees of an organisation).

The initiatives realised by UniCredit Bank for the professional development of human resources are mainly aimed at increasing competitiveness and contributing to the sustainability of the enterprise. These initiatives have been developed in the three areas of HRM: (1) Talent attraction and development; (2) Increase of professional skills and competencies; and (3) Investments in communication and dialogue. With regard to the content of the educational e-courses provided to the staff, it should be noted that the bank has developed training initiatives in three main macro-areas such as technical training, management skills training and language training. The main tool used by the Bank in order to realise online training activities in this sphere is the Intranet digital platform "OneGate" in seven languages, where different types of activities for the personnel have been performed.

The programmes related to the financial education for external users are aimed at supporting the competitiveness of enterprises and the ability of citizens to make economic choices sustainable over time, which consequently will have a positive impact on the sustainability of the Banking Group itself. Regarding the content of the programmes, we can highlight that the topics of the educational initiatives are mainly related to banking and finance, designed for specific needs of different clusters of participants (e.g. individuals, SMEs, non-profit associations) and delivered by highly qualified staff of the bank. These training activities have been provided, at least partially, through the Internet and/or by means of digital technologies. The main technologies used for this type of education were tools such as educational video pills realised and distributed through the official online channels of the programmes' partners; online video seminars with the possibility to interact with the trainers through digital devices – smartphones or tablets; online courses provided in collaboration with some MOOC platforms.

In comparison with other pre-existing studies on e-learning application by banks, this work has the objective to demonstrate the relationship between two essential areas of learning (financial education for external stakeholders and staff training in the internal organisational environment) on which this paper is focused. The conclusion that could be drawn is that only with the development and promotion of an integrated policy of web-based and ICT-assisted training will bank organisations be able to start and success-fully perform the process of innovation and digital transformation, and to become more competitive in the global arena of financial services.

Regarding the managerial implications of the present study, we should emphasize the usefulness of this research in the framework of a possible application of some elements of the "best practice" developed by UniCredit Bank and described above not only by managers of other companies in the banking sector but also by organisations operating in other areas of the service sector.

Some directions for further research in this scientific stream could be as follows:

- more in-depth study and comparison of initiatives on the example of different banking organisations (multiple case studies) within the same country, as well as in different national contexts and economic systems;
- comparative analysis of some organisational practices in the field of internal and external e-learning in other areas of the service sector in order to identify the main similarities and differences, depending on the characteristics and peculiarities of a sector.

References

1. Eisenhardt, K.M.: Building theorise from case study research. Acad. Manag. Rev. **14**(4), 532–550 (1989)
2. Ryan, B., Scapens, R.W., Theobold, M.: Research Method and Methodology in Finance and Accounting. Cengage Learning, London (2002)
3. Yin, R.K.: Case Study Research: Design and Methods. Sage Publications Inc., Thousand Oaks (2003)
4. The importance of economic education and financial literacy. https://goo.gl/dCQsnb
5. Lusardi, A., Mitchell, O.S.: The economic importance of financial literacy: theory and evidence. J. Econ. Lit. **52**(1), 5–44 (2014)
6. D'Alessio, G., Iessi, S.: Over – indebtedness in Italy: how widespread and persistent is it? Bank of Italy Occasional Paper 319 (2016)
7. Gentile, M., Lucarelli, C., Linciano, N., Soccorso, P.: Financial Advice Seeking, Financial Knowledge, and Overconfidence: Evidence from the Italian Market. Quaderno di Finanza 83, Consob, Rome (2016)
8. Mendoza, N.A., Pracejus, J.W.: Buy now, pay later: does a future temporal orientation affect credit overuse? NA-Adv. Consum. Res. **24**, 499–503 (1997)
9. New Internationalist Magazine. http://goo.gl/H8kCOI
10. Fagerstrøm, A., Hantula, D.A.: Buy it now and pay for it later: an experimental study of student credit card use. Psychol. Rec. **63**(2), 323–332 (2013)
11. Studi Economici dell'OCSE Italia. http://goo.gl/TUewvH

12. Annual report 2007 on the European Community's development policy and the implementation of external assistance in 2006. http://goo.gl/e2lHAX
13. OECD/INFE: Policy Handbook – National Strategies for Financial Education. OECD Publishing, Paris (2015)
14. OECD secretary-general report to G20 finance ministers. http://goo.gl/2GafXu
15. Global Financial Literacy Excellence Center. http://gflec.org
16. High-level principles on financial consumer protection. goo.gl/FfJ2BP
17. Disegno di legge AS 1196. http://goo.gl/jV81Yb
18. Ministero dell'Istruzione, dell'Università e della Ricerca. http://goo.gl/O2Rjdq
19. Banca d'Italia. http://goo.gl/lX0ICI
20. Ellis, R.K.: A Field Guide to Learning Management Systems. ASTD Learning Circuits, Alexandria (2009)
21. Sigala, M.: Integrating Web 2.0 in e-learning environments: a socio-technical approach. Int. J. Knowl. Learn. **3**(6), 628–648 (2007)
22. Zimková, E.: E-learning in the banking sector. BIATEC **14**(2), 20–21 (2006)
23. Andreu, R., Jáuregui, K.: Key factors of e-learning: a case study at a Spanish bank. J. Inf. Technol. Educ. **4**(1), 1–31 (2005)
24. Commachio, A., Scapolan, A.C.: La decisione di adozione di un sistema di e-learning. Un'analisi empirica. In: 4th Workshop dei docenti e ricercatori di organizzazione aziendale, Firenze (2003)
25. European Bank Training Network. http://www.ebtn-association.eu
26. Karaaslan, I.A.: The effect of banking personnel's access to e-learning opportunities on their professional achievement. TOJET: Turk. Online. J Educ. Technol. **12**(2), 269–280 (2013)
27. E-front. http://goo.gl/LJsJGf
28. Karadimas, N.V., Rigopoulos, G.: Enhancing IT skills of banking employees through e-learning technology. WSEAS Trans. Comput. **5**(12), 3165–3168 (2006)
29. Prasad, B.D.: Content analysis: a method of social science research. In: Lal Das, D.K. (ed) Research Methods for Social Work, pp. 174–193. Rawat Publications, New Delhi (2008)
30. Unicredit. https://goo.gl/TjgaXJ
31. Unicredit. https://goo.gl/I6OqfB
32. Unicredit. https://goo.gl/WVBLk9
33. Unicredit. https://goo.gl/ISDKRS
34. Economyup. http://goo.gl/ZG2UXf

Fraud Risk Modelling: Requirements Elicitation in the Case of Telecom Services

Ahmed Seid Yesuf[1]([✉]), Lars Wolos[2], and Kai Rannenberg[1]

[1] Deutsche Telekom Chair of Mobile Business and Multilateral Security,
Goethe University Frankfurt, Frankfurt am Main, Germany
{ahmed.yesuf,kai.rannenberg}@m-chair.de
[2] Department of Applied Mathematics and Computer Science,
Technical University of Denmark, Kongens Lyngby, Denmark
lpwo@dtu.dk

Abstract. Telecom providers are losing tremendous amounts of money due to fraud risks posed to Telecom services and products. Currently, they are mainly focusing on fraud *detection* approaches to reduce the impact of fraud risks against their services. However, fraud *prevention* approaches should also be investigated in order to further reduce fraud risks and improve the revenue of Telecom providers. Fraud risk modelling is a fraud prevention approach aims at identifying the potential fraud risks, estimating the damage and setting up preventive mechanisms before the fraud risks lead to actual losses. In this paper, we highlight the important requirements for a usable and context-aware fraud risk modelling approach for Telecom services. To do so, we have conducted two workshops with experts from a Telecom provider and experts from multi-disciplinary areas. In order to show and document the requirements, we present two exemplary Telecom fraud scenarios, analyse and estimate the impacts of fraud risks qualitatively.

Keywords: Fraud risk · Requirement elicitation · Fraud modelling · Service security · Telecommunication · Risk assessment

1 Introduction

Telecom providers are losing billions of dollars every year due to frauds. According to the Communications Fraud Control Association (CFCA), the global revenue of Telecom providers was affected by almost $38.1 billion (USD) in the year 2015 [1] due to frauds. Fraud is the use of a Telecom service to gain money or with no intention to pay. There are many types of Telecom frauds. The top five types mentioned in the 2015 CFCA report are International Revenue Share Fraud (IRSF), interconnect bypass, premium rate services, arbitrage and theft/stolen goods. IRSF costed the Telecom providers almost $10.76 billion (USD) globally in the year 2015 [1], which is almost threefold from the year 2011.

Telecom providers store subscribers usage behaviour - known as call data records (CDR) - in a database. Telecom fraud detection systems (FDSs) usually

© Springer International Publishing AG 2017
S. Za et al. (Eds.): IESS 2017, LNBIP 279, pp. 323–336, 2017.
DOI: 10.1007/978-3-319-56925-3_26

use anomaly-based fraud detection relying on identifying anomalies through comparing the past and current or recent behaviour of subscribers [2–4]. Although a FDS has advantages, it has also a problem of detecting frauds only once a fraudster follows a predefined and/or behaviour-based pattern stored in the detection system – which increases the reaction time if it happens in a larger scale. Given these and the (revenue-loss) figures above, it is not surprising that Telecom providers aim to reduce the damage. Thus, in order to predict fraud risks beforehand and take preventive measures, a fraud risk assessment approach is required as part of Telecom service security.

According to ISO/IEC 27005 [5] a risk assessment process starts from understanding the context and performing risk analysis. In terms of the Telecom domain, the risk assessment process begins by understanding the Telecom service under assessment, identifying the potential risks and estimating the consequences of those risks. We believe this process substantially helps to reduce the level of fraud risks before a given Telecom service is under attack. Basing on the general risk assessment process, the first step for a usable and context-aware fraud risk modelling approach is to elicit requirements necessary to develop the intended approach. The main contribution of this paper is, therefore, eliciting the important *requirements* for a context-aware and model-based fraud risk modelling at different stages of the assessment.

The rest of the paper is organised as follows. Section 2 provides background on the context of Telecom services and fraud risks. Section 4 describes the methodologies used in order to elicit the requirements of fraud risk modelling. Section 3 discusses the related work on security requirements in the Telecom domain. In Sect. 5, two exemplary fraud scenarios are described. We also performed a risk analysis on each scenario to show fraud enablers and document the initial requirements on fraud risk modelling. These requirements at different stages of fraud risk modelling are discussed in Sect. 6. At last, Sect. 7 summarises and concludes the important points in the paper and provides insights for future work on fraud risk modelling approaches.

2   Background: Telecom Services and Types of Fraud

The core elements of Business Process Management (BPM) in a Telecom provider involve "products" (actually services) and the infrastructures to "produce" them [6]. The other elements of the BPM are based on these core elements - the operations, marketing and service delivery management [6]. Telecom services include call services (mobile and fixed-line), roaming services [7], data services, Internet services (e.g. VoIP service) and messaging services.

When taking a closer look, most Telecom services require a complex technical network architecture. This is due to a multitude of different interconnected networks, network operators and service providers. At the same time, in most cases, the underlying business models are tailored to compete in highly competitive markets. This entails opposed financial interests of the participants and very often even more complex services. Furthermore, this proves fertile ground for fraud or misuse.

Telecom frauds affect both fixed line and mobile services which could be provided through either prepaid or post-paid contracts [4, 8–10]. There exist several categories of fraud in the literature, e.g. [11–14]. Some of the known methods that fraudsters use to commit fraud include:

PBX hacking: A Private Branch Exchange (PBX) is a telephone network used within a company to switch calls coming from outside the company rather than installing individual telephone lines to each user in a company - which is very expensive. Fraudsters identify the potential weaknesses of PBXs to forward calls from the outside network, as PBXs provide the capability to remotely connect to a central portal (voicemail) [15, 16]. According to CFCA [1], the damage of PBX hacking in the year 2015 is estimated to $3.93 billion (USD).

Subscription fraud: This is another typical method of committing fraud [3, 4]. It happens through using users' information illegally either through scamming or other forms of social engineering attacks. According to CFCA [1], the damage of subscription fraud in the year 2015 is estimated around $3.53 billion (USD).

Wangiri fraud: Wangiri fraud, originated in Japan, is also known as "one ring and cut" fraud which makes everybody a target by making calls and waiting for a reply [17]. When victims call back, they would listen to advertisements from a premium internet service or calls. This will cause expensive bills to the users who call back to the missed-call.

In the paper, we focus on business-related fraud risks and used the following definitions.

- Telecommunications Service Provider (TSP): A TSP covers the different types of providers of Telecom services, regardless of whether they operate a network by themselves or just (re)sell services.
- Call termination: Call termination is a service of a TSP to route phone calls from another TSP to customers of the first TSP. Call termination could happen for national and international calls.

3   Related Work

Ensuring the security of a system is a fuzzy challenge but through procedural techniques, the level of system or service security can be improved. One way is through "implementing" security requirements using security engineering methods. Unfortunately, traditional risk analysis and security engineering methods have several limitations when applied to Telecom cases [18]. Zuccato et al. [18] proposed a security requirement engineering method - SKYDD - including infrastructure, business and information requirements based on checklists, guidelines and expert knowledge. In continuation to this work, Zuccato et al. [19] proposed an approach which is a step by step process to use service security requirement profiles. They outlined four basic requirements for the approach: economic feasibility, agility, multi-disciplinary and help without the need for security experts. The approach is, therefore, intended to span all kinds of security issues (mostly related to business related risks).

Tian et al. [20] presented a requirement model not for the security domain but for a solution domain in the Telecom field such as an SDP (Service Delivery Platform). This domain requirement model allows producing a model that can be easily navigated through its hierarchical structure to incorporate featured and detailed requirements for stakeholders and developers respectively. In relation to the usage of Telecom services, Krogstie [21] highlighted the areas to focus in order to cope with the technology shifts of mobile information systems. He described the advantage of involving the concept of mobility and people's involvement (the context) in mobile information systems using requirement engineering mechanisms.

In general, even though there is a limited amount of literature in the area of Telecom service security, some of the papers above have some relation with security and requirements engineering. Still, none of the papers focuses on preventive risk modelling requirements to deal with fraud risks. Recently, we showed a risk analysis using a value-based approach [22]. As a continuation to cover identification, analysis and estimation of fraud risks, this paper highlights requirements on fraud risk modelling for the case of Telecom services.

4   Methodology

To elicit the requirements on fraud risk modelling, we have conducted two workshops. In total, eleven experts have attended the workshops. Four of them are experts from a TSP who have experience of Telecom fraud and seven of them are security researchers from inter-disciplinary fields (information security, computer science and mathematics). On first sight the number of involved experts may look small. However due to the high secrecy requirements of anti-fraud measures it is actually quite difficult to conduct workshops like these, where several experts in front of researchers discuss issues and even contribute different opinions in a single session. Therefore, the workshop lead to results, that are novel, even though the quantitative base is not the largest.

The first workshop was conducted with the experts from the TSP (seven participants including the authors for three hours). The workshop was conducted with a moderated discussion where the people from the TSP presented different types of Telecom fraud (revenue-share fraud, roaming fraud, call selling fraud and PBX fraud). In the discussion session, they also discussed the effects of those frauds to the TSP. Through questioning and answering, they discussed their expectations of fraud prevention approaches to reduce the effects of fraud to the TSP. In this paper, we present two of the fraud scenarios to show how we elicited the requirements on fraud risk modelling.

Taking their expectations, in the second workshop – where only security researchers met with eight participants for three hours – we formulated problems of those example fraud scenarios, identified the potential fraud enablers, estimated qualitatively the impacts of the fraud scenarios on the Telecom. The main outcome of this workshop was a list of initial requirements for a risk modelling approach. A risk modelling approach consists of three main components: conceptual model, analysis and estimation. The requirements are covering all of the three components.

5 Analysis of Exemplary Fraud Scenarios and Expectations of TSP

In this section, two exemplary revenue-share fraud scenarios are presented and analysed to show how we have elicited the requirements on a risk modelling approach for Telecom services. We also discussed the expectations of the TSP expressed in the first workshop. We chose the two exemplary fraud scenarios because they show very well: (1) the involvement of several actors and entities to commit a fraud; (2) chain(s) of fraudulent activities in a simplified way; (3) money flows (fraud risks) of fraudulent actors; and (4) the contribution of technological weaknesses to fraud risks. Some other fraud scenarios can be found in (e.g. [22]).

5.1 Exemplary Fraud Scenarios

Customers of TSPs in possession of insider knowledge could identify possible arbitrage scenarios by combining TSP services and products (often involving multiple TSPs). One prominent example is the use of flat-rate tariffs for call termination. Another example is the involvement of weak identity checks for new connections to enable a portfolio of usage scenarios that violate at least the terms and conditions of the TSPs involved, which results in a profit for the fraudster. We are using these two example scenarios to show how we documented the requirements on the risk modelling. For both scenarios, even though the net loss of a TSP resulting from a single customer's usage pattern is not a problem per se for the respective TSP, a large-scale systematic exploitation of such a scenario however will add up to a critical level.

Fraud scenario 1 - Tariff misuse for call termination. Mr. Clever, a fraudster, has (multiple) fixed, mobile or virtual IP connection points with TSP A. These are billed either as flat-rate or in tariff schemes which include capacious minute budgets. Also, Mr. Clever has (multiple) fixed, mobile or virtual IP connection points with TSP B. Call termination fees are paid on a per minute basis by TSP A when calls are delivered from TSP A to TSP B. TSP B passes a part of the received call termination fees to Mr. Clever, thereby providing a payout per minute for incoming calls as incentive to generate as much incoming traffic as possible to the network of TSP B. This process is mostly done by an intermediary company, whose main activity is bridging between the fraudster and fraudulent TSPs. Mr. Clever makes calls from TSP A to TSP B in order to maximise his profit from these payouts. The money flow is shown in Fig. 1. The source of Mr. Clever's profit is the *call termination fee* paid by TSP A to TSP B, which is then partly paid out to Mr. Clever by TSP B (Mr. Clever's costs at TSP A are fixed due to the chosen tariff).

Fraud scenario 2 - fraud involving the false pretence of being willing and able to pay. This scenario can be described in five sequential steps.

1. Mr. Clever obtains a high number of prepaid (pay as you go) SIM cards. These SIM cards are either not (yet) registered or registered using fake or

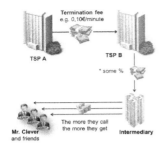

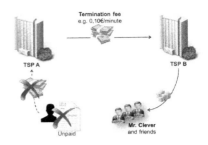

Fig. 1. Money flow for scenario 1   **Fig. 2.** Money flow for scenario 2

stolen IDs. Furthermore, these SIM cards are billed either as flat-rate, have a very low price per minute, or have free minutes (upon activation). In addition to these prepaid SIM cards, Mr. Clever manages to establish one post-paid mobile contract with TSP A using a fake ID or forged identity and bank card credentials. Thus, with respect to the post-paid mobile contract, this scenario is a matter of fraud involving the false pretence of being willing and able to pay.

2. Mr. Clever activates call forwarding on the post-paid mobile connection point to a (foreign) external TSP B. He then makes the highest number of possible parallel calls to that post-paid mobile connection point using the above prepaid SIM cards. All calls will be diverted to TSP B. Call termination fees are paid on a per minute basis by TSP A when calls are delivered from TSP A to TSP B.

3. The fraud detection system (FDS) of TSP A may detect the violation of limits on the post-paid mobile connection contract and disconnect it within the response time.

4. Mr. Clever does not pay the post-paid bill.

5. TSP B passes a part of the received call termination fees on to Mr. Clever, thereby providing a payout per minute for incoming calls. The money flow is shown in Fig. 2.

The source of Mr. Clever's profit is the *call termination fee* paid by TSP A to TSP B, which is then partly paid out to Mr. Clever by TSP B, while outstanding receivables of TSP A will remain unpaid and become a bad debt.

5.2   Fraud Risks: Partial Risks from the Exemplary Scenarios

For the two scenarios presented in Sect. 5.1, we have discussed the (partial) risks identified in the workshop which allow a fraudster to perpetrate the fraud. Risks are named *partial* as their discrete or isolated occurrence would not necessarily be relevant to the company while multiple risks constitute a (relevant) fraud scenario. We have then estimated qualitatively the partial risks based on their contribution to the occurrence of the fraud scenarios. For the sake of simplicity, we decided to use a risk estimation scale with three values: *High*, *Medium* and

Low. A *High* rating has a significant contribution to the occurrence of a fraud and has impact at least on one of the dimensions (e.g. reputation, money, etc.). *Medium* and *Low* means less contribution and impact. The identified partial risks are described below.

R.1 Weak identity checks enable a fraudster to acquire a high number of prepaid (pay-as-you-go) SIM cards using fake IDs: Sound identity and credit checks are costly. Due to high pricing pressure in the Telecom market, the economically reasonable level of effort is limited. This holds true especially for prepaid (pay-as-you-go) SIM cards, leading to a higher risk of identity fraud.

– Impact: *High*
– Initial requirement: the risk modelling approach should model the processes of acquiring different types of Telecom connections and identify fraud risks due to weak identity checks.

R.2 A fraudster manages to mask fraudulent calls to simulate legitimate traffic: Fraud detection systems (FDSs) are operated by TSPs to detect excessive usage of their network resources. In a case of misuse, the FDS should detect the violation of limits, e.g. multiple long-lasting connections of similar length or to the same destination number. Distributing calls using multiple destination numbers and a random duration would mask the fraudulent activity making it much harder to detect.

– Impact: *High*
– Initial requirement: the risk modelling approach should model the possible actions in the existing fraud detection mechanisms, identify the potential risks against such mechanisms and analyse their impact.

R.3 A fraudster acquires a post-paid contract using fake IDs: Stolen identity documents and bank account information enable fraudsters to sign up for a post-paid contract. Factors which facilitate this are: Contract concluded online, sloppy work of retail personnel, acceptance of copied identity documents.

– Impact: *Medium*
– Initial requirement: the risk modelling approach should model the processes how customers or fraudsters acquire different types of connections (including post-paid) and identify weak points that create fraud risks.

R.4 A fraudster manages to find a TSP B which provides payout: There are different TSPs in the market offering payments and incentives to their customers for incoming calls. These incentives need to be usable in a profitable scenario with the products of a TSP A in order for the fraudster to make a profit.

– Impact: *High*
– Initial requirement: the fraud risk modelling approach should incorporate the influences of external services. In reality, this might be difficult to achieve, but with existing data about the external services, it is possible to estimate the damage of the external influence.

R.5 High response time of TSP A's implemented fraud detection system (FDS): Depending on the implementation of the FDS, response times may vary between 15 and approximately 60 min.

- Impact: *Low*
- Initial requirement: besides modelling the possible actions, the fraud risk modelling approach should model the response time of fraud detection mechanisms in order to analyse and estimate the impacts of a fraud.

R.6 A new product thwarts the business model of existing products: A TSP launches a product without properly considering its side-effects and interdependencies with existing products and the resulting user behaviour.

- Impact: *Low*
- Initial requirement: the fraud risk modelling approach should incorporate impact analysis of new products to the existing models.

5.3   Expectations of TSP Experts

New Telecom services usually have to be launched under significant time pressure, e.g. to minimize the time-to-market or to react swiftly to a move of a competitor. This leaves little time and space to account for potential misuse. As discussed in Sect. 5.2, misuses could happen due to sloppy (service) design, underestimation, or misjudgement of the extent of possible exploitation by fraudsters and their motivation to find combinations of different services often involving multiple providers. Based on the discussion in the workshop with the TSP experts, they expect that a preventive approach to:

- identify as many misuse scenarios as possible to find out whether a service is reasonably profitable;
- minimise potential risks associated with the service (and, as a result, potential losses);
- work as a preventive solution or technique for dealing with potential fraud related to Telecom services before it is exploited (in a large scale).

6   Requirements on Fraud Risk Modelling

Based on the outcomes of initial requirements and the expectations of the TSP, we propose a fraud risk modelling approach (cf. Fig. 3). This fraud risk modelling approach involves representation of service, identification and analysis of potential fraud risks and estimation of their impacts. The fraud risk assessment starts with modelling the context of the TSP service under assessment in a *conceptual model* – a model which represents the context of the target of assessment to the intended level of abstraction. Through this conceptual model, a fraud analyst is able to communicate, identify fraud risks and estimate potential damages. Taking this into consideration, the first three subsections in the following describe

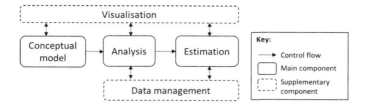

Fig. 3. Components in the fraud risk modelling approach

the requirements of each component in the fraud risk modelling approach (conceptual modelling, fraud analysis and estimation) followed by supplementary requirements (data management and visualisation).

The requirements at each component should generally be evaluated based on the following criteria:

- *context-awareness* – the ability to include internal and external factors in the analysis of a Telecom service;
- *representativeness* – the capability of representing essential elements of the target of assessment;
- *scalability* – the ability to handle complex Telecom cases with a reasonable performance;
- *usability* – the ability to support users efficiently and effectively. This includes the ability to generate the potential fraud risk scenarios and prioritise them. It also includes the capability of using different data formats in the assessment process.

6.1 Requirements on the Conceptual Model

As shown in the example scenarios above, the context of a TSP service includes several independent stakeholders, contract agreements between them, anti-fraud agents, Telecom employees, hardware and software, virtual communications, and network infrastructures.

The model needs to be able to represent a list of relevant properties of the entities (products and services) involved, including ownership of entities and a financial budget of the entities. The TSP service includes at least two parties (from now on we call them actors): subscribers and one or more service providers (TSPs). Subscribers receive a service with a short term or long term agreement provided by a TSP. The model, therefore, should represent contractual agreements and jurisdictional requirements.

For a roaming service, for instance, an additional external TSP is involved to fulfil the goals of the home TSP. Both the subscribers and the home TSP pay service fees. From the home TSP's perspective, a subscriber pays money for the service in different forms: prepaid, post-paid, or with some contractual agreement. The home TSP pays for the service given to its subscribers while roaming.

These transactions between several independent actors create chances for fraud-sters to perpetrate frauds or misuse the service. Therefore, the model should rep-resent relations between different entities, as interrelations between different par-tial risks coincide and enable an attacker/fraudster to carry out an attack/fraud.

Depending on the type of a TSP service, the relevant entities should be iden-tified and represented in the conceptual model, where the model serves as a medium of communication between, for example, the modeller and fraud ana-lysts. Therefore, the model is expected to have formal or semi-formal semantics.

The *representativeness* of the model with respect to the service under assess-ment is in general the evaluation criterion to the requirements on the conceptual model. The following summarises the high-level requirements on the conceptual model. The conceptual model should represent:

- entities and their relevant properties, including ownership of entities and finan-cial budget of the entities;
- the possible actions at each entity;
- relations between different entities, as interrelations between different partial risks may coincide and enable a fraudster to carry out a fraud;
- contractual agreements and jurisdictional requirements; the processes of acquiring different types of Telecom connections;
- the existing fraud detection mechanisms of the Telecom service in the model.

6.2   Requirements on the Analysis

Fraud analysis is the process of identifying the potential fraud scenarios and estimating the related potential economic damage to the TSP. Given the relevant entities in the contextual model of a TSP service, it should be possible to show or generate damaging incidents from the model. Damaging incidents have a damaging effect on the revenue of the TSP when implemented with a large enough level of scale. The analysis has to predict the possible fraud within a small amount of time before it amplifies the damage.

Depending on the type of a TSP service, the analysis process may leverage necessary data in relation to the contextual model - for instance, CDRs to analyse the fraud on fixed line call services. As the goal of the analysis is to predict potential fraud risks and propose preventive measures, it might leverage the usage-data in order to achieve this goal.

The analysis process should also incorporate the contractual and jurisdic-tional requirements on the TSP service under assessment. In the TSP service environments such as roaming, the jurisdictional requirements are crucial. In some countries, the delay in providing service usage files to the home TSP pro-vides a chance for fraudsters to perform the fraud undetected for an extended time span. In some other countries, where Near-Real-Time Roaming Data Exchange is applied, the delay of the data exchange is limited to only four hours - if delayed the visited TSP is responsible for any kind of fraud detected after those four hours. Additionally, there are contract agreements in place between TSPs, covering subjects of interconnection fees and inter-operator charges.

The fraud analysis, therefore, should consider both the contractual and jurisdictional requirements in order to predict the potential fraud risks in a TSP service.

Fraudsters use complex ways to perpetrate fraud in order to hide or legitimise their acts. Depending on the TSP service, the analysis approach should identify complex correlations of partial risks (cf. Sect. 5.2) to estimate the potential damage resulting from the fraud risks. The following summarises the high-level requirements on the analysis component. The analysis component should:

- identify fraud risks by identifying fraud factors from the conceptual model; different ways to identify fraud factors. To mention some: (1) by identifying the weaknesses of the components in the contextual model; (2) by using fraud patterns that fraudsters use as a guideline to identify fraud factors (e.g. due to weak identity checks). Evaluation criterion: *context-awareness* and *scalability*;
- consider contractual agreements and jurisdictional requirements in identifying the fraud risks. Evaluation criterion: *context-awareness*;
- handle complex correlations of partial risks. Evaluation criterion: *usability*;
- collect fraud scenarios identified due to fraud factors in a file format so that other fraud risk modelling components would use it (e.g. the risk estimation component to prioritise those fraud risks). Evaluation criterion: *usability*.

6.3 Requirements on the Risk Estimation

As explained in Sect. 5.3, new TSP products need to be launched under significant time pressure resulting from strong competition in a dynamic market, leaving little time and space to analyse and account for potential misuse of the product. At the same time, there is a (natural) conflict between on the one hand, the (marketing) department – responsible for product development and seeking business opportunities with a tendency to put a gloss on business case figures – and on the other hand, the misuse department – which has its focus on potential misuse and avoiding risks. In many organisations, these two departments act independently from one another and have different target functions.

Due to a persistent increase in product complexity, diversity on the feature side, and margin pressure on mass products, risks cannot be eliminated, but need to be assumed, estimated and accepted with some level of threshold, whilst minimising their possible impact. Hence, the focus of the risk estimation process should be on the question of calculability. The identified fraud scenarios should be estimated from different impact *dimensions*, mainly based on their economic impacts.

Basing on the types of the TSP services, the fraud risk modelling approach should have a framework for estimating fraud scenarios – either a qualitative or a quantitative estimation. TSP risks can qualitatively be represented in different levels of magnitudes. These magnitudes are interpreted based on the consequences of the fraud risk. As an example, High could be the loss beyond $10K, Medium between $10K and $4K, and Low below $4K. In the example scenarios (Sect. 5.2), we used three levels of qualitative measurement – High, Medium and

Low to estimate the partial risks. Risk estimation using contextual models is usually qualitative but when it is supported by statistical and usage data, it could be expressed in terms of quantitative measurements. In general, a preventive fraud risk modelling approach should estimate those risks identified in the previous stage in either qualitative or qualitative measurements.

The following summarises the high-level requirements on the risk estimation. The risk estimation component should estimate:

- the impacts of the identified fraud risks based on different dimensions (e.g. its economic impact). Evaluation criterion: *usability*;
- the fraud risks in either qualitative or quantitative scale. Evaluation criterion: *usability*.

6.4   Requirements on Visualisation

In the course of fraud risk modelling, visualising inputs and outputs is necessary for a usable approach to facilitate decision making. Decision makers (commercial managers at the TSP) are interested in understanding the effect of fraud risks at a larger scale. To satisfy the goals of TSP stakeholders, visualisation requirements on fraud risk modelling are, therefore, important.

The approach should provide sufficient visualisation at different diverging aspects. A model should visually represent actors (economically or technically), assets, existing policies, the possible connections between actors, boundary restrictions and potential fraudulent actors. It should also depict, which part of the TSP service or product is vulnerable to a known type of fraud.

At the stage of fraud analysis, the approach should visually represent the costs of damage produced by the fraud on a given TSP service - for instance, in a graph where the likelihood of fraud is shown while the detection time increases or when fraudulent actors increase in number. In addition to visualising individual effects of some particular fraud, it should also be possible to visualise the correlations of misuse scenarios within a TSP service that make up fraud. Mostly, fraud is committed either through disguising as legitimate users or through leveraging legitimate services. Thus, the approach should clearly depict accepted scenarios, hidden transactions that could possibly happen between fraudulent actors, and potentially unwanted fraud scenarios in a given TSP service.

The evaluation criterion of these requirements is *usability*. The following are the high-level requirements on the visualisation component. In general the visualisation component should visualise:

- the contextual model, the impact of fraud and the correlations of misuse scenarios;
- fraud scenarios of a given TSP service.

6.5   Requirements on Data Management

The core targets of the fraud risk modelling approach are TSPs and similar service providers. A TSP has a lot of usage data of subscribers available for analysis, but their availability for processing may be legally restricted for e.g. data protection reasons. Thus, the fraud risk modelling approach should consider data management requirements.

The approach needs to provide performance scalability in order to be able to cope with massive amounts of data if needed, e.g. billions of CDRs. It should also be flexible to process data coming from different data formats as data sources stem from the domain of the different Telecom partners. The following summarises the requirements on the data management component. It should:

- be scalable to handle huge amounts of Telecom data. Evaluation criterion: *Scalability*;
- handle different data formats. Evaluation criterion: *Usability*.

7 Conclusion and Future Work

Fraud risks emerge due to single or combined individual risks against weakly protected elements of a service. Fraud risk modelling helps to identify weaknesses in the elements of the service, analyse the potential individual risks and estimate their impacts to the service provider. Even though the requirements are elicited from a limited number of fraud cases, they provide a *guideline* to develop a context-aware and usable fraud risk modelling approach besides improving the state-of-the-art on fraud risk assessment. One way of developing such an approach is by realising the requirements and apply the resulting approach to a list of Telecom services one by one and iteratively improve the approach in each development milestone. In this regard, the involvement of service providers is undoubtedly important in the evaluation of the approach.

In future, we plan to use the requirements above to develop a context-aware fraud risk modelling approach and refine it repeatedly based on feedback from Telecom providers, so that the approach can be applicable to one or several types of Telecom frauds. The other future research direction is to find detailed criteria to evaluate the effectiveness and usefulness of the fraud risk modelling approach.

References

1. CFCA: Global Telecom fraud report. Technical report, Communications Fraud Control Association (2015)
2. Yesuf, A.S.: A review of risk identification approaches in the telecommunication domain. In: Third International Conference on Information Systems Security and Privacy (ICISSP) (2017)
3. Farvaresh, H., Sepehri, M.M.: Subscription fraud prevention in telecommunications using fuzzy rules and neural networks. Expert Syst. Appl. **31**, 337–344 (2006). Elsevier
4. Estévez, P.A., Held, C.M., Perez, C.A.: A data mining framework for detecting subscription fraud in telecommunication. Eng. Appl. Artif. Intell. **24**, 182–194 (2011). Elsevier

5. ISO/IEC: ISO/IEC 27005:2011: Information Technology – Security Techniques – Information Security Risk Management (2011)
6. TM Forum: Enhanced Telecom Operations Map (eTOM) The Business Process Framework (2015)
7. Macia-Fernandez, G., Garcia-Teodoro, P., Diaz-Verdejo, J.: Fraud in roaming scenarios: an overview. IEEE Wirel. Commun. **16**, 88–94 (2009). IEEE
8. Yufeng, K.Y., Chang-Tien, C.T., Sirwongwattana, S., Yo-Ping, H.Y.: Survey of fraud detection techniques. In: IEEE International Conference on Networking, Sensing and Control. IEEE (2004)
9. Burge, P., Shawe-Taylor, J.: An unsupervised neural network approach to profiling the behavior of mobile phone users for use in fraud detection. J. Parallel Distrib. Comput. **61**, 915–925 (2001). Elsevier
10. TIR Corporation: The 2015 telecommunications industry review: an anthology of market facts and forecasts. Technical report, The Insight Research Corporation (2015)
11. Cortesão, L., Martins, F., Rosa, A., Carvalho, P.: Fraud management systems in telecommunications: a practical approach. In: Proceeding of ICT (2005)
12. Hilas, C.S., Mastorocostas, P.A.: An application of supervised and unsupervised learning approaches to telecommunications fraud detection. Knowl. Based Syst. **21**, 721–726 (2008). Elsevier
13. Jürjens, J., Schreck, J., Bartmann, P.: Model-based security analysis for mobile communications. In: Proceedings of the 30th International Conference on Software Engineering, pp. 683–692. ACM (2008)
14. Gosset, P., Mark, H.: Classification, detection and prosecution of fraud in mobile networks. In: Proceedings of ACTS Mobile Summit, Sorrento, Italy (1999)
15. The Smartvox Knowledgebase. http://kb.smartvox.co.uk/asterisk/secure-asterisk-pbx-part-1/. Accessed Dec 2016
16. Kuhn, D.R.: PBX Vulnerability Analysis: Finding Holes in Your PBX Before Someone Else Does. US Department of Commerce, Technology Administration, National Institute of Standards and Technology (2001)
17. Yelland, M.: Fraud in mobile networks. Comput. Fraud Secur. **2013**(3), 5–9 (2013)
18. Zuccato, A., Endersz, V., Daniels, N.: Security requirement engineering at a Telecom provider. In: Third International Conference on Availability, Reliability and Security (ARES), pp. 1139–1147. IEEE (2008)
19. Zuccato, A., Daniels, N., Jampathom, C.: Service security requirement profiles for Telecom: how software engineers may tackle security. In: Sixth International Conference on Availability, Reliability and Security (ARES), pp. 521–526. IEEE (2011)
20. Qi, M.T., Xiao, Y.C., Jin, L.P., Ying, C.: Asset-based requirement analysis in telecom service delivery platform domain. In: IEEE/IFIP Network Operations and Management Symposium: Pervasive Management for Ubiquitous Networks and Services, pp. 815–818. IEEE (2008)
21. Krogstie, J.: Requirement engineering for mobile information systems. In: Proceedings of International Workshop on Requirements Engineering: Foundation for Software Quality, Interlaken, Switzerland (2001)
22. Ionita, D., Gordijn, J., Yesuf, A.S., Wieringa, R.: Value-driven risk analysis of coordination models. In: Horkoff, J., Jeusfeld, M.A., Persson, A. (eds.) PoEM 2016. LNBIP, vol. 267, pp. 102–116. Springer, Cham (2016). doi:10.1007/978-3-319-48393-1_8

Mega Sporting Events and Technology: The Role of Social Networks in Co-creating Value for the Olympic Games

Luisa Varriale[1](✉), Giuseppe Perna[2], and Maria Ferrara[3]

[1] Department of Sport Science and Wellness, University of Naples "Parthenope", Naples, Italy
luisa.varriale@uniparthenope.it
[2] Department of Law, University of Naples "Parthenope", Naples, Italy
giuseppe.perna@uniparthenope.it
[3] Department of Economics and Business Studies, University of Naples "Parthenope", Naples, Italy
maria.ferrara@uniparthenope.it

Abstract. This paper investigates the application of new technologies, specifically social networks, in the mega sporting events, adopting the value co-creation perspective. We aim to evidence how, by adopting the co-creation value concept, thanks to the new technologies, specifically the innovative communication channels, the Olympic Games assume a greater value and are more successful in terms of followers, as athletes, fans, participants, and so forth. Thanks to the spread of social networks in the Olympic Games, the general consumers, mainly the fans, are able to give feedbacks, opinions, suggestions, and evidence criticisms about the same mega sporting events, by interacting directly with athletes, organizers and all the actors involved in this very complex managerial and organizing process. This study, conducted through a review of the literature and a comparative analysis of Summer Olympic Games editions during the period 1992–2016, represents a research starting point that allows us to evidence the crucial role played by the social networks in the mega sporting events improving their social value in the co-creation value perspective. This paper may provide interesting theoretical and managerial implications.

Keywords: Technology · Co-creation value · Mega sporting events · Social networks · Olympic Games · Facebook · Twitter

1 Introduction

During the last decades the information technology (IT) and Internet are widely using especially for the services industry, such as the sport field. The sport field has been now recognized as a very significant business with a crucial role, compared to the past when it was not considered relevant in terms of economic contribution to the overall society, and it has been characterized by many changes also thanks to the use of technology, especially the new social media like social networks, creating the conditions for improving its social value, in the co-creation value perspective too. Due the growing relevance of the sport field, scholars and practitioners tend to develop and adopt more interesting analysis perspectives like the co-creation value approach.

© Springer International Publishing AG 2017
S. Za et al. (Eds.): IESS 2017, LNBIP 279, pp. 337–351, 2017.
DOI: 10.1007/978-3-319-56925-3_27

According to the co-creation perspective, suppliers and customers are, conversely, no longer on opposite sides, but they tend to interact with each other for developing new business opportunities. Co-creation can be defined as the added value providing from the collaboration between organizations and customers, both materially and symbolically [1–3]. It is a form of marketing or business strategy that emphasizes the generation and implementation of a corporate value shared with the customers, specifically, and with all the partners involved, in general.

In this paper, we consider the co-creation as a more general concept which includes all the specific aspects able to push business organizations and customers to generate value through interactions [4], but with focus on mega sporting events, that is the Olympic Games.

The phenomenon of co-creation has been seen as a phenomenon of significant change in the corporate culture: a way to share ideas with the different actors, to open the overall specific world, overcoming the limits related to the business or customer world to understand what the real needs are. Nowadays, thanks to the use of more interactive and diversified communication channels and tools, like social networks (Facebook and Twitter), the organizations are getting more and more very close and frequent interactions with the customers, by receiving relevant support by the customers making them the central player: an example of co-creation is the use of user-generated content. The active contribution for the organization derives by people who voluntary participate, perhaps because they are fans of a specific brand, to increase the value and the emotional bond with the company. Therefore, organizations have to attract and engage users, that is customers, relying on their emotional engagement, but also on their artistic and creative skills.

In this study, we aim to investigate the influence of new technologies, in particular, the digital social networks, Facebook and Twitter, for improving and making more successful mega sporting events, the Olympic Games, through the involvement and participation of customers, that is specifically passive participants or visitors. In the recent years, mega sporting events, mainly the Olympic Games, have been receiving an increasing attention and they are held everywhere in the world, thanks to media such as television and internet. Of course, although mega sporting events present a deep message related to the diffusion of equality and social values through sport activities, they aim to generate a profit, above all for the event organizers and all the players involved in the complex and wide planning and managerial process. However, the passion, the excitement, the audience expressed, the involvement play a key role in the implementation of a mega sporting event, mostly Olympic Games, and, therefore, the co-creation of value, involving all the players, public and private actors, and especially the fans (final consumers) acquires more importance. In this frame, the adoption of new technologies, especially digital social networks, can have a significant positive impact making Olympic Games successfully, because they widely facilitate the interactions between the players involved, mainly the customers, the athletes and the sport event organizers.

This exploratory study aims at investigating the Olympic Games, conducting a comparative analysis of the Summer Olympics Editions in a period 1992–2016, for analyzing the role played by the social networks, Facebook and Twitter, in contributing to improve the active participation and collaboration of all the actors, especially fans and athletes, in the co-creation value perspective for Olympics management process.

This paper is structured as follows: in the Sect. 2 the focus is on the specific application area of mega sporting events, also briefly describing the main related challenges in the perspective of co-creation value concept. The Sect. 3 briefly describes the co-creation value concept with focus on Service Dominant Logic evidencing the main contributions on this issue in the literature. The Sect. 4 provides the comparative analysis of the Summer Olympic Games in the period 1992–2016 with focus on the role played by the social networks, mainly Facebook and Twitter, in terms of their use and application and trends during the different editions. Finally, in the Sect. 5 some final considerations are provided about the phenomenon investigated.

2   Mega Sporting Events

The events generally refer to a composite and complex universe and it is not easy to develop a comprehensive definition also because their numerous typologies. Otherwise, in the literature on events there is a certain convergence of three key elements: the events celebrate, communicate and make things and people interact. The dimension "celebratory event", among others, is stressed by Getz [5], one of the most relevant authors in the field, according to his approach the event can be conceived as "*a public celebration theme*", also according to the International Festival and Events Association [6].

Getz [5, 7, 8] argue that "*the festivals, events and civic celebrations are the foundations of those characteristics that distinguish the community of human beings. The development of such communities around the world will depend in part on the existence of events celebratory*" [5: 11–17]. The communicative aspect of the events is underlined, among others "*the event is an activity that allows you to reach a target audience in a specific place and time, a meeting where you address the message and carry out activities recreational*" [5: 11–17]. Van Der Wagen [9], however, introduces the concept of relationships between individuals believing that "the majority of events are in fact reports within a community"; the event is "the thing in relation", that is the object of the subjects [10], it is ultimately human relationships and social communication [11, 12].

One of the first classifications of events has been developed by Roche [13: 11], who distinguishes the events into four categories: *Mega Events*, *Special Events*, *Hallmark Events* and *Community Events*. The *mega events* refer to events with a global involvement which is related to the target/market that the TV coverage; this category includes the Olympics and Expo events. The *special events* are global or national target with an interest of media conveyed mostly by international or national TV. The *hallmark events* include such events or sports tournaments that cater to a national target, while the *community events* have a target, a narrower scope, and catalyze the attention of the local media (TV and print).

This taxonomy has some weaknesses as "*not fit to prestigious events such as film festivals and musical events high-profile recurring*" [14: 7]. The events categories identified are not adequately reflected even more high-level events, such as international summits, or the network initiatives of cities in the network (the "educational city", the "city of art", the "European City of Culture", and so forth) which also move journalists, visitors, media, and play directly or indirectly a function in promoting the city and the overall hosting community. Next types or taxonomies of events proposed by scholars have increasingly integrated the previous categorization and expanded taxonomy of Roche [13]; However, the events can also be classified according to some factors discriminated, such as the participation, timing schedules, profit, the media attention, the organization and the headquarters of conduct [15].

The complexity and the articulation of the events, in fact, show a variety of possible situations; for instance, in relation to the participation of entities recipients, the events can be characterized by an active involvement in case of races (for example, athletes) or by a passive involvement in the case of performances (e.g. spectators). With reference to the nature of the main event organizer, we are able to distinguish between events managed by public bodies, private or mixed; the events will benefit from specially established organizations or stable character structures; their management will be centralized or decentralized.

The events can also establish a distinct relation to the unwinding of the seat event, since it may be held in fixed locations or "carried around" repeating in more or less similar ways in several locations. The organizing and managerial process of events can generate media interest at national, international or local level, involving limited areas, and, in reference to discriminating time, the manifestations may be bound or not to predetermined schedules. Events can also be characterized with respect to duration (one or more days) and the time rate (periodic events or occasional events). In the light of the previous general considerations and focusing the attention specifically on the "mega events", we can define them as "events important, arranged one or more times, of limited duration, which serve to raise awareness, image and economy of a tourist destination short and/or long term." The mega events have three distinctive features: *although the cyclical event is unique and unrepeatable* [13: 11], *moves a large deployment of economic resources* and *exposes the media the place of development*, so called the "the big event space" [16: 861–894].

The complexity of a "Great or Big Event" configures these events (Olympics, but the International Expo Summit also) such as real "systems" that require the presence of several carriers of different objectives actors. A big event may be regarded as an economic and social system, which takes part a multiplicity of actors and, more or less explicitly, a substantial number of interlocutors really interested into the event. As part of the planning of activities and the subsequent control of the results it is essential not to overlook any "event relationship stakeholders". The stakeholders of big events may be divided into primary, i.e. individuals or groups of people without whose support the event would cease to exist, and secondary, namely the actors who, though not directly involved in the event, can seriously influence it or even prevent its success.

The relations established by the organizer/promoter of the big events with all of supra-component systems in which the event is entered, as well as interactions with all the elements and the resources necessary for its operational implementation, it influences the probability of success and, therefore, the added value derived from the event for the overall community. Moreover, one of the tasks the so-called *event manager* is precisely to ensure that the big event able to interact functionally with the elements of the external environment, while ensuring the achievement of the goals considered priority, and ensuring "respect" of the social climate and the communities in which exhibition/event is being organized.

3   Co-creation Value and *Service Dominant Logic* (SDL)

Co-creation is the joint, collaborative, concurrent, peer-like process of providing new value, both materially and symbolically. Prahalad and Ramaswamy [17] introduce the co-creation concept by acknowledging the changing roles in the theatre of the market: Customers and suppliers interact and largely collaborate beyond the price system that traditionally mediates supply-demand relationships. They consider consumer and business markets, as well as downward (customers) and upward (suppliers) relationships. Later, Prahalad and Ramaswamy [1, 18, 19] problematized and articulated the various directions in which co-creation could and should provide benefits for organizations and customers, such as improving consumption and usage experiences [20, 21] and stimulating product and service innovation [22, 23].

From the service science perspective, Vargo and Lusch [24] suggest that organizations should not focus on products, but should just consider their offerings in terms of the services they can offer to the customers. Co-creation is one of the constitutive elements of this theory: it is through customer collaboration that the market offering is realized and the required benefits (activities and services) are generated.

More generally, according to the literature, co-creation is inherent to service businesses in which market offerings (quantity, quality, attributes) are actually created in the service encounter [25, 26]. Recent developments in the service theory are grounded in the co-creation debate and converge in this direction; they claim that the service offered from organizations can be understood as "the application of specialist skills (knowledge or skills) through actions, processes or services in favor of another entity or the entity itself" [24: 1–17, 27: 43–56], which implies recognizing the value of the service produced collaboratively through forms and ties of interactions between the service systems oriented to mutual exchange and benefits. These systems are characterized by a network and/or own resources constellation, made available and integrated with each other [28] and they clearly contribute to the co-creation of value.

In general, the study of service systems and methods of co-creation of value within the scope of the Service Science can be briefly systematized and read thanks to the study conducted by Ostrom and colleagues [29: 5], who define the phenomenon "as an emerging interdisciplinary field of inquiry that focuses on fundamental science, models, theories, and applications to drive service innovation, competition, and wellbeing through co-creation of value".

Through the study and analysis of the services systems, it is possible, on one hand, to enhance and emphasize those factors that enable collaborative and adaptive co-creation of value and, on the other hand, to define balanced framework and marked by mutual logical benefit that allows access to and the sharing of resources. In addition, these systems can be individual (individuals) or groups of individuals (families) who live and evolve through an exchange and use of resources (such as knowledge and skills) to other systems. It is from this perpetual incremental and adaptive interaction between systems that arises effective co-creation of value, the value of which is the opportunity for all the players to survive and make growing the systems. It should also be pointed out that the systems in relation to each other generate a network link that generates, in fact, a macro system open to new ties with other macro systems whose ability to create value is not given by the mere sum of the potential of every single element (quantitative factor), but by the quality of the links and mode/coordination requirements and management of flows in/out (quality factor).

This phenomenon of macro and micro services systems in the perspective of networks is strongly related to *service-dominant logic* (SDL), even if there are critical remarks about the way the SDL approach articulates the relationship between the actors in the co-creation process [30, 31].

Indeed, relationships with customers, intangible property and the co-creation of value with the customers form the basis of the production of services by businesses [24]. As expressed by Dalli [32: 325–339], "undertakings shall make available to consumers the resources necessary for the generation of value. Under certain conditions, the consumers can be considered manufacturers and even themselves suppliers of the companies for the preparation of the necessary resources to provide the services required by the market. Everything on the assumption that consumers do not ask for goods or material objects, but functions, activities and, in general, services: these are services that are the basis of the perception of value". Hence, Vargo and Lusch [27: 43–56] point out that "the customer is always a co-creator of value." According to Grönroos [3], organizations, specifically service providers, only provide the resources or the means to make possible to create value for customers. In this sense at least, when vendors and customers interact, they are involved in co-creating value.

In this approach, the service (rather than the product) is what creates advantage for the customer and, accordingly, the goods are interpreted as simple tools or mechanisms of distribution of service provision. Therefore, the service is the general case, the common denominator of the exchange process; in fact, the service is always exchanged, while the goods when they are used, are the supports for the process of services delivery. The customer is co-creator of value, in an interactive process with the service provider. The latter is not able to create value by itself, but provides it collaboratively using resources and contributes to the value following the acceptance by the customers of such interaction offered.

Ultimately, the paradigm of Service Dominant Logic (SDL) proposed by Vargo and Lusch [4, 24] shows the importance of consumers as "economic and social stakeholders that interact within networks". Thus, it argues that the value is the result of healthy approach application operant resources which can sometimes also be guided through goods or materials on which to act. The value is so co-created by the commitment integrated "systems" such as organizations, employees, consumers, social partners and all the other parties with an interest in operationally share the resources available according to their specificities and needs. When we follow and adopt SDL logic, the distinction between producers and consumers, in fact, disappears and all the actors/participants become active players in the co-creation of value for themselves and for others. It acknowledges the customers as integrated resources (as well as suppliers), in line with the concept of co-creation value.

Considering sport events in the perspective of co-creation approach, they can be described as services that are demanded by different reasons. For example, sport spectators experience good stadium service quality by different dimensions: perceived team performance, stadium service quality and spectator induced atmosphere. Satisfaction is led by fulfilling all three dimensions [10]. Thus, the value is created by different parties. Furthermore, this value depends not only on the origin value creator, but it is spread over several participants. Business drifting integrates the customer into the value creation that leads the customer to an active role of value creation [1]. The interaction leads to a cooperation for innovations within the networks [33]. Successful customer integration needs two aspects: information about customer needs and information about how to best solve these needs [34]. In this direction, in order to collect, manage and share information and data between the players, organizations and customers for sport events, technology play a significant role, because it could significantly facilitate these interactions.

In recent years, digital technology has led to the development of social networks (e.g. Facebook, Twitter) which are an important resource for a wider dissemination of sport events and support of the needed collecting, processing and management of information and data about the same sport events for the players involved. The social channels provide the ability to "take out" the sport event, and given the immense contribution they can make, should not be underestimated. Social media today are a way to create a virtual reality on various platforms through which you can communicate and share interests, opinions and any type of content. The combination of social and sport events now not just about the promotional aspect, but many other factors related to the interactivity of the sport event, and in general, the overall management process, the most direct and immediate involvement of people (even distant) and the ability to extend the life of the event.

Regardless of the choice of communication channels and the type and nature of sport events, it is essential for almost all events to have a small presence or indirect social media. To increase awareness of the public of an event and, therefore, the participation, the sport event organizer should have a strategy and a well-designed communication plan: it must start from the marketing objectives of the event organizer or entity organizer, from the analysis of the target audience to the choice of the media to be used. The numbers, as the amount of "Like" or comments on a Facebook page, do not speak for themselves, but provide important information if used to understand what people think and want [35]. Social media are not vertical, such as advertising or public relations, but they are a horizontal instrument that touches the entire business, from customer service, acquisition or customer retention [36]. With social media you can "co-create value" through direct communication with the public, for example, by taking actions such as: "Respond to every comment and all questions that directly involve"; "Leave space on their pages or those of others to post events that you find useful or interesting"; "Keeping a blog, a section of the site, or a series of posts on themes dear to your target audience". With the power that social media give, people become important to pay attention to small details, so it is crucial for the organizers to manage feedback. If anyone had negative impressions in a sport event for any reason, even from acting independently of the organization, social media can represent the perfect medium to make their voices heard and criticisms.

Ultimately, through social media, the sport event organizers can obtain significant advantages, especially in terms of effective and efficient management of sport events, as a part of one system composed of numerous and unlimited possibilities of communication with participants. Underestimating the potential of social media can be an unforgivable mistake, especially for those organizations involved in event management process [37]. As shown by a survey carried out over 2000 XING event organizers, mainly European actors, trust continues to be strong in social media. Almost all the event organizing companies, especially the sport organizers, as shown in the last sport mega-events like the Olympics, are planning an increase in social media activity for the future. In 2014, in general, the event industry recorded an expansion of the online business, following the trend of recent years. It will be interesting to see if the event organizers are willing to continue to invest more and more resources, and if social media will continue to evolve in this direction, developing new qualities and potential suitable for this area.

4   Empirical Study on Mega Sporting Events and Social Networks in the Co-creation Value Perspective

For the purposes of this study, we focused our analysis on the mega sporting events, more specifically, on the Summer Olympic Games events that took place in the period between 1992 and 2016. In Table 1 the dataset of our analysis has been shown.

The choice to start our investigation in 1992 is not random. The 1992 Barcelona Olympics showed profound changes in the planning and organizing process of the Olympic Games. For the first time, for example, the opening ceremony has a significance that had not previously assumed ever had, becoming an important and almost autonomous event and globally expected. The Olympic Games in Barcelona were those of a profoundly different world from the one that had marched four years earlier in Seoul. The Barcelona Olympic Games were the first since the fall of the Berlin Wall, with a world completely changed. The fall of the Berlin Wall in 1989 led to the unification of Germany, the release of Nelson Mandela in 1990 removed the embargo on South Africa, and in 1991 the Soviet Union and Yugoslavia country ceased to be unitary nations. The XXV edition of the Olympic Games that took place in Barcelona from July 25th to August 9th in 1992 assumed a special meaning by defining a kind of divisions of the games, in fact, it was necessary to take into account the newly established independent states and submit bids for a single team for the two German nations now working together, and the readmission of South Africa. At the 1992 Barcelona Olympics, the independent teams of Estonia and Latvia made their first apparition since 1936 and Lithuania sent its first team since 1928. The other ex-Soviet republics participated as a "unified team", although the winners were honored under the flags of their own republics. The only controversy concerned Yugoslavia, which was the subject of United Nations sanctions because of its military aggression against Croatia and Bosnia-Herzegovina countries. In the end, Yugoslavia was banned from taking part in any team sports, but individual Yugoslav athletes were allowed to compete as "independent Olympic participants". Croatia, Slovenia and Bosnia-Herzegovina competed as separate nations for the first time. These geopolitical changes had positive effects on the number of countries involved (169) and the participating athletes (9356, of which 6659 men and 2721 women).

Table 1. Synthesis of the Olympic Games editions. Period 1992–2016. Source: IOC - International Olympic Committee – Olympics

Event	Year	Nation	Number of participants athletes	Number of participants countries	Media
Olympic Games Barcelona	1992	Spain	9356 (6659 men) (2721 women)	169	13082 Media (written press and broadcasters)
Olympic Games Atlanta	1996	U.S.A	10318 (6806 men) (3512 women)	197	15108 Media (written press and broadcasters)
Olympic Games Sydney	2000	Australia	10651 (6582 men) (4069 women)	199	16033 Media (written press and broadcasters)
Olympic Games Athens	2004	Greece	10625 (6296 men) (4329 women)	201	21500 Media (written press and broadcasters)
Olympic Games Beijing	2008	China	10903 (6294 men) (4609 women)	204	24562 Media (written press and broadcasters)
Olympic Games LONDON	2012	England	10568 (5892 men) (4676 women)	204	21000 Media (written press and broadcasters)
Olympic Games Rio de Janeiro	2016	Brazil	11303 (6213 men) (5090 women)	207	25000 Media (written press and broadcasters)

Games of the XXVI Olympiad took place in Atlanta in the United States of America from July 19th to August 4th in 1996. Countries that participated were 197 and 12 nations of the former Soviet Union participated for the first time as independent states: Armenia, Azerbaijan, Belarus, Georgia, Kazakhstan, Kyrgyzstan, Moldova, Russia, Tajikistan, Turkmenistan, Ukraine, Uzbekistan. Because of that, and also due to the large number of participants, 14 countries won for the first time an Olympic medal. The athletes who took part in this edition of the Olympics was 10318 of which 6806 men and 3512 women. The Atlanta Olympics are also known as the Centennial Olympics, as you perform exactly one hundred years after the first modern Olympic Edition (Athens 1896).

The Sydney Olympics of 2000 are the XXVII edition of the Summer Olympic Games of the modern era. The races are held in Sydney (Australia) from September 15th to October 1st in 2000. 199 nations participated with 10651 athletes, of which 6582 men and 4069 women in 28 sports and 300 disciplines. For the second time in the history of the modern Olympic games, they are organized in the southern hemisphere of the planet, after the Olympic Games in Melbourne (Australia) in 1956. Unlike the previous Australian edition of 1950s, carried out between November and December, these

Olympic Games are held in the Southern Hemisphere spring season to prevent damaging the Western habit of attending the Olympic Games in the summer.

The 2004 Athens Olympics are the XXVIII edition of the Summer Olympics. The races take place in Athens (Greece) August 13th–29th in 2004. 201 countries participated with 10625 athletes of which 6296 men and 4329 women in 28 sports and 301 disciplines, one more than the previous Olympics in Sydney. Athens hosted for the second time the Olympic Games, the Greek capital has already hosted in 1896 the first edition of the modern Olympics. The 2004 Athens Olympics are also the first to take place after the attack of September 11th, 2001 and the first to have a high cost not only for the realization of the event but also to ensure the safety of the public against the risk of terrorism, about 10% of the total construction costs of the Olympiad.

The Beijing Olympics of 2008 represent the XXIX summer Olympic Games of the modern era. The races take place in Beijing (China) from August 8th to August 24th in 2008. 204 countries participated with 11028 athletes of which 4637 women and 6305 men in 28 sports and 302 disciplines. The Beijing Olympics are the third Olympic edition to be held in Asia after the Olympic Games in Tokyo in 1964 and Seoul also those of the edition 1988. The Games brought many tangible and intangible benefits to China, especially in terms of public infrastructure improvements. While some of the positive benefits are immediately apparent, others will emerge in the future after the end of the big event. The Chinese games, well organized but pressed by strict controls, have the symbolic meaning of China's entry among the Great Powers of the world.

The Games of the XXX Olympiad were held in London in United Kingdom, from July 27th to August 12th in 2012. Thus, the British capital has become the first city to have hosted the summer Olympics three times, after the editions of 1908 and 1948. 204 countries participated with 10768 athletes of which 4776 women and 5992 men.

Finally, the games of the XXXI Olympiad were held in Rio de Janeiro, Brazil, August 5th–21st in 2016 and Brazil became the first South American State (and The Second Latin American) to host an edition of the Summer Olympic Games. 207 countries participated (record in the history of the Olympic Games) with 11303 athletes of which 6213 men and 5090 women. This edition of the Olympic Games hosted for the first time in history 10 athletes, including four women, fleeing their country which in fact represented the team of political refugees. One in Rio de Janeiro is the first Olympics in South America, the third in 120 years in the Southern Hemisphere (earlier editions were Melbourne in 1956 and Sydney 2000). The only continent that has never organized an edition of five games circles, summer or winter, remains Africa.

The Games that were held in Rio de Janeiro mark, as already in the case of Tokyo in 1964, Seoul in 1988 and Beijing in 2008, the full membership of a nation not European or Western one of the stars of the international scene, not only in terms of sport, but also in terms of image and its economic and productive relevance. The Rio Games celebrated and showcased the sport, thanks to the beautiful surroundings of the city and a desire to lift event presentation to new heights. At the same time, Rio 2016 was an opportunity to deliver the broader aspirations for the long-term future of the city, the region and the country, the opportunity to accelerate the transformation of Rio de Janeiro into an even greater global city.

The various editions of the Olympic Games have had over the years a growing media interest amplifying the advent in recent editions, and by daily use of social networks. This is confirmed by the following data. For each edition of the Olympic Games, accredited media, including the active role of journalists, photographers and cameramen, had the following trend: 13082 for the edition of the Olympic Games of Barcelona in 1992; 15108 for the edition of the Olympic Games of Atlanta in 1996; 16033 for the edition of the Olympic Games of Sydney in 2000; 21500 for the edition of the Olympic Games of Athens in 2004; 24562 for the edition of the Olympic Games of Beijing in 2008; 21000 for the edition of the Olympic Games of London in 2012; 25000 for the edition of the Olympic Games of Rio de Janeiro in 2016.

The web did not remain indifferent to the charm of the Olympic Games, there are in fact hundreds of sites created to report events in streaming, publish news or provide the results of real-time competitions. In fact, the last three editions of the Summer Olympics were characterized by the use of social networks (mostly Facebook and Twitter) that helped foster an interest and a greater involvement of the public. The usual television broadcast of the Games was supported by a social media storytelling made in person by athletes, commentators, celebrities, and not least by the public.

The Beijing Olympics in 2008 inaugurated the use of social networks in the world of this event. It is known that in China there is a total prohibition about using social networks and those prohibited are: Facebook, Twitter and Youtube. The only available is Instagram. However, on the occasion of the Beijing Olympic Games the access to social networks have been unlocked for hundreds of journalists who have been able to use the network to tell the chronicles of the sport events.

For the first time in the history of the Summer Olympic Games, the International Olympic Committee (IOC) has awarded the broadcast rights to the event also Internet sites. According to a study of the Chinese Government Center, in spite of the massive media coverage and TV, almost 80% of the 207 million users who have followed the games has chosen before the internet to inquire. In addition, nearly 27 million users have never turned on their television or read the newspaper to follow the competitions. This study shows that thanks to the Olympics, internet has been recognized and accepted as a medium for all purposes because it allows you to meet the needs of all target, and also to niche interests, instead it was impossible to achieve these goals for television.

Thanks to the Olympics in London in 2012 there is the true consecration of the use of internet and social networks by sport fans and not only. Two weeks after the end of the Olympics official data were released on the use of Twitter during the competition. Throughout the Olympics there have been more than 150 million tweets divided over the 16 Olympic days. The moments in which Twitter has peaked is not, as you might think, during the awarding of the medals but in the final moments of the accesses and exciting matches. The events with the highest peak have been two, the final basketball between the USA and Spain, and the women's football final competition between the USA and Japan.

From these statistics it is clear what the sports that have generated most of the conversations on Twitter, were primarily football with 5 million tweets and then others are swimming, athletics, gymnastics and volleyball. It is proper to make a separate discussion for the impressive opening and closing ceremonies, also at the center of conversations with very important numbers. In the closing ceremony, one of the most anticipated moments was the exceptional participation of the Spice Girls, the famous singers group, pending all over the world and that of facts helped to create 116.000 tweets per minute. Those presented are important numbers, highlighting the centrality of Twitter during sporting events.

About the Olympic Games in Rio de Janerio in 2016, these have inspired 187 million tweets and generated 75 billion total impressions while the Olympic Games in London had collected though it is proportionally fewer, considering subsequent platform growth (back in 2012, Twitter only had around half the users it had now). Hence, even though Twitter had 146 million monthly active users, there was only an increase of 37 million total tweets. This shows the growth in the popularity of games but the same or may be indicative of less commitment on Twitter over time; in fact, there was a more important use of Facebook, competitive platform of Twitter. Regarding Rio de Janeiro 2016, Facebook reported that there were 227 million posts and comments on the games while, in London 2012, 116 million people interacted.

Between the two platforms, Twitter's "impressions" stat and Facebook's "interactions" are pretty closely related, which would suggest that Facebook came out well on top in overall Olympics conversation volume. Regardless of the competition between the two social networks, the data provided by them demonstrate the meaningful participation and involvement of users. Still with reference to the Olympic Games in Rio de Janeiro, Facebook also reported that Instagram saw 916 million interactions from 131 million users on the games and that more than 15.2 million people used profile frames to show their support for their favorite teams. Of these, frames were more popular among users in India, the Philippines and Pakistan.

Facebook, Twitter, Instagram can be seen as relevant tools able to revolutionize the way people watch the Olympics inaugurating an innovative communicative cut.

Thanks to contents, opinions, and ideas shared by users about events related to the Olympics, the social networks are proposed as "a place to gather," and in which everyone can feel part of this great sporting event regardless of their geographical location. Social accounts of the athletes have millions of followers (Usain Bolt for example, has nearly 18 million fans on Facebook and only the Indian tennis player SaniaMirza more than 11 million) following each shared moment from the preparation to break the race moments. Social media allow all people to immerse themselves in the atmosphere of the Olympic Games to 360° and experience exclusive moments. The data on social networks provide more and more opportunities to witness these offerings at the events and especially the enormous value covered by the contents in real time. The great objective in the organization of future Olympic Games will be to convey the here and now of the Games by making the community share in a collective storytelling, because it is now recognized the active role played by customers (participants, visitors, fans) and all the players in improve and make successfully the Olympics by interacting thanks to social media in the co-creation value perspective.

5   Concluding Remarks

All the users, that is the customers, as specifically visitors or passive participants of mega sporting events, mostly the Olympics, have more freedom to interact with the organizers of sporting events and the same athletes through the use of social networks, especially Facebook and Twitter. Hence, the co-creation of value that determines the mega sporting event positive impact, measured in terms of socio-economic benefits and the number of participants who always show their satisfaction through social media. Therefore, co-creation can be seen as a business strategy even in the case of mega sporting events sub-field that emphasizes the creation of a corporate value shared with the customer/visitor/fan. As already outlined, the idea behind this process is the vision of the market as a place where the various actors involved (companies, consumers, employees, share-holders, suppliers) share, combine and renew together the resources and the ability to create value through new forms of interaction, like the social media.

Thanks to the social networks, customers (visitors, participants, fans) can stay always in touch with all the players directly or indirectly involved in the Olympic Games events, giving continuous feedbacks and transforming themselves into participants-actors. Certainly, the co-creation is not an innovation process that guarantees its success, there is in fact no certainty that the ideas, and especially, the realization of Olympic Games events will be successful, responding to effectiveness and efficiency criteria. But the result is that the thought of the project value increases as it comes from personalized and unique experience for the consumer, as well as a better perception of the Olympic event by the actors. According to co-creation economic perspective, this can lead to an increase in revenues and profitability because the co-created will meet more the fans/costumers' needs and requests in terms of standards and required services, which will increase as a result of their participation, also considering the direct opinions of athletes and active participants of the Olympics.

This process, in which organizations adopt different channels available to consumers to allow them accessing to more direct and engaging experiences, can also be seen from an opposite point of view. It is possible that the phenomenon of co-creation is perceived as a form of free exploitation of ideas and "energy" external businesses. Also, we should support a "mutual growth" for both parties involved, so that the Olympic event organizer and users can develop in parallel, taking advantages of each other but respecting equity criteria. It is important that both parties have benefits, including material as rewards customers/co-creators with free products/services, because the collaboration to create new ideas does not become a form of free exploitation, but an enrichment of mutual value.

This explorative study have several limitations, mostly because of its nature, in fact, it allows only to represent and describe still undeveloped ideas about the phenomenon investigated, briefly describing how the Summer Olympic Games events acquire a more relevant role and have been more successful thanks to the spread of the social networks by all the users, mainly fans and athletes. In the future development of the study, we might conduct a meta-analysis to identify in a wide research design the main variables of the impact of social media, specifically the social networks, on the mega sporting event management process adopting the co-creation perspective, and also considering theoretical frameworks and not easily providing descriptions of the main existing effects. This exploratory study represents a research starting point, in fact, in the future we are working on a mathematical analysis model to evaluate and measure the concrete impact of social media on the effectiveness and efficiency of Olympics and all the other mega sport events by adopting always the co-creation value approach.

References

1. Prahalad, C.K., Ramaswamy, V.: Co-creating unique value with customers. Strategy Leadersh. **32**(3), 4–9 (2004)
2. Prahalad, C.K., Ramaswamy, V.: Co-creation experiences: the next practice in value creation. J. Interact. Mark. **18**(3), 5–14 (2004)
3. Gronroos, C.: Adopting a service logic for marketing. Mark. Theory **6**(3), 317–333 (2006)
4. Vargo, S.L., Lusch, R.F.: Service-dominant logic: continuing the evolution. J. Acad. Mark. Sci. **36**(1), 1–10 (2008)
5. Getz, D.: Special events: defining the product. Tour. Manag. **10**(2), 125–137 (1989)
6. IFEA, International Festivals & Events Association (2002)
7. Getz, D.: Event tourism and the authenticity dilemma. In: Theobald, W.F. (ed.) Global Tourism: The Next Decade, pp. 313–329. Butterworth Heinemann, Jordanhill (1995)
8. Getz, D.: Festival places: a comparison of Europe and North America. Tourism **49**(1), 3–18 (2001)
9. Van Der Wagen, L.: Event Management: for Tourism, Cultural Business and Sporting Events. Prentice Hall, Upper Saddle River (2001)
10. Dalla Sega, P.: Gli eventi culturali. Ideazione, progettazione, markeing, comunicazione. Franco Angeli (2005)
11. Varzi, A.: Parole, oggetti, eventi e altri argomenti di metafisica. Carocci (2001)
12. Diodato, R.: Estetica del virtuale. Bruno Mondadori (2005)
13. Roche, M.: Mega-events and Modernity: Olympics and Expos in the Growth of Global Culture, Routledge (2000)
14. Guala, C.: Per una tipologia dei mega eventi. Bollettino della Società Geografica Italiana, serie XII, vol. VII, 4, (2002)
15. Cherubini, S.: Il marketing per generare valore nel sistema evento. In: AA.VV., Le tendenze del marketing. Convegno de l'Ecole Supérieure de Commerce de Paris, EAP, pp. 184–186 (2009)
16. Dansero, E.: I luoghi comuni dei grandi eventi. Allestendo il palcoscenico territoriale per Torino 2006. In: Dansero, E., Segre, A. (a cura di) Il territorio dei grandi eventi. Riflessioni e ricerche guardando a Torino 2006, numero monografico Bollettino della Società Geografica, pp. 861–894 (2002)
17. Prahalad, C.K., Ramaswamy, V.: Co-opting customer competence. Harvard Bus. Rev. **78**(1), 79–90 (2000)
18. Prahalad, C.K., Ramaswamy, V.: The Future of Competition: Co-creating Unique Value with Customers. Harvard Business School Press, Boston (2004)
19. Prahalad, C.K., Ramaswamy, V.: The new frontier of experience innovation. MIT Sloan Manage. Rev. **44**(4), 12–18 (2003)
20. Gentile, C., Spiller, N., Noci, G.: How to sustain the customer experience: an overview of experience components that co-create value with the customer. Eur. Manag. J. **25**(5), 395–410 (2007)

21. Payne, A.F., Storbacka, K., Frow, P.: Managing the co-creation of value. J. Acad. Mark. Sci. **36**(1), 83–96 (2008)
22. Sawhney, M., Verona, G., Prandelli, E.: Collaborating to create: the internet as a platform for customer engagement in product innovation. J. Interact. Mark. **19**(4), 4–17 (2005)
23. Bitner, M., Ostrom, A.L., Morgan, F.N.: Service blueprinting: a practical technique for service innovation. Calif. Manag. Rev. **50**(3), 66–94 (2008)
24. Vargo, S.L., Lusch, R.F.: Evolving to a new dominant logic for marketing. J. Mark. **68**(1), 1–17 (2004)
25. Solomon, M.R., Surprenant, C., Czepiel, J.A., Gutman, E.G.: A role theory perspective on dyadic interactions: the service encounter. J. Mark. **49**, 99–111 (2008)
26. Bitner, M., Brown, S.W., Meuter, M.L.: Technology infusion in service encounters. J. Acad. Mark. Sci. **28**(1), 138–149 (2000)
27. Vargo, S.L., Lusch, R.F.: Service-dominant logic: what it is, what it is not, what it might be. In: Lusch, R.F., Vargo, S.L. (eds.) The Service-Dominant Logic of Marketing: Dialog, Debate, and Directions, pp. 43–56. ME Sharpe, Armonk (2006)
28. Spohrer, J., Maglio, P.P.: The emergence of service science: toward systematic service innovations to accelerate co-creation of value. Prod. Oper. Manage. **17**(3), 238–246 (2008)
29. Ostrom, A.L., Bitner, M.J., Brown, S.W., Burkhard, K.A., Goul, M., Smith-Daniels, V., Demirkan, H., Rabinovich, E.: Moving forward and making a difference: research priorities. J. Serv. Res. **13**(1), 4–36 (2010)
30. Gronroos, C., Voima, P.: Critical service logic: making sense of value creation and co-creation. J. Acad. Mark. Sci. **41**(2), 133–150 (2013)
31. Gronroos, C.: Service logic revisited: who creates value? and who co-creates? Eur. Bus. Rev. **20**(4), 298–314 (2008)
32. Dalli, D.: Il ruolo del consumatore nei processi di marketing. In: AA.VV., La scuola di Riccardo Varaldo. Relazioni personali e percorsi di ricerca, Pacini, pp. 325–339 (2010)
33. Chesbrough, H.: The Era of Open Innovation (2003)
34. von Hippel, E., Katz, R.: Shifting innovation to users via toolkits. Manage. Sci. **48**(7), 821–833 (2002)
35. Highamy, D.J., Grindrodz, P., Mantzarisx, A., Otley, A., Laflin, P.: Anticipating Activity in Social Media Spikes. arXiv preprint arXiv:1406.2017 (2014)
36. Hougland, C.: Things Fall Apart: How Social Media Leads to a Less Stable World. Knowledge at Wharton, University of Pennsylvania (2014)
37. Safko, L.: The Social Media Bible: Tactics, Tools, and Strategies for Business Success. Wiley, New Jersey (2012)

Sustainability: Service Ecosystems, Environment Control and Transportation

Towards a Proposal for the Sustainability Through Institutions in Public Transport Services in Times of Emergency

Monica Drăgoicea[1]([⊠]), Saber Salehpour[1], Henriqueta Nóvoa[2], and Virginia Ecaterina Oltean[1]

[1] Faculty of Automatic Control and Computers, University Politehnica of Bucharest, Splaiul Independentei 313, 060042 Bucharest, Romania
monica.dragoicea@acse.pub.ro, ecaterina.oltean@upb.ro
[2] Faculty of Engineering - FEUP, University of Porto, Rua Dr. Roberto Frias, 4200-465 Porto, Portugal
hnovoa@fe.up.pt

Abstract. The public transport service system provides safe and secure urban mobility for all citizens. This paper describes work-in-progress regarding its role in times of emergency, such as earthquakes, tornadoes, or terrorist attacks. In order to deliver good and safe transport service, it requires adequate resources such as information, staff, property, vehicle, infrastructure, the set of operational rules, and governance. In this perspective, establishing a sustainable and smart transport model to cope with transport services' disruptions especially for emergency events is vital. A general characterization of critical actions for public transport Emergency Management Planning is provided. A new perspective on designing sustainable public transport services in times of emergency is introduced in order to support transport service company's operational activities based on sustainable institutions principles.

Keywords: Urban mobility · Emergency management · Public transport · Sustainability · Information intensive processes

1 Introduction

One of the main criteria to define smartness in a city is *urban mobility* [1,2]. During the last decades, the subject of urban mobility and sustainable transport has been taken into consideration by experts trying to solve neglected aspects such as providing a clean and peaceful city with fast, secure transport services and effective transportation service systems.

Transportation itself is evolving in the *digital transformation* era that embraces cities and regions in all aspects of cultural, social, technological, and economical changes. It is prone to the case of *information intensive business processes* which are driven by the large scale integration of digital technologies into transportation infrastructure, traffic management systems, and transportation means. Transport 2050 [3], European Commission's initiative to define

© Springer International Publishing AG 2017
S. Za et al. (Eds.): IESS 2017, LNBIP 279, pp. 355–369, 2017.
DOI: 10.1007/978-3-319-56925-3_28

major challenges and key measures to enhance the quality of transport services with a major impact on people's quality of life, acknowledges the role of digital technologies for making transport safer, more efficient and inclusive. For greater potential, these technologies must be well integrated in sustainable mobility concepts for smart governance in the local digital ecosystem [4].

Reliability of the public transport service system needs to be strongly connected to all *public transport service systems' stakeholders* such as government, transport operators, travellers, and public authorities. Reliability, the ability of public transport service system to provide a solid transport service over a longer period of time, is generally defined on several components like travel time, waiting for time, arrival time, and punctuality [5]. Other authors asses reliability as being a key level of service indicators where poor reliability leads to customers' discomfort, dissatisfaction, and anxiety [6]. Several other studies propose on improving service reliability in urban transit networks based on the relevancy of strategic and tactical levels of transit planning on disturbance mitigation. They claimed that changes in demand and supply such as a change in traveller's behaviour, infrastructure quality, availability, and operator's performance may eventually lead to service disruption [7, 8].

The public transport service system is prone to many daily disturbances which impact on delivered service to its users. Besides its impact on customers' satisfaction, it also has a negative influence on companies' financial situation. Public transportation operators' response to disruption differs based on the type of event, time of the event, or location of the event.

Recent literature on public transport network management addresses disruption in several ways. In [7] disruption is mainly categorized in three dimensions such as *event's frequency* (frequent events, semi-frequent events, low frequent events), *event's predictability* (predictable events, unpredictable events), and *event's regularity* (minor quasi-continuous ongoing, major discrete). The transport disruption ontology in [9] approaches a framework for modelling travel and transport related events with disruptive impact on public transport. Aggregate data exposing different variables of travel behaviour highlight disruption patterns leading eventually to transport policy change [10].

Public transport operators are examples of forefront actors to maintain and support citizens' mobility which may be impacted by predictable or unpredictable events [11]. As an inseparable part of public transport, these events have an impact on service running time, service punctuality and service regularity [7].

Besides interruption of local public transport service, history of public transport shows that it also plays a vital role as a quick response to emergency events. The organized response of New York City and Washington DC's public transportation system [11], and failure of New Orleans's public transport system after 9/11 terrorist attack and Hurricane Katrina [12] proves the role of developed and sustainable public transport in case of emergency.

According to the above introduced perspective, this paper presents an approach to define the main characteristics of the public transport service system in times of emergency. In Sect. 2 some successful and unsuccessful transport operators' responses during emergency events are briefly presented in order to understand why some transport operations are prosperous in resource provision, allocation and appropriation. Identifying public transport system's crucial resources such as people, information, and properties help public transport operators to have a better understanding of vulnerable components which cannot be lost by disruption. In Sect. 3 the public transport system is described as a service system which provides for its users mobility, evacuation, and emergency mobility services. Section 4 introduces the role of public transport service before and after emergency events, and describes the application of Elinor Ostrom's principles on sustainable institutions in case of emergency events for public transport services. Section 5 concludes the paper and draws further research directions.

2   Public Transport Response in Times of Emergency

In times of emergency, identifying involved organization, transport agencies and the level of theirs resource is important. Authorities' lack of coordination, uncertainty, and confusion of primary responsibility may lead to avoidable problems. Authorities' action after Katrina and Rita is one of the best studied instances [12]. For this reason, an emergency management plan should be prepared in advance to make sure that all involved parties are protecting beings from threats and hazards in their areas of responsibility.

Emergency Management Planning (EMP) for an organization can be seen under the umbrella of Business Continuity Management (BCM). In BCM, critical parts of a business which cannot be lost in case of disruption are identified before this event occurs [13]. Its primary goal is to prevent or to minimize the impact of the event on scarcity of resources (such as customer, information, staff, properties) and to deliver a proper maintenance service after event [14].

Before designing a BCM action plan to assure sustainability in public transport service delivery, the types of services the transport operators can provide need to be defined for this specific case. This implies evaluating which are the required resources, which kinds of risks threaten public transport activities, and how the transport service can be maintained after a disruption happens.

In this respect, Table 1 introduces a classification of several types of critical actions involved in the public transport emergency management planning. The proposal follows the classification of emergency-based activities in the four-stages of a disaster cycle: mitigation, preparedness, response, and recovery [15–17].

Recent literature reveals insight on *successful* and *unsuccessful public transport operators' response* during emergency events. This type of analysis is useful when trying to understand why some transport operations are prosperous in resource provision, allocation and appropriation.

The role of Port Authority Trans-Hudson Corporation (PATH) after September 11th, 2001 was remarkable. Round-the-clock service provided by PATH, with more than 1000 trains a day, was the result of the development of a new service plan and team working [11]. In the meantime, PATHs prevention plan

Table 1. Critical actions for Public Transport Emergency Management Planning

Phase	Purpose	Public transport action
Mitigation	All actions which public transport does in order to prevent or minimize the impact of emergency events on its assets (People, Information, Infrastructure) before and after emergency event	– Risk reduction strategy – Awareness and Training Programs – Security Awareness – Infrastructure and vehicle maintenance, standard infrastructure and vehicle design
Preparedness	A set of plans which public transport operators make in order to be prepared for human-made or natural emergency events before an emergency event	– Analysis of Hazard and Threats assessment – Communicating about risk – Emergency procedures – Coordinating with stakeholders – Education, Training, and Evaluation
Response	All emergency operations that public transport operators does to save people and animals' life and provide necessary resources such as equipment, expert, authority, policies, foods, and water to threatened location during emergency event	– Preservation and protection – Responsibility of public transport personnel – Service restoration – Inter-agency Coordination
Recovery	All efforts taken by transport operators in order to return the transport service and citizens' life to the safe and normal level and repair transport infrastructure in order to support public transport service	– Transport service continuity – Maintenance of transport service – Support employee involved in emergency event – Return to normal transport service – Learn from experience to build a robust public transport service

to pump water from train infrastructure, exchange place station and tunnel, to prevent infrastructure water damage was exemplary. Cooperation of PATH employees and Port Authority Police officers at a professional level led to a successful evacuation of people in World Trade Centre Station. Transport service to World Trade Centre Station was suspended due to a notification from Port Authority Police. Information staff redirected passengers to a safe exit, PATH station inside the World Trade Centre was coordinated by Police and the last train did not stop at World Trade Centre station, being returned to New Jersey. New York's Metropolitan Transportation agencies' quick reaction after 9/11 such as moving trains to a safe place, discharging the passengers, halting the trips, suspending the subway and rail services and closing all MTA bridges and tunnels based on an emergency plan ended up with no single dead. As a quick and proper response to service demand and travel pattern they adjusted their service to passengers [11].

On the morning of March, 22, 2016, Brussels airport and Maalbeek metro station were attacked by ISIS. This attack, as an emergency event, impacted heavily on public transport service delivery. As a direct response all public transport services were closed due to security purpose, but train service was resumed after 4 pm [18]. As well, (a) all flights were suspended, (b) public transportation network was temporarily closed, (c) tramway and bus services were active after 4 pm, (d) evacuation service was operational for public transport, (e) public transport service disruption was localized. This disruption lasted one day and extra vehicle were used in service to reduce the impact of disruptions. An alarm application was offered by the Brussels Intercommunal Transport Company (STIB/MIVB) to inform citizens about disruption and online data. This attack impacted on another part of country and caused closure of public transport. After attack the public transport service experienced many disruptions in service caused by operation security, clearing the backlog resulting from disruptions caused by the attack, and reparation works [19].

New Orleans's public transport response to 2005 hurricane can be the best example of management failure during emergency response. The scarcity of public transport resources, failure in resource allocation, and lack of establishment of self-government of common pool transport resources led to failure in the emergency plan and transit dependent residents' evacuation from threatened areas.

Todd Litman enumerated inordinate dependence on cars, dearth of fuel and traffic congestion as reasons of failure of evocation plan during Rita [12]. During this disaster hundreds of thousand of people trapped in critical conditions. Before the disaster, according to [20], it was proved that only 60% of citizens are able to move during an emergency event and 40% of transit-dependent residents are reliance on public transport. Authorities were aware of no-drivers statistics, but the lack of adequate information about evacuation plan to transport users and adequate stations to take people to the shelter in congestion with adequate water, food, and medical service led to the humanitarian crisis.

For the point of view of public transport service system, some planning failures during Katrina and Rita can be highlighted [12]: (a) failure of enough resource to fulfil public transport need such as drivers; (b) lack of effective emergency plan especially for evacuation response; (c) failure to consider other beings in rescue plans such as animals and pets; (d) mismanagement in enlisting of other means of transport such as bus or train for evacuation plan; (e) failure in maintenance service such as fuel shortage; (f) lack of free ride for low-income people or who lost their money; (g) failure of fire evacuation operation for people who trapped in treated area; (h) failure of provision of up-to-date information for public transport crews and people.

In Boston, however, METRO's response to Hurricane Rita was exemplary but ad-hoc arrangement left many stranded motorist on the highway [21].

These case studies of information-based response activities in case of emergency shape the idea that transport operators need to gather insight on how citizens' demand behaviour changes after disasters in order to draw a proper supply strategy as an immediate response to keep service sustainable. These responses may vary from a nation to another. Many psychological patterns impact on commuter design-making after a disaster. Giuliano and Golob [22] studied on citizens trip pattern after January 17th Northridge earthquake. The study shows that citizens used their own car but changed the route, schedule or destination instead of using public transport service.

Many elements are engaged in passengers' reaction pattern, such as psychological reasons, lack of sufficient capacity of public transport, road working, reconstruction program, to name but a few. According to passengers' behaviour, suitable strategies can be made to avoid congestion in the first week of disasters such as increasing the fuel price, decreasing transport fare, rescheduling of school, universities or non-work trips in order to convince people to use public transport service. However, several studies show that the main reason for congestion in the highway is the business trip, and several adjustments can be made by authorities, such as: (a) road closure and traffic detour; (b) operating of two one-way parallel lanes in peak hours; (c) opening truck bypass lanes to public; (d) construction program; (e) opening old road; (f) reopening of additional connector; (g) adding additional stations; (h) additional parking space; (i) operating additional train during rush hour; (j) fare discount up to 50%; (k) expansion of service, local bus, and shuttle service; (l) providing express route [12,23].

Insights from these case studies of public transport response in case of disruptive events can be further used in order to characterize the public transport system as a service system in time of emergency.

3 Public Transport Service Systems

The public transport system can be seen as a *service system* (SS) which provides for its users mobility services, evacuation and emergency mobility services, and transportation of emergency resources such as equipment, technicians, and authorities from/to critical locations in the time of emergency. In this paper emergency service refers to above-mentioned activities provided by the public transport service system. In order to provide a better understanding of public mobility service and public transport service system characterization, some fundamental concepts can be identified as detailed in Tables 2, 3, 4, 5, 6 and 7, following a service oriented line of thought [24,25].

Safe, fast, secured, clean, reliable and economical daily trip can be considered as a value proposition offered to a customers' segment to convince people to use this service instead of using personal cars. Another critical value proposition promised by public transport operations can be rescue and evacuation services for citizens and other beings, providing also emergency resources to severed locations. All entities such as public transport operators can interact together in a complex value co-creation process in public transport service delivery.

Table 2. Public transport provider as a SS in times of emergency

Service system description elements for public transport: **Resources**
Physical: Public transport provider, shelter providers, other emergency response providers, emergency events first response providers, local transportation provider, departments involved in public transport (human resource, finance, technical department, safety department, local emergency management team), operating and maintenance companies, transit staff, security providers, passengers, visitors, contractor, vendors, community members, external information providers (state and local agencies, other public and private organization and peer transport agencies), local emergency medical service, employees and customer information systems, vehicle identification systems, storage, passenger, and maintenance facilities, computer system and communication equipment, internet providers, transit threat alert systems, security advisor systems, telecommunication and phone companies, emergency response team, personnel, legal counselling, risk manager and executive staff, local and state emergency operation center, local emergency planning committee, local transport department rail system, freight rail dispatcher center, local hospitals emergency rooms, independent living center, local media, emergency phone service provider, emergency management training company, environment protection agencies
Conceptual: Operating polices, institutional roles, emergency management policies and plan, mutual preparedness agreement, statistics, accident and vehicle maintenance reports, insurance claim, human resource record, staff and passenger records, information provided from social media, news and people, internal and external contact information, local critical information, regular and after-hours contact information, emergency public and sensitive information, incident planning and information, reports, statistics and documentation about activities during and after event

Table 3. Public transport provider as a SS in times of emergency (continuation)

Service system description elements for public transport: **Entities**
Service principal: Minister of transport, Federal transit administration, department of transportation, transport provider, public transport authorities, public transport operators, Emergency Management team;
Service producer: infrastructure maintenance companies, IT provider companies, Technical department, Customer service department, Planning department, safety department, Finance department, Telecommunication provider;
Service provider: Public transport operator, Private transport provider, Bus school provider;
Service customer: Commuter, passengers, citizens, authorities, emergency management team, police officers, fireman, Technician, authorities who needs to be transported to the location of event with their devices, passengers, visitors, contractor, vendors, community members;
Service object: citizens, beings, authorities;

Table 4. Public transport provider as a SS in times of emergency (continuation)

Service system description elements for public transport: **Access Rights**
Open: Emergency transport service, transport of emergency devices and resource, evacuation service;
Owned: Vehicle, infrastructure, assets, public, transport operators' internal roles and policies and protocols;
Leased: Loaned equipment, device and vehicle;
Shared: Infrastructure in common like stations, necessary information shared with other stakeholders and passengers and citizens;
Privilege: Company's internal information

Table 5. Public transport provider as a SS in times of emergency (continuation)

Service system description elements for public transport: **Interactions**
Value co-creation interaction: IT service support interaction, Instructional service interaction, Emergency management team interaction, public transport operator's emergency management team interaction, use of vehicle, equipment and devices
Governance interaction: Staff contract, mutual agreement with other parties, authorities, stakeholders and entities, crew, vehicle and resources schedule during emergency event, collecting information and statistic gathered from emergency events for public transport association and internal use

Table 6. Public transport provider as a SS in times of emergency (continuation)

Service system description elements for public transport: **Stakeholders**
Customer: Community, commuter, passengers, citizens, other beings, other service providers such as fire station, police, security companies, authorities, technician, research centres and institutions
Provider: Public transport operators, bus school operators
Authority: emergency management team, public transport authority, public transport operator's authorities and staff
Competitor: Private transport operators (local/neighbourhood), other public transport operators(local/neighbourhood), personal car owner

Table 7. Public transport provider as a SS in times of emergency (continuation)

Service system description elements for public transport: **Networks**
Local Public transport staff, crew and managers, local public transport technician, emergency management team, emergency management control center, public local transport emergency management control and center, neighbourhood and external public transport operators, passengers

4 Sustainability and Institutions in Public Transport Services

This section proposes a new perspective on designing sustainable public transport services in times of emergency, following the line of thought introduced by Elinor Ostrom on institutions, as seen as *"formal and informal rules that are understood and used by a community"* [26,27].

In a broader view, institution and institutional arrangements are proposed to be introduced as a driver supporting service systems development in the Service Dominant Logic perspective [28–31]. Recent literature reveals also new frameworks for resource allocation in sustainable institutions such as computational justice [32] leading to further investigation of interactions between humans, rules and norms [33,34].

In the previous section, some case studies were reviewed to figure out why some organizations were successful with the allocation of resources and provision of service at a sustainable level after a disruption event occurrence. Considering the given insight in developing economic activities in which *"ordinary people are capable of creating rules and institutions that allow for the sustainable and equitable management of shared resources"* [27], this section further elaborates on the application of the first five from Ostrom's eight principles for sustainable governance [26] in the special case of public transport services.

1. Clearly defined boundaries in CPR evaluation. One of the vital parts of an emergency response based on an emergency plan is stakeholders' collaboration and cooperation. Transportation system's primary task in the time of emergency is evacuation of vulnerable citizens to/from damaged area to safer areas, providing the sustainable transport service in another part of the city which is not impacted by disaster, and transporting equipment, technicians, and resources to damaged locations.

All duties, responsibilities, boundaries should be defined, clarified and classified during all four emergency management phases to save transit company's valuable assets. Public transport operation's employees such as drivers, technician, and other personnel should be clearly informed and educated about their responsibility during hiring, training, and operating processes. Responsibility of stakeholders which are involved in emergency management planning should be identified in the beginning of emergency management process to ensure proper coverage on all phases. Figure 1 shows an example of responsibility chain for emergency management in public transport service systems.

Sets of institutional rules are prepared as emergency plans in time of emergency in developed countries prone to disasters, man-made or natural, in order to clarify the primacy of state and local governments. Responses to disasters are categorized based on a set of rules and operating conditions which define who, where and when and how can appropriate the resources in order to have a quick and emergency response.

The complexity of self-organizing in CPR (Common-Pool Resources) situations increase when the size of CPR increase with many appropriators. Hurricane

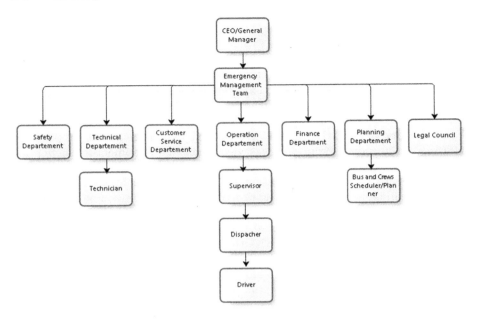

Fig. 1. Chain of Responsibility - emergency management in public transport SS

Katrina and Rita demonstrated that appropriation of public transport resources for the provision of evacuation service is difficult in the metropolitan area. Availability of resources in time of emergency is an important part of an emergency planning. Contract agreement should be made between all organizations (internal and external) to ensure availability of the resources [21].

2. Congruence between appropriation and provision rules and local conditions. Public transport response to emergency events differs per characteristic of the urban area, characteristic of emergency, behavioural characteristic, resource, and characteristic of the transit system. This plan includes [21]: area of evacuation; place of shuttle service; route, destination, and mean for evacuation; identifying location of that citizen; proper arrangement to return evacuee to their home.

Providing public transports service in large and populated areas is a challenging task because evacuation of a large number of the vulnerable population relying on public transport may overwhelm local transport resources with too many transit and bus station especially in limited ridership area. Geographical properties of an area, its public transport network, geographical barriers, the number of route and transit capacity effects on an emergency plan. The most vulnerable places should be covered first.

3. Individuals affected by the operational rules can participate in modifying them. People interacting inside the transport system can have a better vision of a system as a whole. With knowledge on advantages and disadvantages of currentrules, they may make more effective modification on a system then external authorities. The users in CPR who are interacting with each other for a longer time can participate in the modification of current institutional rules to make it more sustainable.

As mentioned, all stakeholders involved in the emergency management process especially public transport system as a first responder are familiar will their local resources, needs and obstacles according to past experiences and learned lessons. In this case, regular meeting is mandatory to modify/enhance local emergency management strategic and operational rules according to current resources, capacity, and local condition.

4. Monitoring. The role of monitoring of public transport resources and infrastructure appropriation in times of emergency is undeniable. Monitoring of operation, maintenance, and tracking of staff and vehicles in order to ensure sustainable usage of public transport resources takes place in control centre by decision-makers. Beside the CCTV and monitoring devices, other strategies should be used by operators such as a fence, guard, and silence alarm to control passengers' flow which lead to ordered and optimized resource appropriation.

However, when the scale and disaster impact is high, transport operations send the managers, information staff and experts to the place of damage to provide control centre with real information and provide them with their assessment. For example PATH control centre was informed by Port Authority Police and senior staff member to stop service into the World Trade Center station after 9/11 attack [11]. After this alarm, those trains which were in the station were rerouted or stopped because of security purpose. Other passengers were assisted in finding the safe direction by PATH passenger information agent and operation examiners.

Monitoring of public transport resource appropriation strongly relies on the computer-based communication system. The risk of failure of internet based communication and of the Internet network itself in the time of the emergency event is high. Building a robust data communication infrastructure and an Internet network in congestion with proper maintenance strategy needs to be taken into consideration.

5. Graduated Sanctions and Penalties. A graduated sanction and punishment strategy prevents appropriators of the public transport CRP from violating rules and government policy to guaranty sustainability against rule's violation. Different kinds of penalties can be defined to ensure that companies will not cross-government rules, will increase customers' satisfaction, and they will provide proper standards for safety of journey especially during emergency events. Besides these rules, there are some internal rules in public transport operators regarding appropriation and provision regime in line with international and government rules.

Provision of public transport service in the time of emergency is one of the most important factors for evaluation of a service performance. For this reason, privately own firms are sensitive of being penalized or sanctioned by higher authority [35]. They penalize their drivers for late or early running, absence, negligence of vehicle maintenance, or non-compliance for working or resting hours [36,37]. In addition, the government punishes private public transport companies who forced their drivers and infrastructure to cross government rules like hours of service rules for drivers and infrastructure. Maintenance of vehicle and other infrastructures for the safety of journey, passenger satisfaction and its impact on the provision of appropriated resources is vital. Mainly in most private transport companies technical crews will be penalized if they do not deliver faulty infrastructures in scheduled time because of facing the failure of resource provision. Infrastructures like roads will be maintained by other authority like a municipality.

In this perspective, the response to disasters may be categorized based on a set of *rules* and *operating conditions* which define *who*, *where*, *when*, and *how* can appropriate the resources in order to have a quick emergency response.

5   Conclusions and Further Research Directions

Several national, policy, or industry initiatives, such as *Transport 2050*, *Industrie 4.0*, *l'Industrie du futur*, *Health 2.0*, or *Digital Europe 2.0*, approached also under the European Commission's industry-related initiative of the *Digital Single Market* package, are proposed today. Under this broader umbrella, addressing societal challenges based on the utilization of the latest digital technologies, the scale and the impact of this changes on citizens, users, or consumers have to be better understood. Transportation, as well as its related evolution context, *Smart Transport* which is seen as an application domain of the *Smart(er) Cities* concept, does not evolve by itself, but in a larger perspective of using digitalisation for productivity and growth in manufacturing and services. The role of the transport service customer as an active actor in triggering information intensive business processes is redefined through process digitization. At the same time, digital innovation in public transport service systems coping urban mobility issues has to be closely aligned to strategic digital transformation in transportation domain.

In this context, this paper presents work-in-progress describing a new approach to support transport service company's operational activities in times of emergency based on sustainable institutions principles. Several research directions may be initiated following these inceptive ideas.

As people's behaviour is heavily dependent on their culture and the location they live, further studies on how this insight can help authorities to make a proper adjustment and improvement of emergency management planning activities based on people's reaction is recommended. Therefore, the definition of the corresponding CPR problem for public transport service delivery and a deeper evaluation of the application of Ostrom's eight principles for sustainable governance in the special case of public transport services are further need. This aspect closely relates to the definition of mobility as a service in times of emergency. This activity requires *information based design* for improved *interaction* in services. Therefore, a thoroughly analysis of information-intensive service processes to answer the hardest question of resource allocation in transport service delivery in times of emergency can be further developed. It would provide a better understanding of public transport service systems by studying complex possible entities interaction episodes and outcomes through improved service business process models in times of emergency.

References

1. Pardo, T., Taewoo, N.: Conceptualizing smart city with dimensions of technology, people, and institutions. In: 12th Annual International Conference on Digital Government Research, pp. 282–291. ACM, New York (2011)
2. Smart cities - Ranking of European medium-sized cities. European Smart Cities Project. http://www.smart-cities.eu/
3. Transport 2050: The major challenges, the key measures. European Commission, MEMO/11/197, Brussels, 28 March 2011
4. A European Strategy for Low-Emission Mobility. European Commission, Brussels, 20.7.2016, COM 501 final (2016)
5. Yatskiv, I., Pticina, I., Savrasovs, M.: Urban public transport system's reliability estimation using microscopic simulation. Transp. Telecommun. **13**(3), 219–228 (2012)
6. Van Oort, N., van Nes, R.: Improving reliability in urban public transport in strategic and tactical design. In: Compendium of Papers TRB 87th Annual Meeting, Washington, DC (2008)
7. Tahmasseby, S.: Reliability in urban public transport network assessment and design. Ph.D. Thesis, TU Delft (2009)
8. Tahmasseby, S., van Nes, R.: Improving service reliability in urban transit networks. In: European Transport Conference. AET Papers Repository (2010). http://abstracts.aetransport.org/
9. Corsar, D., Markovic, M., Edwards, P., Nelson, J.D.: The transport disruption ontology. In: Arenas, M., Corcho, O., Simperl, E., Strohmaier, M., d'Aquin, M., Srinivas, K., Groth, P., Dumontier, M., Heflin, J., Thirunarayan, K., Staab, S. (eds.) ISWC 2015. LNCS, vol. 9367, pp. 329–336. Springer, Cham (2015). doi:10.1007/978-3-319-25010-6_22
10. Marsden, G., Docherty, I.: Insights on disruptions as opportunities for transport policy change. Transp. Res. Part A **51**, 46–55 (2013)
11. America under threat: transit responds to terrorism. American Public Transport Association, Special report, 11 September 2001. http://www.apta.com/
12. Litman, T.: Lessons from Katrina and Rita. What major disasters can teach transportation planners. J. Transp. Eng. **132**, 11–18 (2006)
13. Sterling, S., Duddridge, B., Elliott, A., Conway, M., Payne, A.: Business Continuity for Dummies. Wiley, Haboken (2012)
14. How prepared are you? Business Continuity Management Toolkit, HM Government. https://www.gov.uk/
15. Alexander, D.: Principles of emergency planning and management. Terra, Harpenden, Hertfordshire, England (2002)
16. Schafer, W.A., Ganoe, C.H., Carroll, J.M.: Supporting community emergency management planning through a geocollaboration software architecture. In: Carroll, J.M. (ed.) Learning in Communities, Interdisciplinary Perspectives on Human Centered Information Technology, pp. 225–258. Springer, London (2009)

17. Emergency Management Planning Guide 2010–2011. Emergency Management Planning Unit, Canada. https://www.publicsafety.gc.ca/
18. Haworth, J., Shammas, S., Lubin, R.: Eurostar and ALL public transport cancelled as Brussels residents told 'stay put' after terror attacks, 22 March 2016. http://www.mirror.co.uk/
19. Attacks on Brussels: The Implications and Outlook. A Drum Cussac Global View. http://www.drum-cussac.com/
20. Wolshon, B.: Planning for the evacuation of New Orleans. Inst. Transp. Eng. (ITE J.) **72**(2), 44–49 (2002)
21. Transportation Research Board: The Role of Transit in Emergency Evacuation. Special report 294, Committee on the Role of Public Transportation in Emergency Evacuation. http://www.TRB.org
22. Giuliano, G., Golob, J.: Impacts of the Northridge earthquake on transit and highway use. J. Transp. Stat. **1**(2), 1–20 (1998)
23. Maintaining Communications Capabilities during Major Natural Disasters and other Emergency Situations. Study Group on Maintaining Communications Capabilities during Major Natural Disasters and other Emergency Situations, Final report, 27 December 2011. http://www.soumu.go.jp/
24. Spohrer, J.C., Demirkan, H., Krishna, V.: Service and science. In: Demirkan, H., Spohrer, J.C., Krishna, V. (eds.) The Science of Service Systems. Service Science: Research and Innovations in the Service Economy, pp. 325–358. Springer, New York (2011)
25. Lyons, K., Tracy, S.: Characterizing organizations as service systems. Hum. Factors Ergon. Manuf. Serv. Ind. **23**(1), 19–27 (2013)
26. Ostrom, E.: Design principles and threats to sustainable organizations that manage commons. In: Digital Library of the Commons, Indiana University, Working Paper W99–6 (1999). http://hdl.handle.net/10535/5465
27. Ostrom, E.: Response: the institutional analysis and development framework and the commons. Cornell Law Rev. **95**(4), 807–815 (2010)
28. Vargo, S.L., Lusch, R.F.: Institutions and axioms: an extension and update of service-dominant logic. J. Acad. Mark. Sci. **44**(1), 5–23 (2016)
29. Koskela-Huotari, K., Siltaloppi, J., Vargo, S.L.: Designing institutional complexity to enable innovation in service ecosystems. In: 49th Hawaii IEEE International Conference on System Sciences, pp. 1596–1605. IEEE Press (2016)
30. Koskela-Huotari, K., Vargo, S.L.: Institutions as resource context. J. Serv. Theor. Pract. **26**(2), 163–178 (2016)
31. Vargo, S.L., Wieland, H., Akaka, M.A.: Innovation through institutionalization: a service ecosystems perspective. Ind. Mark. Manage. **44**, 63–72 (2015)
32. Pitt, J., Busquets, D., Riveret, R.: The pursuit of computational justice in open systems. AI Soc. **30**(3), 359–378 (2015)
33. Pitt, J., Diaconescu, A.: The algorithmic governance of common-pool resources. In: Clippinger, J.H., Bollier, D. (eds.) From Bitcoin to Burning Man and Beyond. The Quest for Identity and Autonomy in a Digital Society, pp. 130–142 (2014)
34. Pitt, J., Busquets, D., Diaconescu, A., Nowak, A., Rychwalska, A., Roszczynska-Kurasinska, M.: Algorithmic self-governance and the design of socio-technical systems. In: European Conference on Social Intelligence, CEUR Workshop Proceedings, pp. 262–273 (2014)

35. Stokes, P.: Bus firm pays £ 140,000 penalty for early running. The Telegraph, 18 October 2003. http://www.telegraph.co.uk/
36. Shibghatullah, A.S.: A multi-agent system for a bus crew rescheduling system. Ph.D. thesis, Brunel University, Uxbridge, UK (2008)
37. Brunoro, C.M., Sznelwar, L.I., Bolis, I., Abrahaoa, J.: The work of bus drivers and their contribution to excellence in public transportation. Prod. J. **25**(2), 323–335 (2015)

Digital Services Development Using Statistics Tools to Emphasize Pollution Phenomena

Costin Gabriel Chiru$^{(\boxtimes)}$, Mariana Ionela Mocanu, Monica Drăgoicea, and Anca Daniela Ioniţă

Faculty of Automatic Control and Computers, University Politehnica of Bucharest, Splaiul Independentei 313, 060042 Bucharest, Romania
{costin.chiru,mariana.mocanu}@cs.pub.ro, monica.dragoicea@acse.pub.ro, anca.ionita@upb.ro

Abstract. This paper presents a perspective related to information service integration for pollution awareness evaluation. The proposed methodology is based on indirect information analysis as retrieved from available literature over time. A time series - type analysis highlighting usage of pollution-related terms is employed. The displayed impact of pollution is evaluated based on public awareness, exposed through digitalized available publications. Estimation techniques and tools are also employed in order to evaluate the exact impact of pollution related events on society. The proposed methodology fosters the design of improved environmental monitoring smart services, specifically addressing the development of data processing components in information sub-systems of EISs (Enterprise Information Systems).

Keywords: Digital transformation · Business process digitalization · Digital information services · Pollution events

1 Introduction

Digital services and information-based intelligence define a major trend today that strives towards a strategic transformation of a new generation of business models that use technology paradigms supporting business process automation and digitization [1–3]. The promises of novel utilization modes of existing technologies and the advance of new technologies bring to light increase awareness on the role of "digitalization" as a vector for creating and delivering new value to service customers, as well as improved digital service experiences [4–6].

The advent of digital platforms such as Google, Facebook, Uber, Skype (to name but a few) emphasises the new modes of interactions for work, collaboration, and information management based on mobile, social, cloud, and big data evolution [7]. They foster implementation of a special type of Internet-based applications, the smart services supporting superior digital encounters between humans and technology [8,9]. In this perspective, business process digitization becomes a main driver for value co-creation through customer experience integration in the new digital transformation economy [10].

© Springer International Publishing AG 2017
S. Za et al. (Eds.): IESS 2017, LNBIP 279, pp. 370–382, 2017.
DOI: 10.1007/978-3-319-56925-3_29

Following this research evolution perspective, information-based design has been recently acknowledged as a valuable tool driving IT-enabled innovation in smart services [11,12].

In this respect, this paper addresses the development of data processing components in information sub-systems of EISs (Enterprise Information Systems) supporting the design of improved environmental monitoring smart services. Pollution phenomena is a hot topic, exposed extensively in scientific publications and official documents, at different levels of detail about the phenomena itself and about the underlying effects [13]. Unfortunately, most of the time it is impossible to directly measure the effects of different - predictable or unpredictable - pollution events. Therefore, experts are using various estimations to evaluate the damages that were produced, for example in [14,15].

The working methodology proposed here is general and it can be used for other kinds of events without any major modification. It consist of three main steps that should be undertaken: (a) *identify* the relevant *concepts* for the analysed event using either expert information or information extracted from an ontology; (b) *extract data* as time series of the selected concepts from the available corpora and identify their peaks; (c) perform analysis of the obtained time series in order to *identify* relevant *information.*

In Sect. 2 the relevant concepts for the considered pollution phenomena are identified. For this task, WordNet lexical database was used, and the relationships between concepts were graphically represented with Visuwords. As presented in Sect. 3, for the identified concepts, data is required in order to build their time series. The corpus provided by Google is used for this step, as this is the largest publicly available corpus, comprising information for a large number of concepts over a period of more than 200 years.

The last step of the proposed methodology is presented in Sect. 4. The concept time series obtained in the previous step are analysed for peak detection and the proposed algorithm is described. There are also other methods for peaks detection, such as the one proposed in [16], but these methods are usually more computationally intense and thus take a lot of time to do the required computation. Since the dataset which is used in this work is very large, it was opted out for simpler algorithms in order to obtain the results in reasonable time. Even thought similar methods have been proposed in the literature, they suffer from a series of problems. For example, the algorithm presented in [17] has two main drawbacks: first of all, it does not verify whether the current value is greater than the one of its left and right neighbours. Therefore, it identifies as peaks points that are present on the upward trajectory from one point to another which satisfy the imposed condition. Secondly, due to this condition, it doesn't distinguish between smaller and larger peaks, being able to identify only the extreme ones.

Section 5 describes a perspective related to service integration for pollution awareness and identifies process activities, services, and resources towards future automation of the service integration process. Section 6 concludes the paper and draws possible further directions for improving the proposed working methodology.

2   Identification of Pollution Concepts

In order to identify the relevant concepts for the pollution phenomena, WordNet, a large lexical database for English, is used. It contains parts of speech that are organized into groupes of synonyms, called synsets [18,19]. Between such synsets, relationships can be defined based on lexical criteria, thus creating an extended network of interconnected terms.

A second visual tool, Visuwords [20], was used in order to browse this network. Visuwords is an online graphical dictionary that maps the concepts from WordNet, along with the connections between them. This graph-like representation shows how different words associate.

Using Visuwords tool, several concepts related to pollution events were identified, as presented in Fig. 1, such as synonyms, pollution types and semantic categories to which pollution belongs. For instance, pollution is equivalent to: deterioration, infection, decomposition, dirtying, impurity a.s.o. The study considered water, soil, noise, radioactive, thermal, light, visual, personal and other types of pollution. Nonetheless, it was important to identify the groups of words from the vocabulary, which contain pollution (such as environmental condition, damage).

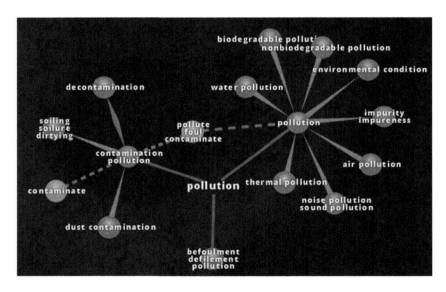

Fig. 1. The concepts related to "pollution" (WordNet representation). (Color figure online)

Using the facilities provided by Visuwords search box, the word of interest is specified by the user and then the concepts connected to it are displayed in the diagram. There are several ways of representation; therefore different types of connections may be present in such a diagram. As can be seen in Fig. 1, in the pollution related diagram, concepts that are not leaves in the graph may be further explored, and they are presented in green. They may be further expanded, leaving the user with the option to explore other elements of interest.

3   Extracting Pollution Data

After the set of concepts related to pollution was identified (as presented in the previous section) the study continued with the search of these concepts within a large volume of articles, books and conference proceedings that have been issued by publishing houses all around the globe. The purpose was to determine how often these concepts appear in these texts and, in this way, to assess the impact carried out by them.

The study becomes even more difficult if we go back in time, first due to the lower availability of publications, and second because, for the historical documents, the access is guarded by security authorizations. Therefore, we used for this purpose the corpus of Google Books N-grams [21], archiving information starting with 2010. First, it was created with the digital versions of more than 5 million books, totaling 5 billion words and about 4% of all the books edited during the centuries. A significant addition appeared two years later, when the corpus overcame 8 million books, covering now about 6% of the entire existing collection of published knowledge [22].

The corpus makes available two kinds of information: (i) how many of the recorded concepts were published per year, and (ii) how many times each of them appeared within the digitalized publications. The former may be used to determine the frequency of a given concept, and to represent its variation in time graphically.

This corpus was successfully used by researchers for a large variety of tasks, such as: analyzing the transformation of words over the centuries [23]; identifying relationships between the history of culture and the correspondent languages [24]; recognizing the modifications of semantics for particular words [25]; identifying the tendency to make use of terms that suggest emotions [26]; indicating clusters for depicting topics important for the humankind and associating them to given time intervals [27].

Although the corpus contains information about all the words that were present in the indexed publications, the inputs used in our study were the pollution concepts previously identified from WordNet. For each of them, the study identifies the number of times of usage in the books published during the specified year.

Figure 2 gives an example of graphical representation, for the 20th century, for the general term called "pollution" - the starting point of our research. The time series describing the pollution events comprises a series of time ordered values representing, for each analysed year, the number of times a concept was present in the publication during the specified time frame.

Each point of the graph in Fig. 2 represents the number of times the selected concept(s) appeared in the documents written during a specific year. On the

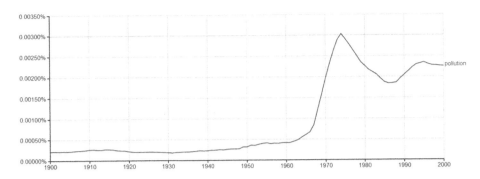

Fig. 2. Example of the "pollution" time series and its visualization.

X axis, years from the selected time range are represented. The Y axis corresponds to the number of times the concept(s) appeared in the documents written during the year corresponding to the values on the X axis.

However, since data is not uniform, as the number of books digitized for each year was growing exponentially with time, instead of representing the concept counts for each year, it was chosen to present the frequency. The appearance frequency of each concept, c, denoted as $freq_{c,y}$, was obtained by dividing the number of times this concept appeared during a specific year, $count_{c,y}$, by the total number of concepts analysed for that specific year, $\Sigma_{c_i} count_{c,y}$ (Eq. 1).

$$freq_{c,y} = \frac{count_{c,y}}{\Sigma_{c_i} count_{c,y}} \tag{1}$$

Therefore, in reality, the Y axis expresses the frequency of the selected concept(s) during each year and this is why most of the times the values are very small. Moreover, since multiple concepts may be chosen for analysis, the time series for each of these words are presented in different colours in order to be easily distinguished.

The following steps describe how such a representation was built, based on Google Books N–grams [21]:

- *Select the concepts to be visualized.* It is possible to visualize the time series of one or more concepts chosen by the user (case-sensitive or not case-sensitive search);
- *Select the time range for visualization.* Although the data can be represented in a 1600–2008 year range, the corpus developers advice the users that more reliable results can be obtained using the data extracted between 1800 and 2000;
- *Select the corpus.* Considering the multi-lingual character of data, the option is of choosing from 22 different corpora in 8 different languages, plus a series of variations (American English, British English, English Fiction, English One Million, a.s.o.);

– *Select the smoothing factor*. Considering the sparse character of data in the corpora, it is advisable to use smoothed values of the data instead of the raw data. The following moving average method was used in order to obtain the smoothed values, using a window of size $(2 * w + 1)$ (Eq. 2):

$$smoothed\_count_{c,y} = \frac{\Sigma_{k=-i}^{i} count_{c,y+k}}{2 * w + 1} \qquad (2)$$

where $count_{c,y}$ denotes the number of times the word c was found in the indexed texts that were written in year y, while $smoothed\_count_{c,y}$ is the value that will be used after smoothing instead of the $count_{c,y}$ value. Since the most important factor in this formula is the range of the smoothing $(2 * w + 1)$, this value has to be chosen by the user.

4 Pollution Events Identification

The idea of this service is to analyse the scientific literature on pollution events along time, to assess the prevalence of concepts having relationship with pollution and to evaluate the risk impact based on time series analysis. Typically, after the normalization, the graphic of number of events vs. time should be horizontal; yet, one notes that, for certain concepts, there are peaks or intervals of higher values, corresponding to the years when the scientific world was more interested about those words.

The interpretation is that their semantics leads to certain pollution events, notorious and with great impact for the given time, thus influencing not only the text of a specific article, but also the number of studies and therefore publications that approached the "hot" topic of the moment. In fact, there is a slight delay between the moment the event actually happened, and that when it started to be the subject of scientific communications.

Table 1 specifies the steps of the peak detection algorithm applied to distinguish the moments when the word for which the graphic was drawn was more frequently used. The parameters referring the time series values that are used in the algorithm are: m, the mean; s, standard deviation; w, window size; h, constant value influencing the application sensitivity for peaks recognition.

Considering **Step 1** of the algorithm, there are multiple ways of computing the peak score for different years. The simplest one is based on finding the points having the shape of an arrowhead upward oriented, and the highest differences between the top of the arrow and its right and left neighbours.

In **Step 3**, only the relevant peaks are considered, both in arrowhead and magnitude forms. Small variations that are not statistical relevant must be eliminated. Therefore, the variable h is adjusted to values ranging between $1 \leq h \leq 3$, in order to adjust the level of statistical relevance, based on the characteristics of a normal distribution. The values ranging from $[m - s, m + s]$ are normal values that cannot be considered peaks, while values for which the distance to the mean is larger than three standard deviations are very improbable, therefore they can be considered to be peaks or valleys. Thus, by imposing that the

Table 1. Algorithm for peak detection of pollution concepts time series

---

Inputs: Normalized counts for the investigated concept $count_{c,y}$
Outputs: Peak(s) from the analysed time series
Score: Vector of scores given to each year denoting how suitable that
year is to be a peak year.

Step 1. For each year y, compute the score, $Score(y)$

Step 2. For each positive value $Score(y)$, compute mean m and
standard deviation s

Step 3. For each $Score(y)$, eliminate values for which the condition
$Score(y) > 0$ and $Score(y) - m > h * s$ and $Score(y) > Score(y - 1)$ and
$Score(y) > Score(y + 1)$ is not fulfilled

Step 4. All $Score(y)$ values are added to the **Output** vector

Step 5. For each consecutive pair in **Output** vector, $Output_i$ and $Output_{i+1}$:
If the two values are too close to each other from the year's point of view
(y_1 for $Output_i$, and y_2 for $Output_{i+1}$ being in the same window of size w),
eliminate the smaller value from **Output** vector

Step 6. Return **Output**, the peaks vector for the analysed time series

---

difference between the current value and the mean to be larger than a number of standard deviations, we are actually retaining only the improbable values (33.33% in the case of one standard deviation - that accounts for smaller peaks - and 3% for three standard deviations - for much larger peaks). Adjusting the value of h in such a way gives the possibility to select less or more important peaks, depending on the requirements.

Finally, the purpose of the **Step 5** is to eliminate closely related peaks in order to retain only the most important ones in a series of consecutive such peaks. For defining how far must be two different peaks in order to be considered independent, a window of size w is used. If the two peaks are not in the same window, they are considered as being independent and they are returned at the output.

4.1   Understanding the Pollution Threat

People have not been much aware of the pollution threat until the late sixties, nor was the available literature oriented towards the study of this issue. This is visible not only on the specific example from Fig. 2, but also in the multiple curves

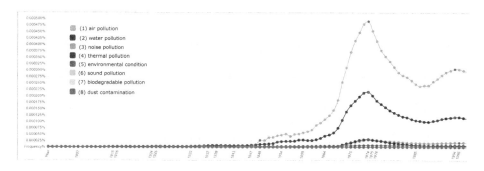

Fig. 3. Frequency of pollution-related concepts in books.

illustrated in Fig. 3, corresponding to noise, thermal, environmental, sound, and biodegradable pollution, as well as to dust contamination.

However, air and water pollution have been hot topics almost two decades earlier. During the time, the awareness progressively embraced more diverse types of pollution. The most prominent peak is in 1974, the year when people became concerned of the global pollution and its effects.

The study of WordNet revealed that pollution has multiple classifications, and the evolution of their presence into the digitalized literature can be determined from the Google N-gram Corpus, by generating the correspondent time series. Figure 3 illustrates them in different colors, in respect with pollution types.

Subsequently, we analyse the evolution of this conceptual diversification from the chronological point of view.

4.2   Conceptual Diversification

Air and water pollution were present in the literature slightly after 1945, under the influence of the World War II and its effects on the environment, but the diversification began after 1965. The difference is also present if one takes into account the next level classifications, like the waterbody types. For instance, the writings record information about river pollution since 1870, and about lake pollution since 1890, proving that there was a serious concern about this topic even in the nineteenth century. In the twentieth century, the attention towards the sea pollution has significantly increased since the 60's, followed a decade later by an increase of the perception of the ocean-related issues (Fig. 4).

A significant rise can be observed in 1974, due to the adoption of relevant and novel regulations, both in UK and in USA. On the one hand, there was the "Control of Pollution Act 1974" [28]; on the other hand, the United States Environmental Protection Agency signalled that Hudson is polluted with polychlorinated biphenyl, therefore fishing was interdicted [29].

Note that, for the same year, we also found multiple peaks in respect with the classification of pollutants (Fig. 5), with a visible appreciation of the public concern for radioactive, hydrocarbon, nitrate, and petroleum pollutions.

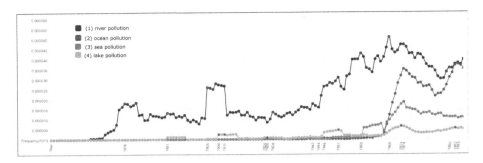

Fig. 4. Evolution of the pollution awareness in respect with the waterbody types.

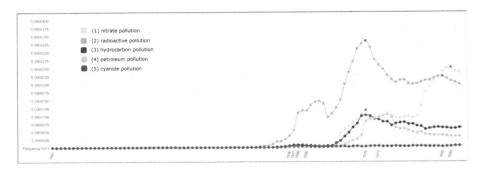

Fig. 5. Evolution of the pollution awareness in respect with the classification of pollutants.

The underground nuclear test performed by India, also in 1974 [30], must also have had its influence on the values obtained in these times series. Afterwards, the interest on this type of pollutant diminished, with a possible cause identified in the ratification of the Treaty on the Non-Proliferation of Nuclear Weapons, progressively followed by the decline of the lobby on this topic.

More recently, in 1990 we also noticed another rise of the water pollution concerns, and we associated it to the nuclear accident from Chernobyl, which happened four years earlier. See Fig. 6, where the time series for water pollution was overlapped with those related to other important terms, such as "war" and "Chernobyl". The frequency values for "war" were diminished a hundred times to make the comparison possible, because the preoccupation for this concept is far more significant all along time.

Since the number of times the word construction "Chernobyl disaster" appeared in documents is extremely small ($15E - 6\%$) compared to the number of times the concept "war" appeared (0.055%), these values were multiplied by 100 in order to be able to visualize them on the same graph.

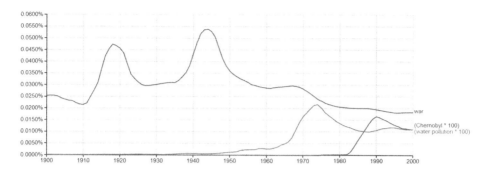

Fig. 6. Relevant peaks of awareness for terms related to the disaster from Chernobyl.

Eventually, the analysis of multiple time series proved that there is generally a delay between the occurrence of an event and its being reported intensively into the literature, especially in books, whose publication cycle is less agile. Nonetheless, the frequency was influenced by different factors, like disasters, war-related activities, lobby effort, national and regional legislation, international treaties.

5   Service Integration for Pollution Awareness

The previous chapters presented tools and statistics algorithms that are used for analysing the available information about pollution and for correlating it to real events. Based on this experience, specific *process activities*, *services*, and *resources* were identified and presented in Fig. 7. They may further serve to enlarge the analysis scope and may stand at the basis of future *automation* of the *service integration process* [31], currently executed in an ad-hoc manner.

For realizing the first process activity, *Identify Pollution Concepts*, one needs services related to a lexical database, lexical relationships, and a graphical dictionary. For the *Extract Pollution Data* activity, it is necessary to integrate services for concept search, but also for smoothing and viewing time series of data related to these concepts. Eventually, for the *Highlight Pollution Events* activity, the required services concern peak detection for the obtained time series, the history of pollution events, and consecrated classifications.

A similar process may be executed starting from another root concept except pollution (as chosen in this study). For example, these services may be reused for the analysis of natural hazards or epidemic diseases. In the present study, a specific choice of resources was made, such as data stores, or search and visualization tools: WordNet, Visuwords, Google Books N-gram Corpus.

Nevertheless, the implementation can be also realized based on other collections of data, or by migrating to other software services tools [32], owned by various agencies for natural resource management and environment protection.

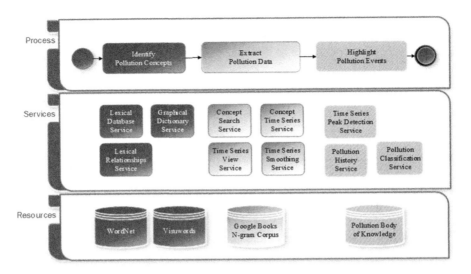

Fig. 7. Service integration for emphasizing pollution phenomena.

6   Conclusions

The current technological paradigm shift highlights a more acute focus toward the orientation on applying Artificial Intelligence (AI) technologies and tools supporting digital transformation of information services along enterprise processes for better assimilation, processing, and interpretation of information.

Dealing with a major concern topic of current years, this paper introduces a working methodology aiming to evaluate the impact of pollution on the society over the years.

The proposed approach highlights an inter-disciplinary perspective, involving several specializations' participation, such as computer scientists, hydrologists, sociologists, service providers, and infrastructure providers. It requires also development of hands-on support for technical skills' formation supporting policies on Digital Transformation, approaching knowledge engineering, requirements management for Systems of Systems (SoSs) integration, and business process digitization.

The main limitation of the current approach is related to the delays that might appear between the moment when an event occurred and the moment when it was reflected in press. Due to this fact, the identification of different events might be delayed by a number of years that is somehow dependent on the importance of that event: the less important events, the more delay they have (important events are reflected almost simultaneous in press).

Another drawback is related to the fact that, depending on the chosen parameters, more or less important peaks might be detected. Thus, it might take a little time to tweak the application in order to respond to the user's needs.

As the current work-in-progress reveals, there are immediate improvements that can be taken into consideration for the proposed approach, such as finding a better way to discriminate between different yearly events, and developing an investigated event model suitable for predictions on the further evolution of that event.

At the same time, it is envisioned that this proposed approach may be investigated in several other applications, such as stock prices, vehicle resell price, gold price prediction, or predicting soccer games outcome.

Acknowledgements. The research presented in this paper is supported by the DATA4WATER Project: H2020-TWINN-2015, Project ID: 690900, Funded under: H2020-EU.4.b. - Twinning of research institutions.

References

1. Markovitch, S., Willmott, P.: Accelerating the digitization of business processes. A McKinsey & Company Report. http://www.mckinsey.com/business-functions/digital-mckinsey/
2. Li, H., Mäntymäki, M., Zhang, X.: Digital Services and Information Intelligence: IFIP Advances in Information and Communication Technology, vol. 445. Springer, Heidelberg (2014)
3. Spohrer, J.C., Bassano, C., Piciocchi, P., Siddike, M.A.K.: What makes a system smart? Wise? In: Ahram, T.Z., Karwowski, W. (eds.) Advances in The Human Side of Service Engineering. Advances in Intelligent Systems and Computing, vol. 494, pp. 23–34. Springer, Cham (2017)
4. Moore, S.: Digitalization or Automation - Is There a Difference? Gartner Report. http://www.gartner.com/smarterwithgartner/
5. Demirkan, H., Spohrer, J.C.: Emerging service orientations and transformations (SOT). Inf. Syst. Front. **18**(3), 407–411 (2016)
6. Scherer, A., Wünderlich, N.V., von Wangenheim, F.: The value of self-service: long-term effects of technology-based self-service usage on customer retention. MIS Q. **39**(1), 177–200 (2015)
7. The acceleration of third platform innovation: here comes the DX economy. i-SCOOP Whitepaper. http://www.i-scoop.eu/the-acceleration-of-third-platform-innovation-here-comes-the-dx-economy/
8. Ostrom, A.L., Parasuraman, A., Bowen, D.E., Patrcio, L., Voss, C.A.: Service research priorities in a rapidly changing context. J. Serv. Res. **18**(2), 127–159 (2015)
9. Pitt, J.: This Pervasive Day: The Potential and Perils of Pervasive Computing. World Scientific, Singapore (2012)
10. Westerman, G., McAfee, A.: The Digital Advantage: How Digital Leaders Outperform Their Peers in Every Industry. MIT Center for Digital Business, Research Report. http://ebusiness.mit.edu/research/Briefs/TheDigitalAdvantage.pdf
11. Böhmann, T., Leimeister, J.M., Möslein, K.: Service systems engineering - a field for future information systems research. Bus. Inf. Syst. Eng. **6**(2), 73–79 (2014)
12. Romero, D., Vernadat, F.: Enterprise information systems state of the art: past, present and future trends. Comput. Ind. **79**, 3–13 (2016)
13. Mortality and burden of disease from water and sanitation. World Health Organization Report. http://www.who.int/gho/phe/water_sanitation/burden/en/

14. Richards, M., Ghanem, M., Osmond, M., Guo, Y., Hassard, J.: Grid-based analysis of air pollution data. Ecol. Model. **194**(1–3), 274–286 (2006)
15. Matějíček, L., Benešová, L., Tonika, J.: Ecological modelling of nitrate pollution in small river basins by spreadsheets and GIS. Ecol. Model. **170**(2–3), 245–263 (2003)
16. Rubner, Y., Tomasi, C., Guibas, L.J.: A metric for distributions with applications to image databases. In: Sixth International Conference on Computer Vision, pp. 59–66. IEEE Press (1998)
17. Palshikar, G.K.: Simple algorithms for peak detection in time-series, Work-in-Progress. https://www.researchgate.net/
18. Miller, G.A.: WordNet: a lexical database for English. Commun. ACM **38**(11), 39–41 (1995)
19. Miller, G.A., Beckwith, R., Fellbaum, C., Gross, D., Miller, K.J.: Introduction to wordnet: an on-line lexical database. Int. J. Lexicogr. **3**(4), 235–244 (1990)
20. VISUWORDS on-line graphical dictionary. http://visuwords.com/
21. Michel, J.B., Shen, Y.K., Aiden, A.P., Veres, A., Gray, M.K., The Google Books Team, Pickett, J.P., Hoiberg, D., Clancy, D., Norvig, P., Orwant, J., Pinker, S., Nowak, M.A., Aiden, E.L.: Quantitative analysis of culture using millions of digitized books. Science **331**(6014), 176–182 (2011)
22. Lin, Y., Michel, J.B., Aiden, E.L., Orwant, J., Brockman, W., Petrov, S.: Syntactic annotations for the google books Ngram corpus. In: ACL 2012 Proceedings of the ACL 2012 System Demonstrations, pp. 169–174. ACM Digital Library (2012)
23. Wijaya, D.T., Yeniterzi, R.: Understanding semantic change of words over centuries. In: International Workshop on DETecting and Exploiting Cultural diversiTy on the Social Web, DETECT 2011, pp. 35–40. ACM Digital Library (2011)
24. Petersen, A.M., Tenenbaum, J., Havlin, S., Stanley, H.E.: Statistical laws governing fluctuations in word use from word birth to word death. Sci. Rep. **2**, Article no. 313. http://www.nature.com/articles/srep00313
25. Mitra, S., Mitra, R., Riedl, M., Biemann, C., Mukherjee, A., Goyal, P.: That's sick dude! Automatic identification of word sense change across different timescales. In: 52nd Annual Meeting of the Association for Computational Linguistics, pp. 1020–1029. ACL Press (2014)
26. Acerbi, A., Lampos, V., Garnett, P., Bentley, R.A.: The expression of emotions in 20th century books. PLoS ONE **8**(3), e59030 (2013). http://doi.org/10.1371/journal.pone.0059030
27. Popa, T., Rebedea, T., Chiru, C.G.: Detecting and describing historical periods in a large corpora. In: 26th IEEE International Conference on Tools with Artificial Intelligence (ICTAI), pp. 764–770. IEEE Press (2014)
28. Control of Pollution Act 1974. http://www.legislation.gov.uk/ukpga/1974/40
29. Hudson River Cleanup. United States Environmental Protection Agency. https://www3.epa.gov/hudson/cleanup.html
30. Laxman, S.: 'Smiling Buddha' had caught US off-guard in 1974. The Times of India, 7 December 2011. http://timesofindia.indiatimes.com/
31. Ionita, A.D., Eftimie, C.-T., Lewis, G., Litoiu, M.: Integration of hazard management services. In: Borangiu, T., Drăgoicea, M., Nóvoa, H. (eds.) IESS 2016. LNBIP, vol. 247, pp. 355–364. Springer, Cham (2016). doi:10.1007/978-3-319-32689-4_27
32. Ioniţă, A.D., Liţoiu, M., Lewis, G.: Migrating Legacy Applications: Challenges in Service Oriented Architecture and Cloud Computing Environments. IGI Global, Hershey (2013)

Service Orientation of Environment Control Processes

Theodor Borangiu[(✉)], Andrei Silişteanu, Silviu Răileanu,
and Iulia Voinescu

Department of Automation and Industrial Informatics,
University Politehnica of Bucharest, Bucharest, Romania
{theodor.borangiu,andrei.silisteanu,
silviu.raileanu}@cimr.pub.ro

Abstract. The paper introduces a framework for designing flexible Environment Control Services (EServices) based on generic sensing, modelling and control process specifications allowing the customization of distributed Facility Environment Control Systems (HFES) in holonic approach. The representation, assessment, configuring and implementation of these EServices by the interacting holons using data structures models in different perspectives their type, features, functionalities, interdependencies and usage: (i) the EService Type perspective used to identify a requested type of service included in an existing service-ontology; (ii) the EService Specification perspective allowing the client (the environment conditioning application) to define the facility's environment needs as service properties; (iii) the EService Profile perspective in which the resources publish their capabilities matching the requested EService Type, exposing them to (iv) the EService Configuring and Implementation perspective. In the resulting Service-oriented HFES (SoHFES), four types of processes: environment monitoring, control, conditioning and production influencing, represented in the service perspective either as simple or composite services according to timing progression, granularity and concurrency of their operations can be standardized into EServices which allows for flexibility, agility, robustness and reusability of control solutions. A generic design framework and experimental results are finally presented.

Keywords: Service oriented architecture · Facility environment control · Holonic paradigm · Environment Services · Multi-agent framework

1 Introduction

System theory in environment control tasks is traditionally concerned with the design of static control algorithms and laws for groups of nonlinear processes such as temperature, pressure and relative humidity, represented by fixed control sequences and timing for sensory data acquisition and command update. Fixed control laws and sequences are selected, considering a priori known nonlinear models of closed space HVAC processes (Heat, Ventilation and Air Conditioning), with stability and precision of set point tracking. This applies for environment parameter conditioning in individual closed spaces.

© Springer International Publishing AG 2017
S. Za et al. (Eds.): IESS 2017, LNBIP 279, pp. 383–396, 2017.
DOI: 10.1007/978-3-319-56925-3_30

In the case of multiple interconnected spaces environment parameter conditioning, like radiopharmaceutical production facilities, there is need to additionally provide flexibility to process control configuring; this means both the capability to dynamically change room parameter ranges, control laws and timing in order to satisfy a global facility environment model such as cascaded pressures, and to switch between control modes and assigned resources (HVAC subsystems, redundant control units for fault tolerance) in response to external conditions such as seasonal conditions or breakdowns.

The holonic approach provides the above mentioned capabilities - openness, flexibility, robustness and agile control in semi-heterarchical architecture in which a centralized layer maintains the global facility model and a decentralized layer of intelligent, autonomous entities - the holons cooperate to maintain local channel set points updated according to a global facility model, in the presence of disturbances. The holonic paradigm has been recognized in industry, academia and research, as providing open, flexible and agile control attributes by means of a decentralized control architecture composed by a social organization of intelligent entities called holons, with behaviours and goals defined by reference architectures such as ADACOR, PROSA, HABPA or CoBASA [1–4]. On the other hand, the service-oriented paradigm defines the principles used for conceiving decentralized control architectures that decompose computational processes into sub-processes called services, and then distribute them among different available resources. Its focus is to leverage the creation of reusable and interoperable function blocks in order to reduce the amount of reprogramming efforts. Due to the fact that: [Holon] ← [Physical Asset] + [Agent = Information counterpart], it becomes possible to treat at informational level four process types: environment monitoring, conditioning, control and production influencing as services in an optimal, while robust and agile to changes mode.

The holonic paradigm has been applied mainly in the manufacturing domain. The main contribution of this research is applying the holonic paradigm to distributed multivariable process control, exemplified here by environment parameter control of interconnected rooms where radioactive materials are handled.

In the Information Systems domain, the service-oriented paradigm provides scalability of computing solutions, systems interoperability and flexibility in reconfiguring tasks by decentralized architectures that use componentization to split computational processes into atomic sub-processes defined as services, which are then recomposed according to particular tasks and assigned to available resources; reusable and interoperable function blocks reduce the reprogramming efforts. The SOA paradigm will be applied in these terms to generic, multiple spaces environment parameter conditioning systems leading to Service-oriented Holonic Facility Environment Control Systems (SoHFES).

The paper introduces a framework for designing flexible Environment Control Services (EServices) based on generic sensing, modelling and control process specifications that allow the customization of a HFES control architecture function of the: (1) facility layout, (2) temperature, pressure and relative humidity models of closed spaces (clean room, production capacity, access space, personnel office), (3) global facility environment model (if any), (4) heat and ventilation channel parameter model

(HVAC process model), (5) imposed performances of automated systems (stability, efficiency, robustness, response time,· a.o.), and (6) control strategy, operating mode and control law (linear, nonlinear, feed forward, observer) of the compensator. The decomposition and encapsulation of process operations, identified as EServices which are an extension of control services [7] provides process flexibility at the entire facility level, because they are related to operations for environment parameter conditioning defined by user-specific parameters and constraints.

The conceptual model of EService and control processes preserves the fractal character of environment process conditioning, making possible the reutilization and composition of sets of various sensing, computing, modelling, control and monitoring operations.

The remainder of the paper is organized as follows: Sect. 2 defines a service-oriented specification of environment conditioning processes. Section 3 introduces the principles of modelling environment conditioning processes through EServices and proposes implementing solutions. Section 4 presents experimental results obtained in the design of a SoHFES providing the necessary environment for the production of radiopharmaceuticals.

2 Using Web Service Technology for Process Control Specification

Process Conditioning Flexibility for environment control needs dual control topologies in which set point values of individual closed space parameters are first computed from the global facility model at centralized, hierarchical layer and then transposed in set points of the resources' conditioning parameters (air and cooling water flows of HVAC units) by a set of active task holons and their information counterparts – the agents updating on the heterarchical, decentralized layer of the distributed control system HVAC process models, operating modes and sub-resource allocation.

According to [4], similarities existing in control class structures translate into certain commonalities in the process domain. We introduce the *Process Classes* concept related to environment monitoring and control applications. Such classes of processes are:

- Heating/cooling coil processes with variable water flow admission;
- Variable fresh air changes/hour in clean rooms;
- Air flow admission (e.g., Variable Air Volume-VAV) and distribution (e.g., Variable Frequency Drives-VFD) of supply and exhaust fans);
- Air humidification, air filtering.

It becomes thus possible to apply the concept of services to environment conditioning process specifications, in the perspectives of: componentisation, representation as services with main attributes: exposing, discovering, negotiating, assigning, recomposing, standard interconnectivity and reusability.

The initiative taken so far to transpose Web Service Technology from the Information Systems domain into industrial ones like process control and to integrate the principles of SOA in the design of distributed control systems and agentification of activities, orders and resources is beneficial for the integration of the client's requirements at business level with the technical planning, control and traceability of processes and products at technical level. Once processes represented as services, it is necessary to define their information counterparts, to classify them according to capabilities, to describe and evaluate in real time the quality of the activities they perform. This requires the definition of a framework for configuring, classifying, discovering and allocating services to create new composite processes by establishing relationships between atomic services in the control context [5].

Radiopharmaceuticals are produced in 4-stage manufacturing cycles: (1) radio isotopes are produced in a particle accelerator (cyclotron), and then (2) transferred into isolators for chemical synthesis followed by (3) portioning (vial dispensing) of the bulk product, and (4) quality control of the final product by conformity tests on multiple parameters.

During the production stages, special conditions must be permanently fulfilled; these conditions are: *radioprotection safety conditions* (radioactivity doses, pressure cascades and number of air changes per hour) and *environment manufacturing conditions* (physical environment parameters) as defined by Good Manufacturing Practice guides (GMP).

The cleanroom is a closed area inside a nuclear facility where radiopharmaceuticals are being prepared and dispensed; the concentration of airborne particles is maintained in strict limits to eliminate contamination. Processes below are strictly controlled:

- Permanent input and filtering of fresh air, air flow recirculation being forbidden;
- Cascades of negative pressure in closed spaces relative to the exterior, in order to evacuate the potentially contaminated air from the production facilities;
- Cascades of positive pressures between adjacent closed spaces preventing the rapid air penetration from one room to another at access doors opening and slowing down the diffusion of air particles in the clean rooms;
- Monitoring temperature, pressure, humidity and radioactivity level in each one of the three rooms: cyclotron, synthesis, and dispensing, with HVAC units [6].

The conditioning processes for key environment parameters are described below:

- *Pressure* (P): a minimum positive difference of 8–15 Pa is kept between any clean room and its adjacent rooms; this imposes a global conditioning process described by a "cascade pressure model". Inside closed spaces surrounding production areas (synthesis, dispensing) a negative pressure (-30–-70 Pa) must be permanently kept to prevent inside air entering the clean-room. In the cyclotron vault, a negative pressure cascade prevents the inside air spreading to the entire facility.
- *Temperature,* T: is kept in the range 18 °C–21 °C, to reduce microbial growth.
- *Relative humidity,* RH: must be kept between 35%–55%.
- *Number of airborne particles,* NP_c: is continuously monitored in the dispensing isolator box; if NP_c exits the imposed cleanness value range, the dispensing process is suspended for a maximum time of 20 min to allow the parameter to re-enter in range; if it does, dispensing resumes, and delivery is delayed with the corresponding time; otherwise production is abandoned and the current production order fails.
- *Radioactivity level* $H_c*(10)$: ambient dose flow caused by radiation in clean rooms.

For this type of production processes, two HVAC units are used inside the nuclear facility to maintain room environment parameters within the range of normal state values: (1) one working in negative-pressure mode inside and around the cyclotron vault to avoid spreading of the radioactive air in the building, and (2) the other working in over-pressure mode inside clean rooms to avoid the penetration of the building air and deteriorating the quality of the purified air inside production and dispensing closed spaces.

The two HVAC units are integrated into a Holonic Facility Environment Monitoring and Control System (HFES), which analyses environment data, and computes accordingly the global facility environment control strategy and parameter set points in centralized mode; the application of this strategy is then decentralized by distributing intelligence in a multi-agent framework which implements: holons collecting environment parameter data (T, P, RH, NP_c, $H_c^*(10)$) from room sensors, holons transposing the global, cascade strategy in HVAC operating modes and control laws and computing the reference values of HVAC parameters (environment tasks), and resource holons executing environment tasks via HVAC process models for air flow and cooling water conditioning [7].

For example, the rooms' temperature conditioning processes are supervised by a number of control processes performed by agents executing on two HFES layers (Fig. 1):

1. The *centralized, hierarchical control layer:* a set of *Expertize agents* compute room temperature set points from values read and pre-processed by local sensors (via room agents) and compared with the prescribed temperature values; significant events, like environment parameters exceeding limits are recorded in logs;

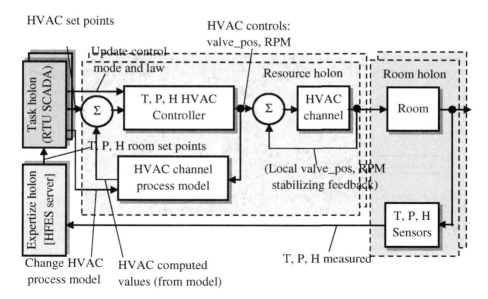

Fig. 1. Generic agent-based control, distributed at room and HVAC channel level, tracking T, P, H room set points computed at global facility environment level by the centralized HFES server

2. The *decentralized, heterarchical control layer:* three sets of agents respectively collect environment data, process the output of Expertize agents providing set points for HVAC channels, and use the channel references to effectively regulate the environment conditioning processes performed by the HVAC controllers.

- A set of *Environment (Room) agents*: process the data measured by the room sensors T, P, RH, NP_c, $H_c^*(10)$ and send it to the Expertise agents;
- A set of *Environment Conditioning (Task) agents*: compute the HVAC channel set points (e.g., mass flow $\dot{m}_{\text{water,r}}$ of the cooling water for temperature conditioning) from the room environment conditioning models, e.g., $T_{\text{room}} = \dot{m}_{\text{water}}$ and send these references to the resource agents;
- A set of *Resource agents*: use the HVAC set points computed by the Task agents and compare them with the HVAC value $\dot{m}_{\text{water,c}}$ computed from the model of the channel conditioning process, e.g., $\dot{m}_{\text{water}} = f(valve\_pos)$ – cooling water is let in the cooling coil area function of the admission valve opening, $valve\_pos$. Resource agents implement model-based channel control (see Fig. 2).

The control architecture in Fig. 1 includes the physical entities of HFES – the holons. HFES has a semi-heterarchical topology, in which some computing processes are carried out at centralized level such as set point calculation of environment parameters and their history, whereas parameter sensing, conditioning, control and computing HVAC channel models are distributed in the heterarchical, service-oriented

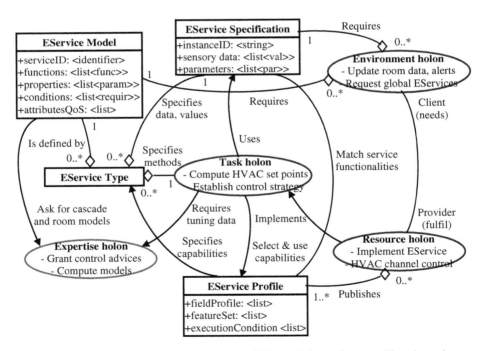

Fig. 2. Modelling the Environment Service (*EService*) from client-provider viewpoint

MAS composed by a network of intelligent and autonomous agents, each one exposing its functionality as services; these services are published in a repository and can be reached via discovery mechanisms.

The HFES paradigm is based on defining a main set of assets: *resources* (technology – HVAC units, sensors, control and computing systems), *orders* (environment requirements reflecting the client's needs) and *control elements* (control laws, algorithms and tasks, operating modes, computational tasks, reflecting the business solutions) – represented by holons collaborating in holarchies according to a set of rules to reach a common goal.

3   Modelling Environment Conditioning Processes with EServices

3.1   Representation and Assessment of Environment Services

Transposing the theory of services to environment conditioning processes and integrating the principles of services into HFES leads to a new type of system: the Service-oriented Holonic Facility Environment System (SoHFES), allowing repeatability and reusability of control operations. By adopting the SOA model for HFES, environment conditioning operations can be standardized into *EServices* exhibiting an unambiguous identification and complete description of both the interactions between the client (who ordered the production of radiopharmaceuticals) and the service provider (the facility ensuring proper environment parameters for production and radio-protection safety), see Fig. 2.

The relational EService Model is subordinated to the concept of holonic control; three basic types of holons interact in a holarchy derived from the PROSA reference one: environment (room) holons, resource holons and environment conditioning (task) holons [2]. Expertise (staff) holons, having a global image of the facility's environment conditions, requirements and control interdependencies process room parameter data according to cascaded room models and assist the basic task holons in accomplishing their activity.

Every oval and rectangle in Fig. 2 which uses the class diagram of UML (Unified Modelling Language) represents an entity type in the system, respectively a holon type and a data structure, whereas each line represents a relation. Depending on the symbol on the line - diamond, full arrow, arrow (not present in Fig. 2) or no symbol, the line refers to a different kind of relation. The *aggregation relation* ("has-a" relation) is indicated by a line with a diamond; for example, any EService Type has associated one model, a specification and a Task holon. The latter is created by selecting from the Profile data the best capability of that resource that fulfils certain conditions (cost, performances or utilisation time) better than any other resource, and best matching the requirements of the EService. The *association relation* ("is-related to" relation) is shown by a normal line; for example, the Environment holon type features a client-provider association with the Resource holon type. The *directed action relation* ("does-a" relation), indicated by a line with a full arrow, refers to a unidirectional action undertaken by one entity relative to another; for example, the Task holon type uses a

EService specification to configure that EService. The cardinality of a relation is specified by numbers placed at both extremities of the line (0…*, 1…*, 0…1, 0, 1, or unspecified), and represents the number of this type of entities that are involved in the relation. The holon classes are defined below:

- Environment (Room) holons, EH: *formulate EService requirements*; they contain the dynamic "safety & quality model" of the production environment as compliance to cleanness and radioprotection. Room holons check against off-line prescribed values the parameters T, P, RH, NP_c, $H_c*(10)$ measured in all closed spaces, send data to the Expertise holon(s) and the request to perform the necessary EServices for the facility.
- Environment conditioning (Task) holons, TH: *configure the control EServices* at HVAC channel (resource) level; the tasks select and update control modes, strategies and laws and compute the references for the HVAC channel controllers that regulate the environment conditioning parameters. Task holons deliver to Resource holons the parameter conditioning knowledge from updated environment data.
- Resource holons, RH: *implement EServices*; hold the knowledge (operating modes and control techniques) to activate the resources to maintain the updated set points.
- Expertise (Staff) holons, SH: *coordinate EServices at global level*; they correlate the set of room parameter references by using their global image about the cascade facility models, and deliver them to the Environment conditioning holons. Also, they provide task holons recommendations on updating control modes (seasonal changes) and tuning.

EServices embody validated operations in fractal mode; they can be reused to integrate different conditioning processes carried out by resources from different manufacturers or by legacy systems, based on the same ontology. In a SoHFES, the service becomes the main element of negotiation and exchange among holons. Environment conditioning and control operations are represented by EServices executed on environment parameters.

In the multi-agent system (MAS) implementing this holarchy, the knowledge exchange is performed in fractal mode based on three classes of reusable operations:

- *Expose a service:* specifying how EServices are presented to the SoHFES in terms of richness of description and identification;
- *Compose a service:* offering strategies and methods to compose process workflows by combining individual services to create new or composite EServices;
- *Consume a service:* making available procedures to access or invoke an EService already created by the Task holon. In the environment control context, this type of action allows implementing and executing simple or composite EServices.

In the centre of the class diagram in Fig. 2, the EService Type identifies a particular type of service included in an existing service-ontology about environment conditioning and control. This data structure lists the properties that characterize an instance of that same service type, to further set up a certain EService Specification: name, category, type of parameters, inputs, outputs. The EService Model instance specifies the components of an EService. An EService Specification, defined by the client (the environment facility needs for the production process) is associated to an EService Type providing information on the values of its properties, and this instance for making service requests. The HVAC resources, as service providers, publish their control capabilities in the EService Profiles data structure, thus exposing their transformation capabilities to the Task holons in view of being assigned services. Hence, the EService Profile of a resource specifies the set of field characteristics, range of parameter values and performances for a certain EService Type that this resource can deliver according to its design, technology, performances, and to its current state, wear, and availability. The selection of the resource's EService Profile best matching a unique set of needed EService Specifications results from a many-to-one comparison of associations between services that can be provided and the requested one.

The setup of an EService, i.e. configuring which resource and how it will be executed - according to the selected EService Profile, is performed by an Environment conditioning holon, whereas the implementation of this EService is done by the Resource holon's physical part (the HVAC channel) controlled by its agent. The Task holon transposes the selected EService capabilities set into the necessary control mode, law and parameters to produce the transformation described by the associated EService Type.

3.2   Configuring and Implementing EServices

The **EService Model** includes the property fields that define environment conditioning or control processes, assessed as services, in the client-provider perspective of global facility environment monitoring and control. The EService structure is represented by the model shown in Fig. 3. This model is specified for any EService Type used in SoHFES tasks developed from environment control service-ontology, by specifying its property fields.

The EService is identified by a *ServiceID* which is composed by four fields: (1) a name, (2) the class to which it belongs (e.g. parameter conditioning, parameter control), (3) its taxonomy, and (4) the service ontology in which it is referred. The second property field describes in text file the *Functions* performed by the EService in terms of: scope, context of execution and tasks to be realized. The third property field, *Service Properties* features the list of parameters which specify completely the EService: the actions or processes to execute, the execution timing (sampling frequency, duration) and the information support used for execution (process models, operating modes, computing results, etc.). *Service Conditions* denote the fourth property field of the EService Model; it accounts for several attributes that must be satisfied by the provider or other agents prior to service execution: (1) prerequisites, such as the existence of updated process models, (2) relational conditions, such as data availability (e.g., set point for HVAC channel parameters already computed by Task holons), and state conditions, such as the provider's operational state. The last of the property fields is *Service AttributesQoS* which lists key performance indicators (KPI) to be initially specified (at service request and offer) or computed at run time to evaluate the foreseen quality of services (QoS), respectively that of delivered ones. For a SoHFES service-ontology, KPIs will be defined both at global (rate of class cleanness

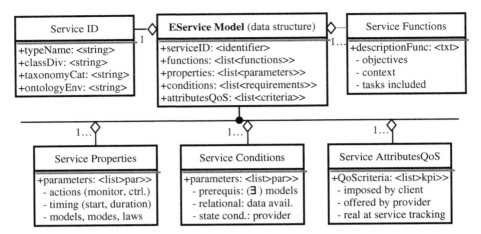

Fig. 3. Components of the EService Model

maintenance during product making, rate of maintaining room environment parameters in normal value range relative to the global model, a.o.) and local (accuracy of HVAC channel parameters matching computed set point values, control loop asymptotic stability, a.o.) level.

The execution capabilities of service providers (HFES resources) are exposed by means of their **EService Profi**le data sets for a requested EService Type that defines exactly the process to be realized as a simple or composite service, with a detailed specification of its elements, characteristics and desired quality. The service-oriented agents of the Resource holons analyse these specifications and, using the EService Profile, write in the currently requested EService Type data structure the sets of their capabilities) best fitting the desired specifications. This set of best available capabilities, called "field profile", depends on the current resource's state too. The model of the EService Profile is shown in Fig. 4.

The EService Profile announces the resource's skills in response to a requested service stored in the EService Type data structure retrieving information from its three lists: (1) a list of field profile sets enumerating all transformation processes that implement the client EService, each one documented with its sets of functional characteristics and parameter values; (2) the performance specification set indicating for each field profile set the list of KPI or attributes used to evaluate the quality of the service delivered by the resource, and the KPI values; (3) the list of conditions needed by the provider to execute the EService according to each field profile: relational conditions, e.g., computation results produced by other information entities (agents), and the real operational state of the provider (RH).

The first list provides information about what a resource can do in its workspace which should ideally include the workspace of the ordered EService; each parameter profile set represents, then, a resource's capability subspace. The second list of specifications defines the properties to be used to evaluate the execution and the performances of a service's method, each attribute set being associated to a field profile set. These properties are first expressed as sets of KPI as criteria to evaluate service

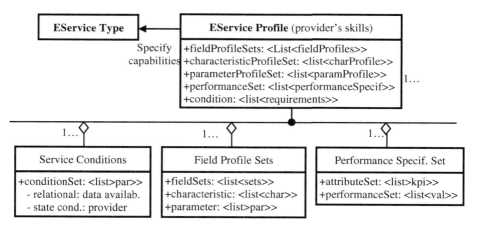

Fig. 4. Components of the EService Profile data structure

accomplishment and effectiveness, e.g. stability and precision of environment parameter control. The quality value of each KPI is then detailed by performance indexes, e.g. steady-state error, overshoot, response time for HVAC channel control.

The third list contains for each field profile set conditions needed by the resource to perform the EService according to a method in a given parameter range; these are either external conditions involving proper functioning of other types of information entities with which the service provider must collaborate to retrieve necessary data (Task agents, Room agents) or internal conditions of proper setup, configuring, a.o.

Although **EService Implementation** is realized only by Resource holons, several other types of information entities participate in EService Model's instantiation stages (Fig. 5):

(1) *Integrating the EService specifications* formulated by the client environment application in the global, cascade environment parameter model of the HFES;
(2) *Matching the functionalities* of the requested service, given by its properties, conditions and attributes, with the capabilities exposed by resources in their EService Profiles;
(3) *Assigning the resource* with the best matching field profile set and confirmed quality of already delivered services to implement the needed EService;
(4) *Reconfiguring the service provider* in the case of breakdown or degraded QoS.

A Task agent is created whenever a client environment application EService request is issued at SoHFES initial configuring, from data progressively stored in the EService Type structure: (i) client application data provided by Room agents, aggregated according to the global model of the facility environment, and (ii) a number of field profile sets provided by Resource agents; these sets could be possibly used to implement the requested EService.

All field profile sets are a priori published by Resource agents in dedicated repertoires of the EService Profile data base, from which, upon request only the subset of profiles possibly matching the client EService is added to the EService Type data

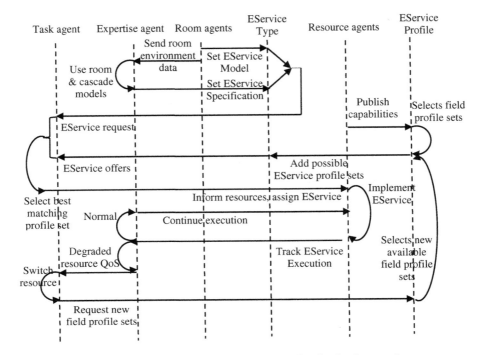

Fig. 5. Agent-based resource selection and EService implementation

structure. Each of these subset profiles contains detailed information on process methods, laws, models implementing the transformations specified by client EService Model and Specifications. *Process methods* describe a sequence of actions that transform the environment parameter specified by that EService Type. A resource may have more than one method achieving the same transformation which its agent publishes in the EService Profile data base.

The Task agent establishes the field profile in the subset offered from the EService Profile best matching the specifications of the requested EService, informs all Resource agents having participated in this bid about the decision taken and configures the selected resource in view of implementing the EService with the chosen process method. The decision support for this selection is granted in centralized mode by one Expertise agent.

The execution of EServices by the assigned resources is monitored by Task agents; if the QoS delivered decreases, then the EService is reassigned to another available resource.

4   Experimental Results and Conclusions

The research objective was to model and manage four types of processes: environment monitoring, control, conditioning and production influencing, in the service perspective as simple or composite services according to granularity and concurrency of their operations set. The modelling technique is based on the holonic paradigm, implemented in multi-agent framework, leading to a service oriented, semi-heterarchical holonic control system, exemplified here by environment facility control SoHFES. The centralized SoHFES layer updates room set points from measured room parameters according to the global facility model, and stores these data during production in the *Environment Data Base* (EDB) server for analysis, reports, audit and historical evidence. In the decentralized layer, room sensors measure environment parameters, local PLCs collect these parameters and compute HVAC channel set points, and HVAC controllers condition the environment parameters. The production management system (MES) is subordinated to the SoHFES.

The Environment Data Base is fed with production data gathered from the MES during execution of a planned client's order, respectively with information about the environment parameters. EDB is organized in 15 parameter value tables: other types of information are computed from data periodically input (max., min., weighted average values over sets of fixed or variable size). The two categories of EDB data: [production + product] and [environment] can be used in separate history reports or combined to correlate production run or product features with the evolution of environment parameters; these are composite *decision* EServices for *production influencing*.

The association of product and environment parameter values below allows evaluating the effect of reconfiguring channel environment parameters of EServices "VFD airflow control for supply/exhaust fan and VAV control of air changes", necessary to bring back the number of particles NP_c in its normal value range for class A cleanness:

Dispensing: [product & process data] + [environment data]

- Volume of bulk product received from synthesis: VBS [ml] dec
- Irradiation level of VBS: IRBS [Mbq/ml] dec
- Time at which IRBS was measured: TMIS [hh:mm:ss] int_string
- Irradiation level of final dispensed product: IRFD [Mbq/ml] dec
- Volume of final dispensed product: VFD [ml] dec
- Planned volume of diluter per vial: VDVP [ml] dec
- Start time of dispensing: TSD [hh:mm:ss] int_string
- Cleanness class imposed in dispensing area: CLDA ["#"] char
- No. of 0.5/5 mm particles measured: NPc_0.5/NPc_5 [#] int
- ID of NPc sensor: ID_ParticleCounter [#] int
- Type of NPc sensor: Type_PC ["#"] char_string (Lasair III 310C)
- NPc sensor location area: NPc_loc ["#"] char_string
- Sample period of sensor reading: SPR [sec] int (e.g., 6 sec)
- Time of cleanness alert at dispensing: TCA [hh:mm:ss] int_string
- Time of cleanness recovery, resume: TCR [hh:mm:ss] int_string
- Remaining volume of final product to be dispensed: VRD [ml] dec
- Irradiation level of bulk product at TCR: IRPR [Mbq/ml] dec
- Post recovery volume of diluter per vial: VDPR [ml] dec
- Real end time of dispensing: REOD [hh:mm:ss] int_string
- Volume of wasted bulk product: VWBP [ml] dec

Future research aims at defining metrics for QoS performed by resources publishing their capabilities, and usage of metrics in optimal resource allocation algorithms based on multicriteria selection of best matching field profiles with requested property sets.

References

1. Leitão, P., Restivo, F.: ADACOR: a holonic architecture for agile and adaptive manufacturing control. Comput. Ind. **57**, 121–130 (2006). doi:10.1016/j.compind.2005.05.005
2. Van Brussel, H., Wyns, J., Valckenaers, P., Bongaerts, L., Peeters, P.: Reference architecture for holonic manufacturing systems: PROSA. Comput. Ind. **37**, 255–274 (1998)
3. Borangiu, T., Răileanu, S., Trentesaux, D., Berger, T., Iacob, I.: Distributed manufacturing control with extended CNP interaction of intelligent products. J. Intell. Manuf. **25**(5), 1065–1075 (2014). doi:10.1007/s10845-013-0740-3
4. Panescu, D., Pascal, C.: On holonic adaptive plan-based architecture: planning scheme and holons' life periods. Int. J. Adv. Manuf. Tech. **63**(5–8), 753–769 (2012)
5. Demirkan, H., et al.: Service-oriented technology and management: perspectives on research and practice for the coming decade. Int. J. Electron. Commun. Res. Appl. **7**, 356–376 (2008)
6. Tang, F.: HVAC system modeling and optimization: a data mining approach. MSc thesis, University of Iowa (2010). http://ir.uiowa.edu/etd/895
7. Borangiu, T., Silişteanu, A., Răileanu, S., Morariu, O.: Holonic facility environment monitoring and control for radiopharmaceutical agent-based production, Chap. 24. In: Borangiu, T., Trentesaux, D., Thomas, A., Leitão, P., Oliveira, J. (eds.) Service Orientation in Holonic and Multi-agent Manufacturing. SCI, vol. 694, pp. 269–286. Springer, Cham (2017)

Service Ecosystems for the Common Good: A Case of Non-profit Network Organization

Sabrina Bonomi[1(✉)], Francesca Ricciardi[2], and Cecilia Rossignoli[2]

[1] eCampus University, via Isimbardi 10, 22060 Novedrate, CO, Italy
sabrina.bonomi@uniecampus.it
[2] University of Verona, via dell'Artigliere 8, 37129 Verona, Italy
{francesca.ricciardi,cecilia.rossignoli}@univr.it

Abstract. Service systems' capabilities to generate (positive or negative) externalities and their impact at the societal level has remained under-investigated so far. This study addresses this gap and explores how an innovative organizational architecture based on (i) a network of competent actors, (ii) an (ICT-enabled) platform serving as a network commons and (iii) a shared institutional logic and worldview enables the emergence and evolution of a service ecosystem with strong positive externalities. It was conducted an in-depth longitudinal study (2013–2016) on an ICT-enabled community of IT professionals, aimed to provide unemployed professionals with employment opportunities while also providing small and micro enterprises and non-profit organizations with affordable, high-level IT services. The case-study shows that the presence of non-profit organizations in the service ecosystem strongly influences the service ecosystem's institutional logic and worldview and facilitates sustainability-oriented self-organizing throughout the ecosystem.

Keywords: Service systems · Service ecosystems · Commons · Self-organizing networks · IT professional employability · Social enterprise · Common good · Italy

1 Introduction

Service science and service-dominant logic provide management scholars with a sound approach to explain and support value co-creation processes [1–3]. So far, this approach has been mainly leveraged to investigate value co-creation in the producer-customer relationship [4].

This special focus on producer- customer relationships led scholars to concentrate on the benefits that service system can offer to actors that play the role of customer and/ or producer in the service system [5]. In other words, service science has so far developed a somehow inward-looking attitude: the impact of the service system is usually measured at the level of the service system itself or at the level of its components, especially producers and customers. As a consequence, service systems' capabilities to generate (positive or negative) *externalities* [6] and their impact at the societal level has remained under-investigated so far.

© Springer International Publishing AG 2017
S. Za et al. (Eds.): IESS 2017, LNBIP 279, pp. 397–408, 2017.
DOI: 10.1007/978-3-319-56925-3_31

This study addresses this gap and explores the organizational features that characterize a service system that systematically generates clearly identifiable positive externalities.

To do so, it has been leveraged recent contributions in the service science and organization studies literatures, and particularly Lusch and Nambisan [5] and Fjeldstad et al. [7].

These studies, in our opinion, converge in identifying the following organizational features as key not only to support collective sense making and innovation, but also to overcome traditional hierarchy-market schemes:

- a network of competent actors who are willing to self-organize as a community;
- an ICT-enabled platform that contributes to provide a commons where resources can be developed, protected and shared by the community;
- an effective architecture of collaborative participation based on a shared institutional logic and worldview.

Following Lusch and Nambisan [5] and Dougherty and Dunne [8], we label systems with these organizational architecture as *service ecosystems*.

Leveraging the theoretical framework synthesized above, this study investigates *how an innovative organizational architecture based on (i) a network of competent actors, (ii) an (ICT-enabled) platform serving as a network commons and (iii) a shared institutional logic and worldview enables the emergence and evolution of a service ecosystem with strong positive externalities.*

Viewing the service ecosystem's ICT-enabled platform as a commons allows us to leverage the rich literature on the commons [9, 10] and new commons [11, 12] to explain the social mechanisms that influence the evolution of the system. This constitutes an innovative contribution to the service science stream.

It has been used this integrated and innovative theoretical lens to understand the longitudinal evolution of an exemplary case, an Italian initiative in the IT consulting industry.

The Italian business structure is highly fragmented, with a significantly higher rate of small and micro enterprises than in other developed countries [13]. These small and micro enterprises are the real backbone of the Italian labor market and business landscape, but they are often specialized in traditional activities and display low innovation attitudes. These firms have stagnated during the 2000s, while larger enterprises enjoyed the positive effects of the digital revolution; in addition, micro and small enterprises have suffered more severely than larger enterprises from the economic regression following the 2008 financial crisis.

The IT consulting market has been negatively impacted by this situation: many small and micro enterprises, severely hit by the crisis, reduced their IT budgets or even renounced to consider the opportunity of ICT-based innovation. Many experienced IT professionals lost their jobs, whilst many young graduates in IT disciplines remain unemployed and are not given the opportunity to gain experience.

The competences of these two categories of unemployed people are highly complementary. Older professionals have a lot of experience of the business world, but they sometimes lack updated knowledge on the most recent technological solutions and approaches Younger IT experts are native of the web-based mobile world, but they lack business experience and relational capital.

The "5020" (FiftyTwenty) organization arises from the idea of transforming this problem into an opportunity. The 5020 initiative was launched in 2013 by integrating the competences of unemployed IT professionals in their 50s and 20s, respectively, into a non-profit organization. The 5020 organization selects, trains and hires unemployed IT professionals, providing them with an ICT-enabled environment for knowledge exchange, collaboration and team building. Thanks to this unconventional approach, 5020 offers high-quality, accessible IT consultancy services to business customers, especially those (such as small and micro enterprises and charities) that are usually excluded from the opportunity to leverage high-quality information systems to improve their effectiveness and efficiency.

The 5020 system generates a community whose key actors are, along with 5020 itself: young and experienced IT professionals, especially unemployed, single or associated; non profit organizations, micro and small enterprises interested in accessible IT services; larger enterprises and government bodies interested in IT services and corporate social responsibility initiatives; system integrators; a cooperative that provides administrative and fiscal services; universities, both as a pool of young people to whom draw from, and as a partner in studies and researches; EthosIT, a network dedicated to promote an ethical approach to IT.

The presence of non-profit organizations and EthosIT in the service ecosystem strongly influences the service ecosystem's institutional logic and worldview and facilitates sustainability-oriented self-organizing throughout the ecosystem.

Since some of the key actors are more interested in value creation at the societal level than value capture [6], strong positive externalities are actually generated by this service ecosystem.

Our explorative study suggests that service ecosystems in which ethically engaged actors play a pivotal role may enable the creation of highly innovative, sustainability-oriented business models, which generate important positive externalities. Our results encourage to further investigating the role of organizational factors in designing service ecosystems for the common good.

2 Background

The Service Science literature views service systems as complex configurations of resources in which one party applies its competences for the benefit of, and in conjunction with, another party, thus enabling value co-creation through (a) interactions between competent and interested actors, and (b) use of the resources exchanged within and across systems [1].

In particular, the service-dominant logic [2, 3] focuses on (ICT-enabled) producer-consumer interaction systems, which should be designed (or innovated) in order to create value both for the provider and the customer. This focus on the producer-consumer relationship is consistent with the disciplinary ground in which the service-dominant logic is rooted, i.e. marketing studies [4].

On the other hand, authoritative authors claim that the service-dominant logic does not limit itself in explaining the pivotal role of customers in value co-creation processes; the service-dominant logic can and should be leveraged not only for business, but also for societal purposes [14]. These claims notwithstanding, the possible role of the service-dominant logic for sustainability and the common good has actually remained quite under-investigated so far. In particular, studies on the possible role of service system to address social problems (such as employability or small businesses' poor innovation capabilities) are almost absent.

Fortunately, recent developments in service science literature are providing novel, useful theoretical tools to leverage the service system approach also for societal challenges. The *organizational* features that are emerging as typical of service systems are particularly promising for scholars interested in identifying new approaches to system sustainability.

Notably, Lusch and Nambisan [5] highlight the importance of overcoming the traditional focus on the producer-consumer divide in service-dominant logic and suggest to develop a broader view of service systems as service *eco*systems. Service ecosystems are understood as *communities* of interacting actors that are capable of mutual value creation thanks to common organizational structures and institutions that allow coevolving capabilities and roles, resource integration and service exchange. Thanks to these common organizational structures and institutions, the actors interacting in the community can co-evolve adaptively and the ecosystem can enable the pursue of the actors' goals [15]. According to this view, the key interactions are not necessarily those occurring in the classical provider-customer relationship: the key interactions for value co-creation could also occur, for example, within a network of voluntary organizations, governmental bodies, and citizens.

According to Lusch and Nambisan [5], some key organizational and institutional features are essential to a successful service ecosystem:

1. Effective actor-to-actor network, where all actors serve as resource integrators and collectively create their environment or the service ecosystem through reciprocally enabled learning and knowledge exchange;
2. Effective digital platform, maximizing resource sharing, mobilization and integration;
3. Effective architecture of participation: shared institutional logics [16], i.e. the shared rules and ethical beliefs that actors leverage to coordinate their value co-creation actions, even without a strong command and control structure (as typical of actor-to-actor networks), along with a shared worldview, i.e. a common set of cognitive tools, assumptions, methods and frameworks enabling the service ecosystem survival.

This view of service ecosystems is strongly consistent with some very recent advancement in organization and management studies. For example, Fjeldstad et al. [7] observe that new, successful organizational designs oriented to innovation and value co-creation "are based on an actor-oriented architectural scheme composed of three main elements: (1) actors who have the capabilities and values to self-organize; (2) commons where the actors accumulate and share resources; and (3) protocols, processes, and infrastructures that enable multi-actor collaboration" (p. 734). This framework highlights the role of commons, understood as common environments where the actors develop, share and use common resources to co-create value. This "architecture of collaboration" proposed by Fjeldstad et al. [7] is highly compatible with Lusch and Nambisan's [5] view of service ecosystem: in both models, coordination and control are emergent capabilities that are accomplished primarily via self-organizing mechanisms, i.e. direct interaction among competent actors that share rules, values, beliefs and tools, rather than by hierarchical subordination or market forces.

In this light, the correspondence between the two models invites us to understand Lusch and Nambisan's [5] *digital platforms* as a possible solution to provide the network of competent actors with Fjeldstad et al.'s [7] *commons*, i.e. environments for co-creating, protecting and/or distributing value.

Self-organizing architectures are particularly effective in enabling adaptive and participated innovation, including sustainability transitions. It is not surprising, then, that these complex self-organizing ecologies of competent actors are growingly viewed as the emerging organizational solution to address societal challenges [8].

3   Method

The research question proposed by this study focuses on how an innovative organizational architecture based on (i) a network of competent actors, (ii) an (ICT-enabled) platform serving as a network commons and (iii) a shared institutional logic and worldview influences the emergence and evolution of a service ecosystem with strong positive externalities.

A qualitative research approach is the one most suited to handle this type of enquiry [17], not only because this study revolves around a 'how' question, but also because scientific research on the role of service systems in addressing societal challenges is still in its infancy.

More specifically, a longitudinal, in-depth case study seems particularly suited for addressing the research question [18].

It has been selected a representative success case [19] of a service system explicitly designed to create three important positive externalities:

- dissemination of sustainable, ethical IT culture;
- up-skilling and employment of IT professionals that otherwise would have remained excluded from the labor market; and
- access to advanced ICT-based innovation on the part of micro and small profit and non-profit organizations, which in traditional market conditions would have remained excluded from this opportunity.

Longitudinal data collection on this case took place from 2013 to 2016. During this period, this researcher had regular contact with all of the initiative's stakeholders and participated in some relevant meetings for the development of related projects.

During the period of data collection, 6 semi-structured interviews (about 40 min each) were conducted with the social entrepreneur who conceived the service system (two interviews) two IT professionals working in the service system (one interview each), the owner of a small enterprise that has implemented ICT-enabled innovation thanks to the service system (one interview) and a manager of the non-profit organization that embeds the service system (one interview). The interviewees represent the entire range of this initiative's stakeholders. It has been also collected relevant documents, web pages, reports and press clips to triangulate the data.

Through group work and discussion, it has been selected the most interesting and relevant contents from the collected materials and transcribed these contents with a word processor in order to build a homogeneous archive for the analysis. The resulting archive of the contents deemed relevant to this study includes roughly 110 pages.

These selected contents were coded using the Atlas.ti software and three types of codes stemming from Lusch and Nambisan's [5] framework, in the light of Fjeldstad et al.'s [7] organizational view: (a) actor-oriented architecture; (b) ICT platform as a commons; (c) architecture of collaboration. Finally, all of the coded parts were further coded for date (year) in order to facilitate the longitudinal analysis of this initiative's evolution.

4   The "5020 (FiftyTwenty)" Case

5020 is the brainchild of a group of IT managers, who worked in this area, for several organizations, for over forty years; they realized that many ex-colleagues had lost their jobs or could work only for a few days a month and they decided to study the phenomenon. They soon realized that the Italian context is particularly unfavourable, because the "body rental" model through large consulting companies results in high prices for the customers, but low wages and employability for IT professionals. After data analysis and especially observing the negative externalities of unemployment in younger and older IT professionals, they created a start-up, with the help of a project manager with specific expertise in social entrepreneurship, to curb the loss and waste of skills, transforming them in a generative resource.

The project consists, in summary, in an "intergenerational pact for work"; it is based on the creation of IT specialist teams, each composed of people in their 20s - who often are not recruited after qualifications because "inexperienced" - and people in their 50s-old people, prematurely ejected from the labour market because of the crisis. In this way, working together, young people can make available the latest knowledge gained in university and the older people can share the experience that comes from working career, combining management experience and innovation. Working together, younger people's and older people's knowledge can be transferred and exchanged, thus enriching both categories of workers and generating positive social impacts in terms of development of knowledge, innovation, employment and their consequences [20]. The 5020 team firstly analysed ICT market trends, indicated by Gartner (world leader in strategic consulting, research and analysis in the field of Information Technology), and then defined the thematic areas on which focusing: business process management, Big Data, Agile Methodology and ad hoc developments (Java and Web).

People were recruited, selected, trained and employed thanks to a dedicated IT platform that allows them do not move from the places of residence, for those who want it, thus also resulting in environmental sustainability. If necessary, 5020 provides certified on specific technologies; people are chosen not only if they already have skills about these technologies (they are still growing and scarcely available in the market) but also selections are based on the individual's potential, and people are trained to increase their skills. Vocational education is also conducted in "5020" logic, therefore always involves mixed sessions between fifty-years-old and new graduates.

After education/training, 5020 allocates IT specialists at the customers' establishments, suited to the demands. Finally organizes training sessions, based on real prototypes developments that may fit the market. 5020 pays particular attention to people who have been expelled from the workplace who possess out-dated skills or master niche technologies, difficult to 'resale', and organizes follow-up meetings, among people already working at customers and those just formed, to facilitate knowledge exchange and sharing, also in training.

Companies that need IT services, can choose 5020 as a supplier; besides getting a qualified service, they pay even less of its value, given that the team members, in order to work and be trained, can accept lower remuneration than that requested by the market (Fig. 1).

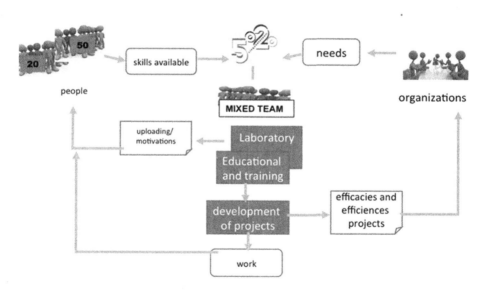

Fig. 1. The 5020 project. (Source: Authors)

The project is a "win-win" model, because all the parties involved have an improvement:

- people in their 50s can be reintegrated into labour market, can have an income and upgrade and enrich their knowledge, adapting behaviours, tools and working methods to the continuous technological innovation;
- people in their 20s can practice, trying a salaried work and especially learn something by working in team with more experienced people;
- small and medium enterprises or non-profit organizations can access innovative IT services with investments that can afford;
- big companies can purchase IT services making a concrete action of corporate social responsibility;
- the government can obtain, from people's employment, positive externalities, especially in health care and social inclusion (such as confronting depression and flaking of family ties, the increase in disposable income and so on).

The interaction among all players continues to produce new knowledge and generates continuous innovation.

The activity is organized as a non-profit; many operators, including the founder, are volunteers and all profits are reinvested in the development of the activity or divided among the employees in terms of higher salaries. The project has always been self-funded.

In these last years, the project has evolved from several points of view about both strategy and dimension, also thanks to the continuous exchange of information. In 2013, the foundation year, it was a simple model of cooperation between seniors and juniors who addressed to SMEs who wanted to implement their processes and therefore their information systems. It aimed at combining expertise and innovation to avoid the waste of resources (i.e. the skills of junior and senior workers), thanks to capable and motivated people, to an effective recruitment model, to a lean organization with "virtual" workplaces, "local" for people. The organization consisted of two founders (volunteers) who practice all functions; six people have been trained and one of them employed in an activity of business process review and improvement, through Appian, of a multinational company.

In the following year, in 2014, a business plan was developed, based on some presentations to companies in Verona and Milan's area (North of Italy), thanks to received references. Other businesses were added to the first activity: a work of Java (programming), to the Veneto Region that financed a software for a network of non-profit organization operating in a project of food waste, a web-site for a social promotion and management, also on Java (programming), for a cooperative company. Then, the 5020 project was embedded in the activities of the same cooperative, in order to have administrative and financial support. Partnerships with software companies were consolidated. Fifteen people have been trained and six of them were employed; they could share and enrich their knowledge and skills.

In 2015, the project had the purpose to enhance the involved people's horizontal competence (customer relationships, clarity of argumentation, balance of activities, project management, etc.), through the organization of dedicated training opportunities, starting from the real cases that they are facing in that period. Eighteen people were trained, seven employed. Compared to the initial platform, there were a lot of changes. Originally, the platform provided mixed senior/junior teams, access to services on all platforms, a dynamic recruitment through "ads" published on 5020 platform, with targeted questions to assess candidate's correspondence to the needs of the project that have to be developed, designed for people who meet the requirements even at a distance, mainly training on the job, and customers founded through sponsorship in the Verona's area. Today, the platform confirms, if possible, the mixed senior/junior team, although the number of unemployed in the over-50 group is greater than the junior's and there is also a significant presence of intermediate (40s); the fields of service are defined by focusing on those which ask more employees and constantly growing (BPM and Business Analytics), to provide more employment opportunities and development to the people; for the recruitment, ads are published on 5020 web-site, on the electronic bulletin boards of the universities, on LinkedIn and other social networks, requiring CV and then interviewing people. In this way, it was possible to better understand the real situation of the job market and the people's problems, analysing their mistakes, especially in job-search, and trying to improve their situation through dedicated training sessions. Thanks to knowledge exchange, it was clear that people's potential capabilities are more important than their vertically specialized knowledge; proximity is preferable to the venue of the projects. Customers are mainly found on Milan through the network of the coordinator of the 5020-project.

In 2016 the project, thanks to the continuous exchange of knowledge, focused more on solutions/tools of Business Process Management, Business Analytics, Digital Communication and to specific customers' needs. For them, 5020 selects the most suitable ICT platforms to their needs, based on the criteria of innovation, easy integration with existing systems, reliability, adaptability, easy development and affordability, mainly Appian and QlickView. The employed team are 6, one for each customer; 4 people are indefinitely employed. 5020, with other partners, created a network focused on ethics in IT and the assumptions, for 2017, is to turn the project into a social enterprise.

5   Discussion

5020 is a project created to fight waste of knowledge and skills and unemployment of 50 and 20 yearsold people, but also to fight the lack of IT investments on the part of SMEs and Social enterprises, which then suffer from decreasing effectiveness, efficiency and competitive advantage.

This project co-creates value by leveraging the resources exchanged within and across service systems, according to Vargo's et al. theory [1]. In particular, thanks to the dedicated ICT platform and instruments, in this case there are two kinds of interactions that co-produce and use resources: the interaction between service provider (5020) and their customers (enterprises), that creates value both for the provider and the customer, who obtain a good service at a low price; and also the interaction between the two groups of people that work to produce those services, 50s and 20s. By sharing knowledge effectively, people that compose the teams improve their expertise, thus resulting in innovation, on the one side, and increased employability, on the other side.

If 5020 is observed through the lens proposed by Lusch and Nambisan [5], this case is a successful service ecosystem. In fact, the 5020 system is an effective actor-to-actor network, where all actors serve as resource integrators and collectively create their environment or the service ecosystem through reciprocally enabled learning and knowledge exchange. I addition, there is a digital platform, that maximizes resource sharing, enables recruitment, training remote work and integration. Finally, there is an effective architecture of participation: shared knowledge, procedures, rules and ethical beliefs are key factors to coordinate the co-creation of value, and the only kind of command and control is the loose coordination by volunteers or in team, emerging from a bottom up process based on shared principles. In the light of Fjeldstad et al.'s work [7], 5020 is based on 50s and 20s people who have the capabilities and values to self-organize. Employment is a common good that people want to safeguard they can accumulate and share resources through the digital platform. Protocols, processes, and the same platform enable collaboration between the two groups for people involved in teams but also between 5020 and their providers and customers to achieve common objectives. The role of the platform, resulting from a wide range of IT tools used for recruitment, education, training and work, is a common environment where the actors develop, share and use common resources to co-create value. Coordination and control are emergent capabilities that are accomplished primarily via self-organizing mechanisms, i.e. direct interaction among competent actors – 50s and 20s on the one side, and 5020 and its providers and customers on the other side - that is self- organized, through the sharing of rules, values, beliefs and tools, rather than by hierarchical subordination or market forces.

This project, in this light, is a service ecosystem that may be a powerful engine to generate positive externalities:

- introduces people in their 20s and reintegrates people in their 40s–50s into the labour market, giving them an income and upgrade and enriching their knowledge, adapting behaviours, tools and working methods to the continuous technological innovation;
- encourages knowledge transfer and learning, by letting people work in team with more knowledgeable and/or experienced professionals;
- provides SMEs and social enterprises with the opportunity to access innovative IT services even with low investments, thus innovating and adapting their instruments and their knowledge to continuous change;
- provides large companies with the opportunity to purchase IT services by making a concrete action of corporate social responsibility;
- provides the entire community with the opportunity to take advantage from people's employment, and its positive consequences, especially in health care and social inclusion fields.

6   Conclusions

Our research question was how an innovative organizational architecture based on (i) a network of competent actors, (ii) an (ICT-enabled) platform serving as a network commons and (iii) a shared institutional logic and worldview influences the emergence and evolution of a service ecosystem with strong positive externalities. This explorative study suggests that service ecosystems in which ethically engaged actors play a pivotal role may enable the creation of highly innovative, sustainability-oriented business models, which generate important positive externalities.

Our results, emerging from the analysis of the interviews and the occupational and profitable data of the 5020 project, shows that this project can be considered a service ecosystem for common goods.

Further studies are certainly needed to better investigate the issue, but we think that the case presented by this study is particularly suitable for illuminating the logic and extending the relationships of the key organizational constructs identified. Moreover, the direct involvement of one of the authors enabled direct access to data, and then a particular thickness of the analysis.

This encourages us to further investigating the role of organizational factors in service systems conceived as self-organizing networks with strong positive externalities. Next steps may include quantitative monitoring thorough questionnaires and structured interviews, longitudinally repeated throughout time to all people involved: volunteers, 50s and 20s professionals, service providers, customers, associations, schools, universities, and governmental bodies.

References

1. Vargo, S.L., Maglio, P.P., Akaka, M.A.: On value and value co-creation: a service systems and service logic perspective. Eur. Manag. J. **26**, 145–152 (2008)
2. Vargo, S.L., Lusch, R.F.: Service-dominant logic: continuing the evolution. J. Acad. Mark. Sci. **36**, 1–10 (2008)
3. Spohrer, J., Maglio, P.P.: The emergence of service science: toward systematic service innovations to accelerate co-creation of value. Prod. Oper. Manag. **17**, 1–9 (2008)
4. Vargo, S.L., Lusch, R.F.: Evolving to a new dominant logic for marketing. J. Mark. **68**, 1–17 (2004)
5. Lusch, R.F., Nambisan, S.: Service innovation: a service-dominant-logic perspective. MIS Q. **39**, 155–175 (2015)
6. Santos, F.M.: A positive theory of social entrepreneurship. J. Bus. Ethics **111**, 335–351 (2012)
7. Fjeldstad, O.D., Snow, C.C., Raymond, E.M., Lettl, C.: The architecture of collaboration. Strateg. Manag. J. **33**, 734–750 (2012)
8. Dougherty, D., Dunne, D.D.: Organizing ecologies of complex innovation. Organ. Sci. **22**, 1214–1223 (2011)
9. Ostrom, E.: Governing the Commons: The Evolution of Institutions for Collective Action. Cambridge University Press, New York (1990)
10. Dietz, T., Ostrom, E., Stern, P.C.: The struggle to govern the commons. Science **302**, 1907–1912 (2003)
11. Hess, C., Ostrom, E. (eds.): Understanding Knowledge as a Commons: From Theory to Practice. MIT Press, Cambridge (2007)
12. Hess, C.: Mapping the new commons. In: Governing Shared Resources: Connecting Local Experience to Global Challenges, The Twelfth Biennial Conference of the International Association for the Study of the Commons, pp. 1–76 (2008)
13. De Gregorio, C.: Micro enterprises in Italy: are ICT an opportunity for growth and competitiveness? Riv. di Stat. Uff. **1**, 79–120 (2003)
14. Maglio, P.P., Spohrer, J.: Fundamentals of service science. J. Acad. Mark. Sci. **36**, 18–20 (2008)

15. Iansiti, M.R.L.: The Keystone Advantage: What the New Dynamics of Business Ecosystems Mean for Strategy, Innovation, and Sustainability. Harvard Business School Press, Boston (2004)
16. Lounsbury, M., Crumley, E.T.: New practice creation: an institutional perspective on innovation. Organ. Stud. **28**, 993–1012 (2007)
17. Bryman, A., Bell, E.: Business Research Methods. Oxford University Press, Oxford (2011)
18. Pettigrew, A.: Longitudinal field research on change: theory and practice. Organ. Sci. **1**, 267–292 (1990)
19. Yin, R.K.: Case Study Research: Design and Methods. Sage publications, Thousand Oaks (2013)
20. Bonomi, S., Za, S., Marco, M., Rossignoli, C.: Knowledge sharing and value co-creation: designing a service system for fostering inter-generational cooperation. In: Nóvoa, H., Drăgoicea, M. (eds.) IESS 2015. LNBIP, vol. 201, pp. 25–35. Springer, Cham (2015). doi:10.1007/978-3-319-14980-6_3

Customer Satisfaction from Inner-City Services: A Case Study

Rafi Zagorie, Shai Rozenes$^{(\boxtimes)}$, and Yuval Cohen

Afeka Tel Aviv Academic College of Engineering,
Mivtsa Kadesh 38, 69988 Tel Aviv, Israel
{rozenes,yuvalc}@afeka.ac.il

Abstract. This paper examines the quality of municipal services within inner-city services. It identifies the most important service quality dimensions that determine citizen satisfaction. System dynamics approach is used to model and analyze ways to improve citizen satisfaction. For that purpose, we developed questionnaires based on ServQual. 634 questionnaires were distributed to respondents in a town neighborhood of about 16,000 citizens. The relevance of the ServQual dimensions was validated. Three of the ServQual dimensions: reliability, empathy and responsiveness, significantly predicted citizen satisfaction. The paper discusses ways to guide practitioners to improve quality attributes and enhance inner-city customer satisfaction.

Keywords: ServQual · Customer satisfaction · Service · Quality · Inner city

1 Introduction

It is well recognized that providing high quality services has a beneficial effect on the performance of the organization. There is a an evidence indicating that providing top quality services enhances profitability, improves productivity, spreading positive word-of-mouth, increases market share and return on investment, and reduces costs [1]. Municipal authorities face various kinds of responsibilities than those of the private services in their endeavor to deliver services that meet customer expectation.

Service quality is a critical to both business competition and customer satisfaction. Efficient and effective service provision system become the main focus of the public service in many countries due to society requirements for local authorities service level that will meet quality standard as expected by the taxpayers [2].

The most important step in defining and delivering good quality service understands precisely the citizen needs. The citizen feedback provides valuable insight into how well the municipal authority meets their customers' needs. The information can be used in quality improvement programs to help municipal management to improve services, and the creation of a positive perception among citizens. Service quality can also identify gaps between citizens' perceptions and expectations of local services.

This paper represents a preliminary investigation of municipal service quality and its influence on citizen satisfaction. In particular, the study attempted to investigate how satisfaction changes can be identified and analyzed by using a system dynamics approach. Finally, the municipal representatives need to understand the important roles

© Springer International Publishing AG 2017
S. Za et al. (Eds.): IESS 2017, LNBIP 279, pp. 409–420, 2017.
DOI: 10.1007/978-3-319-56925-3_32

played by the quality of their services from the customers' point of view. Quality assessments of achieving good reputation within the relevant population can be used as a strategic tool for marketing the municipal services.

The paper structure is as follows: Sect. 2 reviews the relevant literature on service quality and customer satisfaction. A description of the methodology adopted in this study is discussed in Sect. 3. Section 4 presents the results. Section 4.1 presents the system dynamics model, and conclusion appears in Sect. 5.

2 Literature Review

2.1 Service Quality

Quality is a major concern of any organization as well as any consumer as it presents a measure of value that consumers gain for their money, time, and endeavor, and since it owns a direct impact on the short and long-term survival of any governing body. In fact quality is recognized as a strategic tool for attaining operational efficiency and improved business performance [3]. Despite its significance, defining quality is quite a mission. For example, Bitner et al. [4] define quality as a product's ability to satisfy a customer's needs or requirements. Crosby and Stephens [5] define it as 'conformance to requirements', while other definitions include fitness for use. Quality may summarily be viewed as "the totality of features and characteristics of a product or service that bear on its ability to satisfy stated or implied needs" [6]. Any organization should strive to support his customers with quality service to match or exceed customers' needs most of the time. In other words, there should not be any gaps or between customers' expectation and the organization's deliverables.

Service quality is a concept that has attracted considerable interest and a lot of discussion in the scientific literature because of the difficulties in both defining and measuring quality. One of the commonly used definitions of service quality is the ability of the organization to meet or exceed customer expectations. It is the result of the comparison that customers make between their expectations about a service and their perception of the way the service has been performed. If expectations are greater than performance, then perceived quality is less than satisfactory and hence customer dissatisfaction occurs.

Most of the recent work on service quality in the service sector can be credited to the pioneering work of Parasuraman, Berry and Zeithaml. In a seminal research study, Zeithaml, and Berry [7] identified ten dimensions of service quality based upon a series of focus group studies. These dimensions are: (1) tangible features, (2) reliability, (3) responsiveness, (4) communication, (5) credibility, (6) security, (7) competence, (8) courtesy, (9) understanding, and (10) access.

From this primary research, developed a service quality instrument called SERVQUAL that consisted of 22 pairs of statements which measure customer expectations and perceptions of service delivered on a seven or nine points Likert scale. For each pair of statements, the gap difference between the two scores is calculated. The idea is that the service is good if perceptions meet or exceed expectations and problematic if perceptions fall below expectations. The scale combined ten components into five generic dimensions of service quality.

2.2   The SERVQUAL Scale

The foundation for the SERVQUAL scale is the gap model proposed by Parasuraman, Zeithaml and Berry [8, 9]. The gap or difference between customer 'expectations' and 'perceptions,' service quality is viewed as lying along a continuum ranging from 'ideal quality' to 'totally unacceptable quality,' with some points along the continuum representing satisfactory quality.

SERVQUAL is based on a set of 22 variables/items covering five different dimensions of service quality namely:

- *Reliability*: The ability to perform a promised service dependably and accurately
- *Responsiveness*: A willingness to help customers and to provide support services
- *Assurance*: The knowledge and courtesy of employees and their ability to inspire trust and confidence
- *Empathy*: The caring, individualized attention a firm provides its customers
- *Tangibles:* The physical facilities, equipment, and appearance of personnel.

2.3   Customer Satisfaction

Customer satisfaction has been recognized as one of the most important elements of contemporary marketing thought, particularly in the case of service sectors and one of the main goals in marketing. Because satisfied customers tend to maintain their consumption pattern or consume more of the same product or service, customer satisfaction has become an important indicator of the future behavior [10]. Due to its centrality, various theories and models have been developed in an effort to define the construct and explain satisfaction in different products/services and consumption stages [11].

Some researchers believe that service quality leads to satisfaction [12] while others think otherwise. Empirical studies regarding this issue support the assertion that service quality is the cause of satisfaction [13]. Within this causal ordering, satisfaction is described as a post-consumption evaluation of perceived quality. It is an emotional state that results from experiencing a service quality encounter and comparing that encounter with what was expected. Quality is one of the service dimensions factored into the consumers' satisfaction judgment. As a consumer perceives a widening gap between expected and desired levels of service, a feeling of dissatisfaction will increasingly develop [14]. One of the suggestions for improving satisfaction is an emphasis on empathy in relation to the service providers [15].

The social side of service satisfaction is considered particularly relevant where there are high levels of customer-employee contact [16].

Research about using the ServQual methodology within municipality citizens was conducted in several countries. For example, the United Arab Emirates conducted such a study in Al-Ain Municipality. The findings indicated that The customer satisfaction indices and scores of customers' trust were in the mid-eighties, indicating high levels of satisfaction and client trust [17]. Another study was carried out on 256 participants who live in Canakkale, Turkey, and pay their bills to the municipality. Overall, the quality and satisfaction levels are average. The lesson learned for managers and service personnel is the main proposal for the improvement should be focused on consumer satisfaction. Using the ServQual instrument was also reported in the Eskiåžehir Municipalities [18] as well as in the Tehran municipality of Iran and in the municipalities of São Paulo state [19, 20]. It can be noticed that the reports of using the ServQual instrument within municipalities are across the world [21–24].

The review of the literature so far has provided support for the idea that there is a strong direct link between service quality and satisfaction. In this paper, we analyze the relationship between the variables of municipal service quality and citizens' overall satisfaction [25].

The review of the literature so far has provided support for the idea that there is a strong direct link between service quality and satisfaction. In this paper, we analyze the relationship between the variables of municipal service quality and citizens' overall satisfaction.

3 Methodology

3.1 General Approach

The data in this study were collected by using a structured questionnaire, which consists of two parts. Part A was designed to gather information about respondents' contact to the municipal service call center, an overall satisfaction with the services provided by the municipal service center and 22 statements related to measuring respondents' expectation towards the service quality of municipal service center. The SERVQUAL scale that was designed by Parasuraman *et al.* 1988 was used in this study [26]. The respondents were asked to rate their level of agreement with each statement using five-point scale ranging from 1 (*Strongly Disagree*) to 5 (*Strongly Agree*). Part B consists of 22 statements related to measuring respondents' perception towards the service quality of municipal service center with the same scale as part A. Section C gathers information about the respondent's demographic background, such as gender, age, educational level, way of contact (Telephone, Website, Mobile app.), complaint category and residential neighborhood.

3.2 Data Collection and Analysis

The questionnaires were distributed to 634 respondents in a local government contains a population of about 16,000 citizens. A total of 520 respondents out of the 634 contacted the municipal service center. The goal was to choose independent respondents from different neighborhoods in order to avoid imitation or repetition of views. Descriptive statistics, t-test, Pearson correlation and regression were conducted to analyze the data. A system dynamics model was designed to simulate the quantity of incoming complaints vs. how many of them were solved within normal time and how many delayed or not solved at all. The benefit of using such model is a way to perform sensitivity analysis depends on statistic results.

3.3   Perception Score as a Measure of Satisfaction, Loyalty and Retention

For the purpose of this study, the perception score of 3.7 out of 5 ($\sim 75\%$ of the maximum) as shown is the minimum score needed to measure true satisfaction for the combined dimensions as well as each sub-category of the dimensions. For example, a SERVQUAL respondent's perception score of 4 in Responsiveness would indicate an acceptable level of satisfaction for that category. The same would be true for each of the subcategories of that dimension as long as each achieved at least a 3.7 perception score. Conversely, a SERVQUAL respondent's perception score of 3.5 in Tangibility would indicate a non-acceptable level of satisfaction for that category [27]. Manufacturing firms improve their service quality and customer satisfaction to gain and increase the benefits for customer retention and customer loyalty, therefore, it is influenced by various factors involved in the process of service delivery to customers, such as physical facilities and behavior and quality of the service personnel [28]. A common approach to identify and operate appropriate quality improvement strategies to enhance customer satisfaction is to divide the service process into attributes and to elicit the opinion of customers on the performance with respect to each of these SERVQUAL dimensions.

The relationship between customer satisfaction and loyalty can be classified into three groups [29]. The first comes from service management literature and studies the relationship at an aggregated, company-wide level. These studies consider satisfaction to be an antecedent of customer loyalty, which in turn influences firms' profitability [3, 27]. The second focuses on the individual level and has mainly studied customer retention by customer repurchase intension which in this case study is the citizens' will to use the service again [30]. The third focuses on the satisfaction –loyalty link on an individual level with real service purchasing [31]. However, overall satisfaction seems to be a better predictor of citizens' intentions and behaviors.

3.4   System Dynamics Simulation Model

This paper uses System dynamics (SD) as one of its main approaches. SD is a mathematical modeling technique to frame, understand, and discuss complex subjects and troubles. The following references are excellent sources for learning the SD methodology [32–35]. Currently the SD methodology is being employed throughout the public and private sector for policy analysis and invention [36–39].

The simulation model used in this research proceeds from the SERVQUAL results to an analysis on the action that the municipality leaders should take to improve the service quality. The model simulates the results of the survey and shows the overall service quality dimensions that determine citizens' satisfaction and ways to improve it by sensitivity analysis. The modeling and simulation was performed by Vensim PLE software.

3.5   SERVQUAL Results

The following description presents this research SERVQUAL results.

Tangibility. The average unweighted gap score (P-E) for the tangibility dimension of customer satisfaction is 0.028. When applying the tangibility weight score of 18.2% to the gap score, the weighted gap score jumps to 0.501. The survey results show that the perception of the local citizens exceeds their expectations of the tangible appearance aspects of the municipal service center.

The average SERVQUAL perception score is 3.827 out of 5 shows high (>75%) satisfaction ratio in the tangibility dimension.

Reliability. The average unweighted gap score (P-E) for the reliability dimension of customer satisfaction is −0.043. When applying the reliability weight score of 22.4% to the gap score the gap score jumps to −0.966. The survey results show that the perception falls below their expectations of the reliability aspects.

The average SERVQUAL perception score is 3.237 out of 5 shows low (<75%) satisfaction ratio in the reliability dimension.

Responsiveness. The average unweighted gap score (P-E) for the responsiveness dimension of customer satisfaction is −0.083. When applying the responsiveness weight score of 19% to the gap score, the gap score increases to −1.569. The survey results show that the perception falls below their expectations of the responsiveness.

The average SERVQUAL perception score is 3.409 out of 5 shows low (<75%) satisfaction ratio in the responsiveness dimension.

Assurance. The average unweighted gap score (P-E) for the assurance dimension of customer satisfaction is 0.004. When applying the responsiveness weight score of 20.4% of the gap, the gap score increases to −0.086. The survey results show that the citizens' perception exceeds their expectations of the assurance.

The average SERVQUAL perception score is 3.890 out of 5 shows low (>75%) satisfaction ratio in the assurance dimension.

Empathy. The average unweighted gap score (P-E) for the empathy dimension of customer satisfaction is −0.026. When applying the reliability weight score of 20% to the gap score, the gap score increases to −0.523. In both cases, the survey results show that the perception of the Oregon HIDTA drug task force supervisors and investigators fall below their expectations of the empathy aspects of the Oregon HIDTA ISC Analytical Unit's customer satisfaction level.

The average SERVQUAL perception score is 3.651 out of 5 shows low (<75%) satisfaction ratio in the empathy dimension.

3.6   Weighted and Unweighted Gap Scores

Figure 1 presents the overall perceived SERVQUAL score for the five dimensions of customer satisfaction is 3.588. This indicates lower score than 75% as a minimum value of satisfaction.

The overall service quality gap Q = f(E,P) is defined as the difference between customer's expectations of a service provision and the perceptions of the delivered service. To determine if there is a gap we use the Eq. (1):

$$OSQ = \sum_{j=1}^{k} (Pj_n - Ej_n) \qquad (1)$$

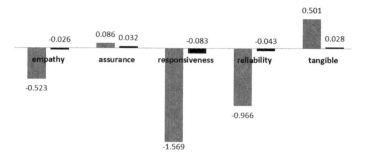

Fig. 1. Weighted and unweighted gap scores

OSQ indicates the Overall Service Quality for the local municipal service center; K = number of service items (which is 22); P = overall perception of respondents with respect to the performance of a service; E = overall service quality expectations; n = number of respondents. Therefore, we get the result is Eq. 2:

$$OSQ = \sum_{j=1}^{22} (Pj_{520} - Ej_{520}) = -0.125 \tag{2}$$

This result indicates that the quality of service provided by the local municipal service center is negative, lower than what customers expect and desire. Overall therefore, there is a gap (−0.125) in the service offered.

With an overall mean of 3.588 for perception and 3.713 for expectation it is evident that quality of service falls below citizens' expectations (<75%).

Statistical analysis was performed on the SERVQUAL. The results show a paired sample correlation results with coefficient of 0.56, indicates a statistically significant medium relationship between the overall score for expectation and that of perception.

4   The Simulation Model

System dynamics, as mentioned previously, is a methodology to understand the non-linear behavior of complex systems in various domains [40]. Researches of municipal systems using system dynamics method show a ways of improvements of processes within the municipal [41]. In this paper a SD simulation model was implemented to examine the citizen's satisfaction. Figure 2 shows a system block diagram which is used to demonstrate the SD simulation that was used. The inputs to the system block diagram are the survey results and all the data collected during the research about citizens' complains. The data contains number of complaints via phone, email or mobile application. The total number of complaints sorted into three channels of which they were being resolved vs. predefined respond time. The first channel is the resolved complaints within the predefined respond time, the second is the complaints that were not resolved within the predefined respond time by the in – house departments, and the third channel includes is the complaints that were not resolved within the predefined

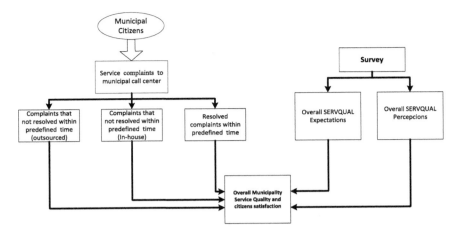

Fig. 2. System block diagram

respond time by the outsourced departments. These inputs are being calculated to determine citizens' satisfaction and ways to improve it by using sensitivity analysis.

4.1 Simulation Results

The simulation results are presented in Fig. 3. Y-axis in Fig. 3a represents the number of citizens' complaints exponentially increasing in relation to X-axis that represents the time frame which is 36 months. As seen there is a relatively growth in the complaints from citizen due to city's population growth. Figure 3b is the result of overall service quality by the survey where Y-axis represents the quality of service that includes the gap between perceptions and expectations from service and results of the SERVQUAL survey. The results show a negative satisfaction from service given by the municipality service center.

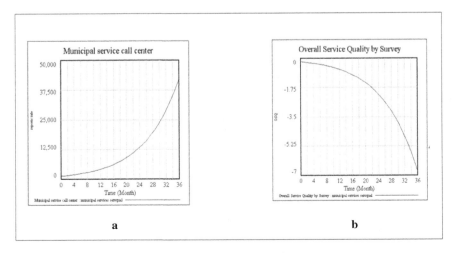

Fig. 3. Simulation results

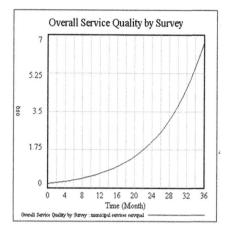

Fig. 4. Overall service quality improvement

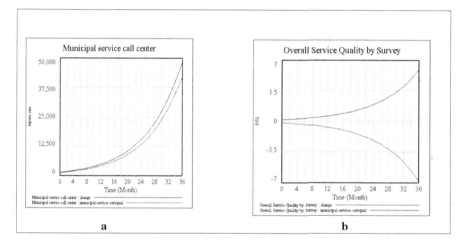

Fig. 5. Simulation results

Three service quality dimensions, namely reliability, empathy and responsiveness significantly predicted citizen satisfaction and as a result the overall service quality is negative as shown in Fig. 3b. However if we change these dimensions the overall service quality can be improved and will influence on the citizens' satisfaction. Figure 4 represents the improvement of overall service quality (OSQ) if the three dimensions of reliability, empathy and responsiveness will increased with future survey where the other two dimensions will remain with the same values.

Figure 5a shows the change of the number of service complaints in a relation to municipality population growth within 36 month time frame where Y-axis represents the number of complaints and X-axis represents the time in simulation. Figure 5b shows the difference between the overall service quality results described by the gap between the perceptions and expectations with the survey result and a change from a negative value of −6 to a positive value of +7 in the overall satisfaction which can lead to a conclusion that using a system dynamics simulation tool can show changes in a nonlinear system.

5 Summary and Conclusion

This study examines the quality of municipal services provided by local government. For that purpose, the 22-item SERVQUAL instrument was used to measure citizen perceptions of municipal service quality. This suggests that basing quality measurement efforts on SERVQUAL is appropriate for the assessing municipal services. The study has also attempted to identify the most important service quality dimensions that determine citizen satisfaction with municipal services. This study provides managerial implications for municipal councils by offering them practical guidelines to improve quality attributes that would increase citizen satisfaction. From a strategic perspective, municipal officials can determine the relative importance of the five service quality dimensions in predicting citizen satisfaction. By doing so, municipal officials can determine which service quality dimension(s) they should pay most attention to.

It is recommended that service quality be assessed on an annual basis. The initial assessment provides a baseline for comparison with future assessments. This comparison is essential for monitoring the effectiveness of service quality improvement efforts and identification of service quality trends as they emerge. The assessment would also provide a broader picture of interest to local governments who would then be aware of the potential areas of interest to their citizens, as an annual assessment can help measure progress and would allow enough elapsed time between assessments for improvement efforts to have an impact, while at the same time still be frequent enough to identify emerging customer service trends.

This study also had certain limitations that must be considered when assessing the outcomes of its findings and implications. The results pertain only to the respondents and generalizations to a wider population or service sector should be done with caution. Additional studies comparing service quality of citizens from different regions of Israel or across different public service settings might produce interesting findings. Furthermore, the study could have been improved by conducting qualitative interviews with individual customers to ascertain other influence factors not identified in this analysis. Future work can also focus on developing a richer model that incorporates other potential variables beyond the two used in this study.

References

1. Sachdev, S.B., Verma, H.V.: Relative importance of service quality dimensions: a multisectoral study. J. Serv. Res. **4**(1), 93–116 (2004)
2. Zahari Wan Yusoff, W., Ismail, M., Newell, G.: FM-SERVQUAL: a new approach of service quality measurement framework in local authorities. J. Corp. Real Estate **10**(2), 130–144 (2008)

3. Anderson, E.W., Fornell, C., Rust, R.T.: Customer satisfaction, productivity, and profitability: Differences between goods and services. Mark. Sci. **16**(2), 129–145 (1997)
4. Bitner, M.J., Booms, B.H., Mohr, L.A.: Critical service encounters: the employee's viewpoint. J. Mark. **58**(4), 95–106 (1994)
5. Crosby, L.A., Stephens, N.: Effects of relationship marketing on satisfaction, retention, and prices in the life insurance industry. J. Mark. Res. **24**(4), 404–411 (1987)
6. Kotler, P.: Marketing Management, Analysis, Planning, Implementation, and Control, 8th edn. Prentice Hall, New Jersey (1994)
7. Zeithaml, V.A., Berry, L.L., Parasuraman, A.: Communication and control processes in the delivery of service quality. J. Mark. **52**(2), 35–48 (1988)
8. Parasuraman, A., Zeithaml, V.A., Berry, L.L.: A conceptual model of service quality and its implications for future research. J. Mark. **49**(4), 41–50 (1985)
9. Miao, L., Mattila, A.S., Mount, D.: Other consumers in service encounters: a script theoretical perspective. Int. J. Hosp. Manag. **30**(4), 933–941 (2011)
10. Schneider, B., Bowen, D.E.: Personnel/human resources management in the service sector. Res. Pers. Hum. Resour. Manag. **10**(1), 1–30 (1992)
11. Erevelles, S., Leavitt, C.: A comparison of current models of consumer satisfaction/dissatisfaction. J. Consum. Satisf. Dissatisfaction Complain. Behav. **5**(1), 104–114 (1992)
12. Zeithaml, V.A., Berry, L.L., Parasuraman, A.: Communication and control processes in the delivery of service quality. J. Mark. **52**(2), 35–48 (1988)
13. Mokhlis, S., Aleesa, Y., Mamat, I.: Municipal service quality and citizen satisfaction in southern Thailand. J. Public Adm. Gov. **1**(1), 122–137 (2011)
14. Oliver, R.L.: Satisfaction: A Behavioral Perspective on the Consumer. Routledge, New York (2014)
15. Moletsane, A.M., de Klerk, N., Bevan-Dye, A.L.: Community expectations and perceptions of municipal service delivery: a case study in a South African Municipality. Mediterr. J. Soc. Sci. **5**(3), 281 (2014)
16. Rod, M., Ashill, N.J., Gibbs, T.: Customer perceptions of frontline employee service delivery: a study of Russian bank customer satisfaction and behavioural intentions. J. Retail. Consum. Serv. **30**, 212–221 (2016)
17. El-Bassiouni, M.Y., Madi, M., Zoubeidi, T., Hassan, M.Y.: Developing customer satisfaction indices using SERVQUAL sampling surveys: a case study of Al-Ain municipality inspectors. J. Econ. Adm. Sci. **28**(2), 98–108 (2012)
18. Filiz, Z., Yilmaz, V., YagÄzer, C.: The measurement of service quality in municipalies by servqual analysis: a case study in Eskiåžehir Municipalities. Anadolu Univ. J. Soc. Sci. **10** (1), 59–76 (2010)
19. Cukier, R., da Costa, M.A.M.: Competitive dysfunction: analysis the level of the gaps, the servqual model in compounding pharmacies the municipalities of São Paulo state. Rev. Científica Hermes. **8**, 70–91 (2013)
20. Darvish, H., Shirsavar, F.: A study on the effects of remote working on quality of services: A SERVQUAL survey on central office of Tehran municipality. Manag. Sci. Lett. **3**(6), 1615–1620 (2013)
21. Alex, O.T., Ondiek, A.B.: Applicability of SERVQUAL/rater model in assessment of service quality among local authorities in Kenya. **2**(1), 1–16 (2014)
22. Appiah, M.K.: Effects of services quality on customer satisfaction: a case from private hostels in wa-municipality of Ghana. Br. J. Mark. **4**(1), 48–54 (2016)
23. Moletsane, A.M.: Community expectations and perceptions of municipal service delivery: a case study in the Emfuleni local area. Ph.D. dissertation, North-West University (2012)

24. Seng, V.: Service quality in One Window Service Office (OWSO): reflection from Takhmao Municipality, Cambodia. In: Public Administration in The Time of Regional Change (ICPM 2013) (2013)
25. Gumus, M., Koleoglu, N.: Factor analysis on service attributes of Canakkale municipality. TQM Mag. **14**(6), 373–375 (2002)
26. Parasuraman, A., Zeithaml, V.A., Berry, L.L.: SERVQUAL A multiple-item scale for measuring consumer perception of service quality. J. Retail. **64**(1), 12–40 (1988)
27. Heskett, J.L., Sasser Jr., W.E., Schlesinger, L.A.: The Service Profit Chain: How Leading Companies Link Profit to Loyalty, Satisfaction, and Value. Free Press, USA (1997)
28. Murali, S., Pugazhendhi, S., Muralidharan, C.: Modelling and Investigating the relationship of after sales service quality with customer satisfaction, retention and loyalty–a case study of home appliances business. J. Retail. Consum. Serv. **30**(1), 67–83 (2016)
29. Bodet, G.: Customer satisfaction and loyalty in service: two concepts, four constructs, several relationships. J. Retail. Consum. Serv. **15**(2), 156–162 (2008)
30. Chatzoglou, P., Chatzoudes, D., Vraimaki, E., Leivaditou, E.: Measuring citizen satisfaction using the SERVQUAL approach: the case of the "Hellenic post". Procedia Econ. Finance **9**(4), 349–360 (2014)
31. Segoro, W.: The influence of perceived service quality, mooring factor, and relationship quality on customer satisfaction and loyalty. Procedia-Soc. Behav. Sci. **81**(5), 306–310 (2013)
32. Forrester, J.W.: Industrial dynamics: a major breakthrough for decision makers. Harv. Bus. Rev. **36**(1), 37–66 (1958)
33. Forrester, J.W.: The impact of feedback control concepts on the management sciences. In: Foundation for Instrumentation Education and Research (1960)
34. Forrester, J.W.: Principles of Systems: Text and Workbook. Wright-Allen Press, Cambridge (1968)
35. Forrester, J.W.: Urban Dynamics. MIT press, Cambridge (1969)
36. Dangerfield, B.: System dynamics applications to European healthcare issues. In: Mustafee, N. (ed.) Operational Research for Emergency Planning in Healthcare: Volume 2, pp. 296–315. Springer, London (2016)
37. Lane, D.C., Monefeldt, C., Rosenhead, J.: Looking in the wrong place for healthcare improvements: a system dynamics study of an accident and emergency department. In: Mustafee, N. (ed.) Operational Research for Emergency Planning in Healthcare: Volume 2, pp. 92–121. Springer, London (2016)
38. Langroodi, R.R.P., Amiri, M.: A system dynamics modeling approach for a multi-level, multi-product, multi-region supply chain under demand uncertainty. Expert Syst. Appl. **51**(3), 231–244 (2016)
39. Sterman, J.D.J.D.: Business Dynamics: Systems Thinking and Modeling for a Complex World. McGraw-Hill, Irwin (2000)
40. Sterman, J.D.: System dynamics modeling for project management, Cambridge (1992). Unpublished Manuscript, http://web.mit.edu/jsterman/www/SDG/project.pdf
41. Winkler, T.J., Ziekow, H., Weinberg, M.: Municipal benefits of participatory urban sensing: a simulation approach and case validation. J. Theor. Appl. Electron. Commer. Res. **7**(2), 101–120 (2012)

Using Ethological Approaches to Understand Skiers' Behavior in Cable Cars Queues in Order to Improve Overall Satisfaction: An Empirical Study Conducted in the Swiss Alps

Emmanuel Fragnière[1,2], Valentine Gaillet[1], Benjamin Nanchen[1(✉)], and Randolf Ramseyer[1]

[1] Service Design Lab, HES-SO, Le Foyer - Techno-Pôle 1, 3960 Sierre, Switzerland
{emmanuel.fragniere,benjamin.nanchen,randolf.ramseyer}@hevs.ch,
gailletvalentine@gmail.com
[2] University of Bath, Bath, BA2 7AY, UK
http://www.hevs.ch/servicedesign

Abstract. In a service quality perspective, the animal behavior of humans (e.g. human ethology) in queues has, to our knowledge, never been observed. This paper provides an empirical exploratory enquiry with the scope to understand skiers' behaviors in cable cars queues in order to improve their overall satisfaction. We carried 82 immersions and 43 semi-directed interviews in the Swiss Alps (Valais), during the scholar vacations of February 2016. Along with the literature review, this research provides hypotheses to better understand the interface between a human queue and a mechanical transportation system. To adjust to the rigid system of the cable cars, our results show that a queue's regulation is mostly based on ethological behavior.

Keywords: Queuing theory · Human ethology · Exploratory study · Human behaviors · Customer satisfaction · Emotional temperature · Preventive control · Cable cars management

1 Context of Research

Many physical services are related to a queue system (e.g. counters, entry points, customs etc.). To our knowledge, no research in queue management has assessed the possibility of early detection of queue-jumping due to such emotional factors as anxiety, nervousness or restlessness. Similarly, very little is done from a practical point of view to defuse conflict while individuals wait in queues. On the one hand, quantitative approaches coupled with Monte-Carlo simulation techniques now allow computers to calculate a series of indicators to manage the queue from essentially an operational point of view. When, for example, waiting time in the queue is considered too long, these models can accurately assess the number of additional counters that need to be available to return to a reasonable waiting time. On the other hand, qualitative approaches make it possible to understand behavioral aspects of the queue. The authors of this paper are

© Springer International Publishing AG 2017
S. Za et al. (Eds.): IESS 2017, LNBIP 279, pp. 421–430, 2017.
DOI: 10.1007/978-3-319-56925-3_33

particularly interested in these qualitative approaches and especially in the studies of human ethology (the animal behavior of humans).

Ethology [1] is the study of the behavior of living beings (animals or humans) in their "natural ecosystem." This discipline is concerned with all factors that induce a certain behavior (e.g. stimuli, innate, acquired, etc.). We are convinced that a human ethology approach can enable us to better understand queues of consumers and users. Human ethology can allow us to study innate human behaviors specific to queues, such as nervousness, impatience and even boredom.

We therefore asked ourselves whether an ethological field survey could be used to better understand skiers' queuing system at cable cars (e.g. ski lifts, gondolas, etc.) in order to develop a tool for managing inconveniences and for prevention of queue-jumping. To answer this question, this article draws on ongoing research in Valais dealing with management of queues in its ski resorts.

This paper is organized as follow. First, a brief literature review on the topic is presented. Then, the methodology is explained. In Sect. 4, the results of our study are summarized. Different hypotheses are discussed in Sect. 5. Finally, we conclude and indicate further research.

2 Literature Review

2.1 Queuing Management

It was at the beginning of the 20th century that the Danish engineer Agner Krarup Erlang (1878–1929) developed queuing theory. His mathematical models were applied for the first time to handle the very long queues of white-collar men and women who wanted to take the skyscraper lift to their offices in downtown Manhattan. Erlang was not only interested in quantitative models to assess waiting time and queue length, but also in the perception of time spent in the queue (qualitative aspects—such as talking to a colleague or reading one's journal—may give the impression that the perceived waiting time is shorter than the actual waiting time). The field of queue studies did not remain solely with mathematical considerations and calculations. Indeed, many researchers have analyzed the marketing and psychological aspects related to queues. Below, we present briefly some research whose contributions deal with the behavioral aspects of queues and provide a better understanding of why this exploratory study of queues at cable cars uses human ethology methods.

Maister [2] is one of the first authors to talk about the psychology of queues. In his paper, he examines how waiting is perceived and experienced by clients, and arrives at the observation that the experience of waiting is specific to the context of the queue. By analyzing this context and improving it, it is then possible to have a significant impact on customers' satisfaction.

Naumann and Miles [3] emphasized the importance of the control process and the announcement of waiting times in order to maintain a sense of justice in the queue.

Nie [4] proposed various stress reduction mechanisms, such as providing clients with a waiting time forecast, or offering clients fast-paced opportunities. These practical suggestions can help managers reduce perceived waiting times, improve the wait-and-see experience of clients, and improve overall queue management.

Baker and Cameron [5] demonstrated that some elements, such as music or the structure of the queue, if properly managed, could have a positive impact on perception of expectation and satisfaction. Their paper concludes with a model that uses all the elements that the authors believe will positively affect the perception of expectation.

McDonnell [6] showed that music or perfume could reduce the level of discomfort during a wait. The overall satisfaction level of the service thus increases by the insertion of external stimuli while in a queue.

Rafaeli et al. [7] highlighted the importance of the relationship between the structure of the queue and the attitudes of clients. The authors of this article questioned the perceived anxiety, whether the service would be delivered according to expectations, and what questions could be asked of clients waiting in the queue.

Choongbeom and Atul [8] developed an econometric model to explain wait times in services. They identified how some independent variables, such as the human factor and visual elements, have a significant influence on the perception of expectations of clients.

Van Riel et al. [9] described waiting as a psychological experience. The authors of this paper found that the traditional queue could produce a sense of injustice, even if from an objective point of view there is no inequality.

Whiting and Donthu [10] examined the difference between real wait time and perceived wait time, based on a case study. It emerged that there were different ways to reduce this difference and that, depending on the emotional state of the client, the perceived waiting time may be longer or shorter.

It is important to note that with the advent of new technologies and the Internet, qualitative research has more recently focused on the perception of waiting for online services. For example, Nah [11] studied the wait time tolerated when consulting websites.

2.2 Human Ethology

Etymologically, ethology means the study of manners. The naturalist Geoffroy Saint-Hilaire coined this term in 1856. Konrad Lorenz and Niko Tinbergen [12] developed ethology as a scientific discipline at the beginning of the 20th century. Human ethology is rarely used by management researchers. As Boris Cyrulnik [13] remarked in an article published in the *Express* on September 30, 1993, "Man is a species that is part of the living world. It has inevitable behaviors that can be made observable thanks to hypotheses and methods derived from naturalist observations."

Human ethology offers definitely interesting potential to study managerial phenomena. Unfortunately, this discipline is not very much employed in management science. Maybe this is due to the fact that it considered to be a natural science and not a social science. It is often called the biology of human behaviour [14]. Nevertheless, it is more and more recognized that in organization research, there is a need to better understand how humans behave, think, feel and react with each other [15, 16]. As such, over the last years there has been a growing number of studies in management relying on human ethology. For instance, Otubanjo et al. [17] have developed a theoretical corporate framework of corporate identity based on ethology. Human ethology has also been recently used to study resource consumption [18]. Closer to our approach, Jones et al. [19] have developed an ethogram to analyze the interactions in the operating rooms (OR).

Concerning the ethological aspect (animal behavior: innate versus learned) of the queues, to our knowledge, no research of this type has yet been conducted.

3   Methodology

To carry out this study, the central focus was an ethological field research. There are two types of approaches in ethology to collect data: naturalistic observation and experimental manipulation. The first is conducted in a natural environment or in a reconstitution of it, while experimental manipulation is carried out in a laboratory environment. In this research, a naturalistic observation was chosen because it is well suited to an exploratory approach with the aim of generating new research hypotheses.

In order to conduct the fieldwork, an immersion guide (commonly called ethogram in ethology) has been developed. According to Weisfeld et al. [20], "a human ethogram is a list and description of the basic observable whole-body behaviors […]. It provides a framework for the study of human behaviour". This ethogram is composed of several sections, including the gestures and behaviors of the skiers, the means of attenuations put in place by the provider (lift operator), and facial expressions. To allow the investigators complete freedom of observation, this ethogram intentionally outlined general elements. Furthermore, an interview guide containing four open-ended questions was used to do the semi-directed interviews. The latter provided additional linguistic elements facilitating the interpretation of the ethograms.

Data collection took place from February, 8th to March, 4th 2016. During that period, Swiss, French, English, Dutch, German, and Austrian skiers spent their holidays in the Alps. A total of 24 days were spent at eight different stations in Switzerland. This research was conducted at the following ski resorts: Crans-Montana, Grimentz-Zinal, Nax and Vercorin (for the Swiss French-speaking regions) and Belalp, Bellwald, Bettmeralp and Saas-Fee (for the German-speaking regions). Eighty-two immersion episodes lasting approximately a quarter of an hour were carried out, and 43 semi-directed interviews were collected.

The ski resorts were selected geographically (near the Service Design Lab in Sierre). A schedule was established and respected in order to cover all the parts of the holiday period and the relative meteorological hazards.

The data were analyzed with RQDA software [21], and summary of the main results is presented below. This synthesis, as well as the various readings carried out on the subject of queue management, allowed us to generate our own hypotheses.

4 Summary of Results

RQDA software was used to analyze the immersions and the interviews. There are four essential findings relating to human behavior which serve as a basis for generating hypotheses of ski queues at cable cars. The four main themes are (1) the notion of proxemics (2) the typical elements of queues and group structure (3) waiting time and (4) the principle of first-in/first-out (FIFO) in relation to queue-jumping or not respecting the queue.

In order to fully understand all the terms used in this synthesis, it is important to differentiate the turnstile from the entry gate. The first is the structure that a skier must go through by scanning his or her ski pass before boarding the cable car, while the latter is a small barrier that opens automatically just before boarding.

4.1 The Notion of Proxemics in the Queue

Proxemics [22], in the words of its creator Edward Hall, means "the interrelated observations and theories of man's use of space as a specialized elaboration of culture." During the observations in the queues, two distinctive waiting moments were identified when skiers approached one another: first before passing the turnstiles, and secondly at the start of the cable car facilities. It can therefore be concluded that the closer the skiers are to the turnstiles and or the lifts, the closer they are to each other. For example, during the same immersion, the distance between skiers ranged from 30 to 40 cm before the turnstiles, and 50 to 60 cm after. This occurred even when few people were present or when skiers were well spaced at the beginning of the line. During the immersions, several other measures were taken, unrelated to the tightening of the end of the line: when the line was compact, only 25 to 30 cm separated the impatient skiers. When skiers knew each other or were in a group, the distance was approximately 30 to 40 cm between people. Thus, groups were relatively tighter and stood out from each other in the queue. In addition, there was no contact between different groups and individuals. It should also be noted that family members stood closer to each other than within groups of friends.

Finally, queuing can induce jostling and (physical) contact between different people. Some then develop techniques to avoid such contact, for example by detaching themselves from the queue. It should also be noted that proxemics varies according to affluence: the more people, the more the skiers tighten. In addition, proxemics can also be influenced by a disruptive element, such as the opening of a barrier that forces people to tighten.

The notion of proxemics in the queue is therefore quite variable according to the situation and affluence levels. It can lead to contact between skiers that some people will try to avoid. This phenomenon affects negatively the skiers' experience.

4.2 The Structure of the Queue and Groups

In introduction to this topic, it is necessary to explain, following the immersions, how a queuing system is generally organized. First, at the end of the ski slope, a funnel is built with physical elements (e.g. ropes). The funnel gets narrower in front of the turnstiles. After the turnstiles, the skiers find themselves in a free space in front of a gate. The end of the queue is at the gate. The latter opens automatically to allow skiers to board the cable car.

In most of our observations, skiers often waited side by side in order to hold conversations. The rows before taking the chairlift were formed automatically, either before the gates or at the beginning of the queue. There is thus an aspect of automation although at certain facilities, the very structure of the queue forces skiers to form a row using the barriers installed before the gates. The queues are well-organized, two by two or in groups, as if it were an innate movement for skiers to line up before taking the chairlift. However, sometimes disruptive elements, such as poorly organized groups or queue-jumping, break this organization.

When skiers are in group, members try to organize themselves to get on the chairlift together, and need to communicate to make that happen. Often the members of the group are separated by the turnstiles and regroup once past. Two types of groups of skiers taking the lift were clearly identified: those who do not want to mix and those who will mix with other skiers if they can remain grouped. Indeed, the separation of a group can become stressful or annoying depending on the situation. The important thing for these groups of skiers is to be together and for this, communication plays an essential role. Communication is similarly important when groups get off the chairlift. Skiers arrive and depart in groups, even if they have to cut in front in order to stay together and not slow down the rest of the line.

The structure of the line is therefore fairly linear: skiers often put themselves side by side or in a very hermetic group, for which staying together is a priority.

Then, arriving at the cable car, the skiers form of themselves columns to adapt to the turnstiles. However, disruptive elements can occur, such as poorly organized groups or queue-jumping, and will therefore break this organization.

4.3 The Waiting Impression

The perception of time is a key factor in the management of queues. The immersions revealed that most individuals have a limit to what they perceive as an acceptable wait time. When skiers reach that limit, they begin to have negative feelings. For the majority of skiers observed, a wait of less than 5 min was suitable and did not cause any negative feelings. After 5 min, individuals began to experience impatience, nervousness, anxiousness, exasperation, and even apprehension, and complained about the length of the line. Some expectations are caused by technical problems and others are driven by human origin. However, it is interesting to note that when expectations about wait time are driven by human factors (e.g. someone falling several times) nobody gets irritated, unlike when the wait time is driven by technical issues (e.g. a cable car breakdown). In other words, the cause of expectations has a strong influence on resulting behavior.

The impression of wait time, as well as the actual wait time, are therefore important in this study because wait time is one of the main causes of dissatisfaction. Some skiers take waiting time at cable cars as a decisive criterion when choosing a ski resort.

4.4 Queue-Jumping Management

Several factors influence queue-jumping: the structure of the lane (e.g. funnel, barrier free space, etc.), the type of cables cars (e.g. cabin, chair lift, etc.) and the arrival area (e.g. uphill, in a turn, several possible arrivals according to the chosen track, etc.).

Certain queue-jumping behaviors are made on the basis of a tacit or explicit agreement, such as when a skier passes a group to get into a cabin that is too small for the group to join, or when a skier lets another skier pass out of politeness, even though it may irritate the individuals behind them.

When skiers were observed jumping queues without tacit or explicit agreement, skiers also in line did not say anything. Some forced a smile but in most cases, there were few reactions, although these incivilities are suspected to be irritating. Only one skier said something to the queue-jumping skier. People are therefore rather stoic when queue-jumping takes places.

Intentional queue-jumping appears to be poorly perceived and displeasing to individuals. There are, nevertheless, explicit or tacit agreements between skiers to allow passing under certain circumstances. However, the more people there are, the fewer agreements there are.

These four elements presented above are the central points that emerge from the synthesis of immersion transcripts and interviews, and characterize the observed skiing queues in the Canton of Valais.

5 Discussion

Based on the synthesis of the different immersions and on the literature review, we generated four hypotheses directly linked to ethological elements observed in waiting queues. A discussion follows each hypothesis.

5.1 Hypothesis 1: In Order to Stay Together, Skiers are Prepared to Wait Longer Without Affecting Their Perception of the Quality of Service

For groups of skiers, staying together is essential. Because the number of seats on cable cars are fixed while group sizes vary, groups have to adapt to this technical constraint and are willing to let other skiers go in front of them. This intentional passing, which causes a longer waiting time for groups, has no effect on global satisfaction of group members. Indeed, the value of remaining with the group is higher than the value of a shorter waiting time.

5.2 Hypothesis 2: A Queue Where Skiers do not Stop Moving Forward Reduces Negative Feelings and Behaviors

In ski queues, several mechanical elements like turnstiles interfere with skiers' progress through the queue. Although these elements have a control and a statistical (count of skiers) function, skiers perceive turnstiles and gates as barriers that slow down the queue, bring skiers closer together (cf. discussion about proxemics in the results section), and possibly separate groups. These three factors generate negative feelings and behaviors. To keep the essential functions of this barrier while suppressing the negative effect that this blockade causes, the structure of the turnstile could be modified: for example, by removing the metal blades but keeping the component that scans the ski pass. Any modification of this sort must be powerful enough to read the ski pass without requiring the skier to stop.

5.3   Hypothesis 3: A Structured and Fair Waiting Queue has a Positive Impact of the Perception of Wait Time

Currently, the structure of ski queues varies for each ski resort and cable car facility. Elements like gates or turnstiles are common, but they are not installed everywhere. This lack of cohesive structure creates perceived injustices characterized, for example, by passing (not respecting the line). Consequently, a consistent queue structure respecting the first-in/first-out principle would likely reduce this negative impact. This reorganization would require several ski resorts to install gates, waiting spaces, and some would need a complete change of queue structure. For example, cable cars could turn to a new model inspired by amusement parks with a serpentine queue to prevent passing. New technologies could also better regulate ski queues.

5.4   Hypothesis 4: Separating Groups has a Negative Impact on Behaviors and Group Members' Feelings

Cable cars have a fixed throughput that groups of skiers have to learn. There is a conflict between this rigid, industrial system and humans' animal instincts that encourages skiers to stay together in their group. When groups have to separate in order to follow the throughput, this animal instinct surfaces and leads to irritation and other negative feelings.

6   Conclusion

This exploratory research was performed through 82 immersions and 43 semi-directed interviews in Valais Ski in February 2016. This empirical research on human ethology in waiting queues in cable cars is, to our knowledge, the first one in this field. Indeed, most of qualitative techniques in queuing management do not distinguish innate human behavior from acquired behavior. In this research, we revealed that most skiers' reactions aren't driven by cultural traits, but rather by common ethological denominators. Specifically, this study demonstrates how to detect and evaluate the skiers' reactions in order to regulate them.

We have managed to generate new waiting queue hypotheses for skiers' behaviors on an ethological basis. In future research, these hypotheses could be addressed to improve the queuing management of cable cars. For instance, a regulation based on ethological behaviors happens naturally in the queue in order to comply to the cable car's logistic. A well-managed procedure of the waiting queue (e.g. physical barriers) would certainly help this phenomenon to happen.

These hypotheses have to be validated by statistical inferences in order to define the typical skiers' behavior in the cable car waiting queues. This research could be completed by using experimental manipulation techniques applied to a random sample.

Thus, in future research we plan to develop an IT Process to control the "emotional temperature" of waiting queues. By "emotional temperature", we imagine a measurement of the state of a waiting queue who could become heated or nervous in case of a special event like an unpredicted closing of a cable car. This process will be similar to a Statistical Process Control (SPC). This process is based on five steps:

1. Definition of the normal behavior of the waiting queue within tolerance based on the ethogram
2. Definition of the deviation of the normal behavior based on the ethogram analysis
3. Alert given by the IT Process
4. Diagnosis from human expertise
5. Mitigation of the situation through human intervention based on the diagnosis to return to normal behavioral conditions

To move from evaluating to treating dissatisfaction of skiers in waiting queues, we suggest to set up preventive controls [23] in a three steps process:

1. Setting up of the objective (e.g. fluidity of the queue)
2. Analyze indicators coming from the IT Process
3. Launch of mitigations in relation to the level of the indicators (e.g. supply of tea or hot chocolate in order to reduce the "emotional temperature")

Several preventive control options based on human ethology already exist and should be considered. For example, a camera automatically analyzes a waiting queue in order to detect individual behaviors like anxiety, anger, or impatience. Then an employee will be alerted and will immediately take in charge the skier. Numerous possibilities exist to aid in managing waiting queues and are necessary in order to prevent a decline of the general satisfaction of skiers. Indeed, reducing waiting time in queue will improve skiers' satisfaction.

Acknowledgements. We would like to thanks Christelle Deillon, Michael Kambly, and Lionel Emery for their commitment to this project, especially for their data gathering.

References

1. Eibl-Eibesfeldt, I.: Human Ethology. Aldine De Gruyter, Hawthorne (1989)
2. Maister, D.: The psychology of waiting lines. In: Czepiel, J.A., Solomon M.R., Surprenant C.F. (eds.) The Service Encounter: Managing Employee/Customer Interaction in Service Businesses. D. C. Heath and Company, Lexington Books, Lexington (1985)
3. Naumann, S., Miles, J.: Managing waiting patients' perceptions: the role of process control. J. Manag. Med. **15**(5), 376–386 (2001)
4. Nie, W.: Waiting: integrating social and psychological perspectives in operations management. Omega Int. J. Manage. Sci. **28**(6), 611–629 (2000)
5. Baker, J., Cameron, M.: The effects of the service environment on affect and consumer perception of waiting time: an integrative review and research propositions. J. Acad. Mark. Sci. **24**(4), 338–349 (1996)
6. McDonnell, J.: Music, scent and time preferences for waiting lines. Int. J. Bank Mark. **25**(4), 223–237 (2007)

7. Rafaeli, A., Barron, G., Haber, K.: The effects of queue structure on attitudes. J. Serv. Res. 5(2), 125–139 (2002)

8. Choongbeom, C., Atul, S.: Assessing the relationship between waiting services and customer satisfaction in family restaurants. J. Qual. Assur. Hospitality Tourism 13(1), 24–36 (2012)

9. Van Riel, A.C.R., Semeijn, J., Ribbink, D., Bomert-Peters, Y.: Waiting for service at the checkout: negative emotional responses, store image and overall satisfaction. J. Serv. Manage. 23(2), 144–169 (2012)

10. Whiting, A., Donthu, N.: Closing the gap between perceived and actual waiting times in a call center: results from a field study. J. Serv. Manage. 23(5), 279–288 (2009)

11. Nah, F.: A study on tolerable waiting time: how long are Web users willing to wait? Behav. Inf. Technol. 23(3), 153–163 (2004)

12. Burkhardt, R.W.: Patterns of Behavior: Konrad Lorenz, Niko Tinbergen, and the Founding of Ethology. ISBE Newslett. 18(2), 38–41 (2006)

13. Sylvie, O.: Boris Cyrulik: L'affectivité nous façonne. L'Express, Paris (1993)

14. Eibl-Eibesfeldt, I.: Human ethology: concepts and implications for the sciences of man. Behav. Brain Sci. 2(1), 1–26 (1979)

15. Markóczy, L., Goldberg, J.: Management, organization and human nature: an introduction. Manage. Decis. Econ. 19(7/8), 387–409 (1998)

16. Guercini, S.: New qualitative research methodologies in management. Manage. Decis. 52(4), 662–674 (2014)

17. Otubanjo, O., Amujo, O., Cornelius, N.: The informal corporate identity communication process. Corp. Reputation Rev. 13(3), 157–171 (2010)

18. Low, B.: The behavioral ecology of resource consumption: why being green is so hard. Hum. Ethology Bull. 29(2), 3–26 (2014)

19. Jones, L.K., Jennings, B.M., Goelz, R.M., et al.: An ethogram to quantify operating room behavior. Ann. Behav. Med. 50(4), 487–496 (2016)

20. Weisfeld, G.E., Goetz, S.M.M.: Applying evolutionary thinking to the study of emotion. Behav. Sci. 3(3), 388–407 (2013)

21. Huang, R.: RQDA: R-Based Qualitative Data Analysis. R Package Version 0.2-0 (2009)

22. Hall, E.T.: The Hidden Dimension: Man's Use of Space in Public and Private. The Bodley Head Ltd., London (1966)

23. Fragnière, E., Sullivan, G.: Risk Management: Safeguarding Company Assets. Thomson, Boston (2007)

Author Index